LAKATOS' PHILOSOPHY OF MATHEMATICS

A HISTORICAL APPROACH

STUDIES IN THE HISTORY AND PHILOSOPHY OF MATHEMATICS

Volume 3

NORTH-HOLLAND

AMSTERDAM · LONDON · NEW YORK · TOKYO

LAKATOS' PHILOSOPHY OF MATHEMATHICS

A HISTORICAL APPROACH

Teun Koetsier

Department of Mathematics
Vrije Universiteit Amsterdam
Amsterdam, The Netherlands

1991

NORTH-HOLLAND

AMSTERDAM · LONDON · NEW YORK · TOKYO

ELSEVIER SCIENCE PUBLISHERS B.V.
Sara Burgerhartstraat 25
P.O. Box 211, 1000 AE Amsterdam, The Netherlands

Distributors for the United States and Canada:
ELSEVIER SCIENCE PUBLISHING COMPANY INC.
655 Avenue of the Americas
New York, N.Y. 10010, U.S.A.

Library of Congress Cataloging-in-Publication Data

Koetsier, T.
Lakatos' philosophy of mathematics : a historical approach / Teun Koetsier.
p. cm. -- (Studies in the history and philosophy of mathematics ; v. 3.)
ISBN 0-444-88944-2
1. Mathematics--Philosophy. 2. Mathematics--History. 3. Lakatos, Imre. I. Title. II. Series.
QA8.4.K64 1991
510'.1--dc20 91-34119
CIP

ISBN: 0 444 88944 2

Printed in The Netherlands

For my parents, Mira and Bogdan

Preface

This book, inspired by Lakatos' s philosophy of mathematics, records my efforts over the past years to draw conclusions with respect to the nature of mathematics by studying its actual historical development. I am very grateful to the Department of Mathematics and Computer Science of the Vrije Universiteit in Amsterdam for supporting this project. Moreover, I would like to express my gratitude to all individuals who helped me in the course of the preparation of the study. I am particularly indebted to Henk Bos and Peter Kirschenmann for reading various versions of the manuscript. Their detailed comments have led to considerable improvements. I greatly appreciated the valuable critical comments on (parts of) an earlier version of the text by Maarten Maurice, Detlef Spalt, Bas Jongeling and Eduard Glas (who on certain points greatly disagrees with my reading of Lakatos). I am much obliged to Sally Kuiper for linguistic improvements and to Andries Pomper for drawing the figures.
The computer illustration on the cover was done by Evert Wattel. It represents the great icosahedron {3,5/2}, a (generalized) regular polyhedron whose vertex figures are pentagrams (Cf. [Coxeter,1973], p.96).

Amsterdam,
July 4,1991

Teun Koetsier

Contents

Chapter III "Cauchy and the continuum"

Chapter IV A meta-methodological intermezzo

Chapter V Two reconstructions of developments in mathematics by means of the Methodology of Scientific Research Programmes

Chapter VIII Two research traditions in analysis

Chapter IX The history of the interchangeability theorem for partial differentiation from Nicolaus I. Bernoulli to Schwarz

Chapter X Some concluding remarks

Bibliography

Index

Introduction

The present monograph deals with the methodology of mathematics. This does not, however, mean that the working mathematician should expect it to solve his[1] methodological problems. The study also deals with the history of mathematics, but here again the historiographer should not expect major contributions to the historiography of mathematics. This study is a contribution to the philosophy of mathematics and like most philosophical treatises it neither addresses itself to any practical problems, nor does it solve them.

In a paper published in 1981 the mathematician Mac Lane encouraged philosophers to renew the study of the philosophy of mathematics, a subject which he described as being "dormant since about 1931".[2] Other authors have expressed themselves in a similar way. For example Putnam, in a paper published in 1979 on recent trends in the philosophy of mathematics, came to the conclusion that none of the existing views on the nature of mathematics were valid.[3] And in 1979 Goodman argued that the four major views in the philosophy of mathematics - formalism, intuitionism, logicism and platonism - arise from "an oversimplification of what happens when we do mathematics".[4] From Goodman's point of view a more adequate philosophy of mathematics had yet to be formulated.[5]

Tymoczko, editor of the 1986 anthology *New Directions in the Philosophy of Mathematics*[6], strikes a more optimistic note in his introductions to the different parts of the book. From his point of view the papers collected in the anthology contain the outlines of a more adequate philosophy of mathematics which he calls *quasi-empiricism.* Tymoczko states: "This anthology delineates quasi-empiricism as a coherent and increasingly popular approach to the philosophy of mathematics".[7]

What is quasi-empiricism? Tymoczko uses a rather loose definition. For him quasi-empiricism is a philosophical position, or rather a set of related philosophical positions, that attempts to re-characterize the mathematical experi-

---

[1] When I refer to mathematicians in general the reader should read "he or she" and "his or her" wherever I write respectively "he" and "his".

[2] [Mac Lane,1981], p.462

[3] [Putnam,1979]

[4] [Goodman,1979], p.540

[5] [Goodman,1979], p.550

[6] [Tymoczko,1986]

[7] [Tymoczko,1986], p.xvi

ence by taking the actual practice of mathematics seriously.[8] Tymoczko writes:

> "If we look at mathematics without prejudice, many features will stand out as relevant that were ignored by the foundationalists: informal proofs, historical development, the possibility of mathematical error, mathematical explanations (in contrast to proofs), communication among mathematicians, the use of computers in modern mathematics, and many more".[9]

Usually the word 'quasi-empiricism' is used in a more restricted sense. As far as I know, the word was first used by Lakatos in the middle of the sixties and also by Putnam later in the seventies. In 1967 Lakatos distinguished two different kinds of theories: *quasi-empirical theories* and *Euclidean theories*. He defined Euclidean theories as theories in which the characteristic truth flow inundating the whole system goes from the top, the axioms, down to the bottom. He defined quasi-empirical theories as theories in which the crucial truth flow is the upward transmission of falsity from the "basic statements" to the axioms. Attacking the foundationalist illusion that there exists a means of finding a foundation for mathematics which will be satisfactory once and for all, Lakatos argued that mathematics is not Euclidean, but instead quasi-empirical. Carried away by his own reasoning and wishing to show the fallibility of mathematics in the sense of Popper's falsificationism, Lakatos exaggerated. There is indeed an upward transmission of falsity in mathematics, but it is not the crucial truth flow. Yet Lakatos had a point. Later, in two papers published in 1975 and 1979, we also find Putnam defending the point of view that mathematics is quasi-empirical.[10] In the 1975 paper Putnam argued that:

> "mathematical knowledge resembles *empirical* knowledge - that is, that the criterion of truth in mathematics just as much as in physics is success of our ideas in practice, and that mathematical knowledge is corrigible and not absolute".[11]

---

[8]Recently *Philosophica* devoted two volumes to issues in the recent philosophy of mathematics ([Van Bendegem, 1988 &1989]). The different contributions clearly show that something interesting is going on in the field. The existing gap between the classical philosophy of mathematics and mathematical praxis obviously has led to new perspectives on the problems in the philosophy of mathematics.

[9][Tymoczko,1986], p. xvi

[10][Putnam,1975] and [Putnam,1979]

[11][Putnam,1975], p. 51; the emphasis is Putnam's.

In the 1979 paper Putnam presents his *quasi-empirical realism* as a modification of Quine's holism. Roughly speaking, Quine's holism consists of the view that science as a whole is one comprehensive explanatory theory, justified by its ability to explain sensations. Mathematics and logic are part of this theory, differing from natural science in the sense that they assume a very central position. Since giving up logical or mathematical truths causes great upheaval in the network of our knowledge, they are not given up. Yet, in principle, mathematics and logic are no different from natural science.[12] Putnam's quasi-empirical realism of 1979 consisted of Quine's view, but with two modifications. First, Putnam added combinatorial facts (for example, the fact that a finite collection always receives the same count no matter in what order it is counted) to sensations as elements that mathematical theorems must explain. Secondly, Putnam required that there be agreement between mathematical theory and mathematical "intuitions" whatever their source (for example the self-evidence of the Comprehension Axioms in set theory). Both Lakatos and Putnam considered mathematical theories to be interrelated sets of statements that are considered to be true. The quasi-empirical element in their positions is the fact that they reject the view that, in principle, mathematics could be described in a Euclidean way in Lakatos's sense of the term. Both Lakatos and Putnam argue that, to a certain extent, mathematical theories always possess a hypothetical status. In that respect mathematical knowledge resembles empirical knowledge.

This study is an essay in quasi-empiricism in every sense of the word. It is an attempt to look at the history of mathematics from a Lakatosian point of view. The actual historical development of mathematics is taken very seriously in a way inspired by Lakatos's unfinished quasi-empiricist philosophical work on mathematics. The first two chapters are devoted to a critical discussion of Lakatos's views and in the remaining chapters the emphasis is on the development of a notion of mathematical rationality that agrees with the historical facts.

Lakatos's position can be summarized in the form of two theses:

I. Lakatos's *fallibility thesis*: Fallibility is an essential characteristic of mathematical knowledge.

---

[12]Quine defended this view of mathematics and logic on numerous occasions, for example in his well-known paper from 1951 entitled "Two dogmas of empiricism" (Reprinted in [Quine,1953]).

Most philosophies of mathematics are infallibilist. Infallibilists argue that, although in practice mathematicians make mistakes, mathematical knowledge is essentially infallible. In fact Lakatos's quasi-empiricism consists in the fallibility thesis. Although their subject matter is different, mathematical theories and empirical theories have in common the fact that they are fallible.

II. Lakatos's *rationality thesis*: In spite of its fallible character the development of mathematical research is not completely arbitrary, but possesses its own rationality. Fallible mathematical knowledge is replaced by other fallible knowledge in accordance with certain norms of rationality.

Most of Lakatos's work with respect to mathematics concentrates on the rationality thesis. A major tool that Lakatos uses is the *rational reconstruction* of developments in mathematics. A rational reconstruction is a reconstruction that is explicitly based on a particular methodology. In Lakatos's work one can distinguish two periods. During the first period he attempted to apply a Popperian brand of falsificationism to mathematics. In this period the *Methodology of Proofs and Refutations* (MP&R) dominated his work. In the second half of the sixties, when Lakatos developed his *Methodology of Scientific Research Programmes* (MSRP), the second period started. During this period he wrote primarily on natural science. In Chapter I, I concentrate on two of Lakatos's rational reconstructions of historical developments in mathematics from the first period. The first reconstruction concerns the history of Euler's formula for polyhedra, and the second reconstruction concerns the history of the notion of uniform convergence. In order to create clarity about their status the reconstructions are confronted directly with the actual historical development to which they refer. One conclusion of Chapter I is that, although the rational reconstructions on the basis of the MP&R are quite counter-factual when taken as reconstructions of historical developments, they reveal important heuristic patterns. Another conclusion is that neither Lakatos's reconstructions nor the corresponding actual historical developments exhibit more than the *weak fallibility* of mathematical knowledge.

In Chapter II, I discuss Lakatos's paper "A Renaissance of Empiricism in the Recent Philosophy of Mathematics", written in the middle of the sixties which clearly shows the problems that Lakatos's early falsificationist views of mathematics encountered. Why is it so difficult to identify potential falsifiers in mathematics? Why is the illusion that mathematics is an infallible science so strong? Some of these problems were, in principle, solved by Lakatos's MSRP, which Lakatos considered applicable to mathematics although it was developed for natural science as an immediate reaction to Kuhn's work on the development of natural science. The conclusion of Chapter II is that the MSRP needs to be modified in order to be applicable to mathematics, and that it would be worthwhile to investigate the validity of the following picture of mathematical rationality which results from such a modification.

On the micro-level (roughly speaking, the level of conjectures and theorems), the overall mathematical methodology is the MP&R (viz: conjectures, attempts at proofs, refutations, critical proof-analysis, etc.). Dramatic refutations, however, are rather rare, particularly when the mathematicians involved are cautious and do not prematurely consider plausible conjectures as theorems. On the micro-level, conjectures, theorems and proofs are fallible in the sense that they can be replaced by other conjectures, theorems and proofs along the lines of the MP&R. On the macro-level (roughly, the level of whole fields, or even mathematics as a whole) the situation is different. There, the developments can no longer be described in terms of the MP&R. The development of mathematics shows that, in practice, coherent clusters of considerable numbers of theorems become established. They are no longer subject to refutation, but form the core of a theory or a set of theories that is being developed further. Such developing theories or sets of theories can be called "research programmes". With respect to such "research programmes" one could also define a form of refutation. The refutation of a "research programme", however, is not a logical refutation but an expression of the fact that the programme has lost ground in competition with another programme. On the macro-level, research programmes are fallible in the sense that they can be replaced.

Chapters III through IX of the book are primarily devoted to an evaluation of the above picture of mathematical rationality. Only in the last chapter do I return briefly to the fallibility thesis.

In Chapter III, I discuss Lakatos's paper "Cauchy and the Continuum: the Significance of Non-Standard Analysis for the History and the Philosophy of Mathematics". This paper is interesting because, although it was written a year before Lakatos introduced the notion of research programme, it in fact contains a reconstruction of two competing research programmes in mathematics. In Chapter IV, I discuss an attempt by Spalt to develop these ideas further and to describe the history of analysis explicitly in terms of competing research programmes. From the failure of his own attempt at reconstruction, Spalt drew the conclusion that all attempts to characterize mathematical rationality in terms of the MSRP or related models are bound to fail. I am less pessimistic. There are circularity problems to be solved, and, in order to avoid vicious circularity, attempts to apply (a modified version of) the MSRP to mathematics require a careful "meta-methodological" approach. The outline of such an approach is also given in Chapter IV.

In Chapter V, I turn to two papers by Giorello. A critical examination of these suggests that we have to modify Lakatos's MSRP as follows. Mathematicians always work on *research projects* that are characterized by their heuristic unity. Research projects usually belong to a research tradition. *Research traditions* concern a particular fundamental mathematical domain and are characterized by the entities that are being studied in that domain and by assumptions about the appropriate methods by which to prove the properties of those entities. A fundamental mathematical domain consists of the most general mathematical entities that still function mathematically in a certain period of time. A good example of a mathematical research tradition is the dominant deductive geometric tradition, represented, for example, by Euclid's Elements. I call that tradition the Euclidean Tradition. The fundamental domain of that tradition is Euclidean space. Another example of a research tradition is the Structuralist Tradition that dominated the twentieth century. I propose a *Methodology of Mathematical Research Traditions* (MMRT) in Chapter VI. I follow a suggestion by Marchi and make mathematical theorems analogous to facts in natural science. The *progress* of research projects and research traditions is defined in terms of the weight of the conjectures and theorems that are being generated by a project or tradition.

In Chapters VII and VIII, two rational reconstructions on the basis of the MMRT are given. The first reconstruction, in Chapter VII, shows that one can argue on the basis of the existing historiographical material that in Greek mathematics the Euclidean Tradition was preceded by another tradition - the Demonstrative Tradition. Although any reconstruction of the development of pre-Euclidean mathematics will be rather speculative owing to a lack of primary sources, my speculations indicate that one must, in mathematics, distinguish deductive traditions from non-deductive traditions. The Demonstrative Tradition was a non-deductive tradition which was superseded by a deductive tradition, namely the Euclidean Tradition. Unlike the Demonstrative Tradition, the Euclidean Tradition possesses an explicitly defined "hard core" (analogous to the hard core that Lakatos takes as one of the defining elements of a research programme in natural science) that was formed in a process of repeated proofs and refutations.

In Chapter VIII, I describe the transition from eighteenth century analysis to nineteenth century analysis in terms of the MMRP. In my reconstruction, eighteenth century analysis was dominated by a particular tradition - the *Formalist Tradition* - which, to a considerable extent, was non-deductive. That tradition was superseded by a deductive tradition - the *Conceptual Tradition*.

The rational reconstructions in Chapters VII and VIII show that the MMRT, within the limits that inherently accompany such models, satisfactorily characterizes the rationality of mathematics.

In Chapter IX, the history of a particular theorem in analysis - the interchangeability theorem for partial differentiation - gives us a good view of the internal development of the two traditions that were described in Chapter VIII. It becomes clear that mathematical research traditions are not rigid, but that, on the contrary, they are subject to change with time.

In Chapter X, I draw some conclusions. Mathematics is fallible, but the rational reconstructions in Chapters I through IX show that its fallibility is very weak. It is remarkable that the question: "How should we explain this weak fallibility?" brings us back to the classical questions in the philosophy of mathematics. I argue briefly that a realist position with respect to mathematics most naturally explains the development of mathematics seen from the point of view of the MMRT.

Chapter I On two Lakatosian rational reconstructions of developments in mathematics

I.1 Introductory remarks: fallibility and rationality

I.1.1 Lakatos's fallibilist philosophy of mathematics

One frequently meets, either implicitly or explicitly, a very narrow view of the methodology of mathematics. This view boils down to the following: the method of mathematics is the deductive method and the methodology of mathematics studies this method.[1] Aristotle, who already described the deductive method, required that the axioms taken as the starting point of a theory "must be so obvious as to be accepted by everybody without any previous justification" ([Beth,1959], p.82). Beth and others replaced this requirement with the condition that every system of axioms should be (1) complete, (2) independent and (3) consistent. Beth's definition of completeness is as follows: "the axioms a_1, a_2, a_3,...,a_n for a deductive theory T must provide us with a sufficient basis for rigorous proofs of every hitherto recognized theorem of T" ([Beth,1959], p.82). This definition still reveals that the systems of axioms are not arbitrarily chosen. Yet the view that the methodology of mathematics should primarily deal with the deductive method inevitably leads to a strong emphasis on mathematical logic and meta-mathematics. In fact Beth and others restricted themselves in their methodological considerations to the final form which, in their opinion, a mathematical theory ought to have. Ignoring the way in which mathematical theories arise may lead to a statement such as: "a mathematical theory is nothing more than a logical theory, determined by a system of axioms".[2]

---

[1]E.g. [Tarski,1964], p.126

[2]The full quotation is: "une théorie mathématique n'est pas autre chose qu'une théorie logique, déterminé par un système d'axiomes [c'est-à-dire de relations construites à partir des relations élémentaires, et posées comme vraies]: les êtres de la théorie sont définis *ipso facto* par le système d'axiomes, qui engendre en quelque sorte le matériel auquel vont pouvoir s'appliquer les propositions vraies: définir ces êtres, les nommer, leur 'appliquer' les propositions et les relations, c'est en cela que consiste la partie proprement mathématique de la théorie logique" ([Cartan,1943], pp. 8-9). In the second part of the quotation Cartan seems to hint at a different picture of mathematics. Of course, Beth, Cartan, Tarski e. a. knew that mathematics is much more than logical deduction from a set of axioms. Probably the emphasis on logic and meta-mathematics in their methodological work is related to the fact that those subjects appeared to be accessible to a precise mathematical treatment yielding many exciting mathematical problems, whereas the other "heuristic" side of mathematics is

Although the view that the methodology of mathematics deals with mathematical logic dominated discussions during the first half of the twentieth century, there is one important exception, namely *Georg Polya* (1887-1985). Throughout his career Polya published on "plausible reasoning" in mathematics. In 1953 he defined it as follows: "In plausible reasoning the principal thing is to distinguish a guess from a guess, a more reasonable guess from a less reasonable guess" ([Polya,1954], p.vi). Plausible reasoning leads to the discovery of proofs. Finished mathematics, however, appears as purely demonstrative. In Polya's view plausible reasoning and demonstrative reasoning complement each other. Concerning demonstrative reasoning he wrote: "In strict reasoning the principal thing is to distinguish a proof from a guess, a valid demonstration from an invalid attempt" ([Polya,1954], p.vi). Polya stated several patterns of plausible inference formally and tried to systematize them. In the course of what follows I shall mention some of them.

In the second half of the twentieth century an attack on the view that the methodology of mathematics deals with mathematical logic was launched by *Imre Lakatos* (1922-1974) under the influence of developments in the philosophy of natural science. Popper had argued that scientific theories are fallible, and that at heart they always possess the status of a conjecture, a guess. From Popper's point of view the search for justification, in the sense of the justification of the claim that a theory is true, has to be abandoned. He writes: "*All theories are hypotheses*; all *may* be overthrown" ([Popper,1981], p.29). In his work Polya still adhered to a simple cumulative view of the development of mathematics: once a theorem has been proved, it is finished and will not undergo any major changes in the future. Lakatos fundamentally disagreed. In attempting to apply Popper's ideas to mathematics, Lakatos defended a *fallibilistic* philosophy of mathematics - mathematical theories are never final, they are guesses and as such refutable. Studying history from this perspective, by means of "rational reconstructions" of developments in mathematics, Lakatos discovered non-deductive patterns of mathematical thought different from Polya's plausible reasoning. That work actually forms the starting point of the investigations that are found in this book. Lakatos's position with respect to the nature of mathematics can be summarized as follows in the form of two theses:

---

not. This is in accordance with the methodology of mathematical research programmes that I will develop in Chapter VI.

i) Contrary to common belief, mathematics is not an infallible, cumulatively developing science. Instead, *all mathematical theories are hypotheses,* and all *may* be overthrown. I call this *Lakatos's fallibility thesis* with respect to mathematics.
ii) Mathematics is fallible but its development is not arbitrary. Its development is in general in accordance with certain norms of rationality. I call this thesis *Lakatos's rationality thesis* with respect to mathematics.
The fallibility thesis is an epistemological thesis concerning the status of mathematical theories. The rationality thesis concerns the methodology of mathematics. In Lakatos's work the two theses are intimately connected. Following Popper, Lakatos defended a *falsificationist methodology*: Lakatos took fallibilism seriously, and in his methodological considerations there is great emphasis on the possibilities of falsifying a theory in one way or another. In Lakatos's view not only may mathematical theories be overthrown, but attempts to overthrow them by falsification are a *necessary condition for progress.*
In Lakatos's work we can distinguish two periods. In the beginning of the sixties the emphasis is on the philosophy of mathematics. In that period Lakatos defended both the fallibility thesis and the rationality thesis by means of rational reconstructions on the basis of the *methodology of proofs and refutations.* At the end of the sixties the emphasis shifts to the philosophy of natural science. His work is then dominated by the *methodology of scientific research programmes.* It is clear that in spite of the emphasis on rational reconstructions of developments in natural science, Lakatos was at that time convinced that the methodology of scientific research programmes was also applicable to mathematics.

I.1.2 Rational (re-)constructions: some definitions

(Re-)constructions
I will first give some definitions. In this study I restrict myself to mathematics which is sufficiently explicit, i.e. mathematical truths and falsities which are explicitly recognized as being truths or falsities. This means that I exclude certain forms of what one could call intuitive mathematics from my considerations. For example, with geometric art one could argue that, although no theorems are formulated, the artists show through their behaviour that they are to a certain extent aware of mathematical regularities.

Moreover, in the course of what follows, mathematics is considered as a *process*: in other words, I concentrate on successions of *mathematical acts*. A description of a particular succession of mathematical acts is in my definition a *construction of a mathematical development*. If such a construction refers to a succession of mathematical acts which actually took place in history (this history may be very recent), the construction is by definition a *reconstruction of a mathematical development*. Reconstructions are special constructions, but not all constructions are reconstructions. For example, pedagogical presentations are often constructions, but they are not usually reconstructions of historical developments.

Rational (re-)constructions

A construction or a re-construction need not be based upon some explicit methodology, i.e. the acts involved need not explicitly be described as instances of known rules. Mathematicians generally do not act in an arbitrary manner. They set goals either for themselves or for other mathematicians; they choose possible approaches to reach the goals, including the setting of (sub)goals; they evaluate the results of those approaches; they change the original goals. And often these mathematical acts can be considered as instances of implicit rules. The methodology of mathematics deals with such rules.[3]

A set of such rules forms by definition a *(partial) methodology of mathematics* or a *(partial) notion of mathematical rationality. I call a (re-)construction of a mathematical development rational if it is explicitly based (at least partially) upon some methodology.*

In general a rational (re-)construction has the following structure. It consists of a sequence of mathematical acts and the results of those acts (including negative results). The rules involved are generally such that they recommend a particular mathematical behaviour pattern on the basis of a preceding series of mathematical acts together with their results (occurring in the construction) and also usually on the basis of certain other factors, for example mathematical results not occurring in the construction.

---

[3] I shall always try to express methodological rules in terms of *recommendations*. Such recommendations, however, will merely be explicitations of rules that working mathematicians implicitly use. The reader should not expect any new methodological rules. As I remarked in the Introduction, this study does not solve the working mathematician's methodological problems.

Regarding reconstructions, there is another useful notion that I will introduce here - their *factuality*. The extent to which a reconstruction captures a sequence of mathematical acts which actually took place in history, constitutes its factuality. And therefore a reconstruction, rational or not, can in various degrees be *factual or counter-factual*. Obviously the degree of factuality of a reconstruction will often be a matter of discussion.
One should distinguish between notions of mathematical rationality in the sense defined above, and broader notions of rationality. In this study I am primarily concerned with testing notions of *mathematical* rationality by means of rational reconstructions in the history of mathematics. The methodological rules that I deal with concern mathematical behaviour that is or might be adequate in a given *mathematical* problem situation and they should be distinguished from psychological heuristic rules.

Deductive logic versus tactics and strategics
The work of Polya and Lakatos opens up the possibility of a broad notion of the methodology of mathematics. After all, the process of mathematics has at least two aspects:
(a) proofs (and refutations),
(b) tactical and strategic considerations.
Research papers in mathematics and textbooks on mathematics contain constructions of particular mathematical developments in the sense that the reader who reads them from beginning to end participates in a succession of mathematical acts. Such constructions are generally rational in our sense of the word: they are rational constructions on the basis of a notion of rationality that is restricted to the rules corresponding to (a), the rules of deductive logic. In this book, however, we are primarily interested in tactical and strategical considerations and in rational (re-)constructions of mathematical developments on the basis of rules corresponding to (b), tactical and strategic rules. I am using here the notions of tactics and strategics approximately in the sense in which they are used by André Weil, who writes:

> "By tactics I understand the day-to-day handling of the tools at the disposal of the scientist or scholar at a given moment; this is best learnt from a competent teacher and the study of contemporary work. For the mathematician it may include the use of differential calculus at one time, of homological algebra at another [...]"

and he continues as follows:

"In contrast with this, strategy means the art of recognizing the main problems, attacking them at their weak points, setting up future lines of advance. Mathematical strategy is concerned with long-range objectives: it requires a deep understanding of broad trends and of the evolution of ideas over long periods" ([Weil,1978], p.436).

The micro level versus the macro level

Tactical considerations dominate at what I call the *micro level* in the development of mathematics, and strategic considerations dominate at the *macroevel* in the development of mathematics. This distinction is important.
The *micro level* is the level of the individual researcher. His considerations concern the short range. He poses himself problems and tries to solve them. The results appear in pre-prints and papers. The mathematical behaviour of the individual researcher is limited by all kinds of factors: he is objectively limited in several ways in his possibilities to do mathematics. He lacks overview, his knowledge and talent are limited. His personal situation may restrict his options, and his subjective personal preferences may also impose limits in various ways. Nevertheless he aspires to produce good mathematics within his limitations.
The *macro level* is the level of the mathematical community as a whole or a major part of it. When we consider trends in the development of mathematics as a whole, we study this level. Here the long-range development of mathematics is important. Survey papers written by top researchers, and handbooks etc., reflect developments on this level. There is a wider overview, greater knowledge and more talent involved here. Subjective personal factors play a lesser role. Internal mathematical factors, and in general factors that transcend the individual and about which, as far as consciously present, there exists consensus, begin to play a more important role. The macro level contains a selection of the results and the possible directions of development at the lower levels. Journal editors, "opinion leaders" etc. act as filters. On the other hand, there is also feedback from the macro to the micro level: for example the individual researcher will not generally want to isolate himself completely from the macro level. The strategic considerations of the opinion leaders in this way play a role on the micro level.[4] Yet at the micro level the considerations are predominantly of a tactical nature.

---

[4]Cf. [Jongeling,1988]

It is relatively easy to identify mathematical acts at the micro level (in particular on the level of the individual mathematician) and it is also relatively easy, at least in principle, to give a more or less complete description of those acts. At the macro level a mathematical act consists in the choice for a particular theory, a particular approach and it will have to be described inevitably in more global terms. That is why, as we will see below, the identification of a mathematical act on the macro level depends more on our theoretical conceptions concerning the way in which mathematics develops. At the macro level the word "act" often no longer refers to an act by an individual but concerns the preferences of the entire mathematical community or a considerable part of it.

Purposes of rational (re-)constructions

There are at least four possible purposes that rational (re-)constructions can have.

(A) A paedagogical purpose: the (re-)construction is meant to teach others a particular methodology of mathematics.

(B) A strictly methodological purpose: the goal is to defend a particular methodology. In this case one shows that the rules applied in a particular construction are effective. One can, however, also aim at recognition. Then one hopes that the working mathematician will say: Right, that is how we work! Thirdly, one can show that (great) mathematicians acted (in the past) in accordance with a particular methodology. In this case we inevitably need a rather factual rational reconstruction of a development.

(C) A philosophical purpose which goes beyond methodology and which concerns the nature of mathematics as a whole. This purpose is related to (B). An argument could run like this: By means of certain rational constructions one defends a particular methodology of mathematics. Then one shows that the methodology or the reconstruction is more in accordance with view V_1 on the nature of mathematics (for example mathematics as a human construction) than with view V_2 (for example platonic realism).

(D) An historical purpose: the goal is purely descriptive or perhaps it is to reveal "laws of development" by means of rational reconstructions.

In this book the emphasis will be on the use of rational reconstructions in the sense of (B) and (C).

I.1.3 A remark on Lakatos's view of methodology

The above given definitions are mine and not those of Lakatos. I will, however, use them to study some of Lakatos's work in the philosophy of mathematics. This requires a remark on Lakatos's usage of notions like heuristics and methodology.

There is an important difference between Lakatos's work dominated by the methodology of proofs and refutations and his later work dominated by the methodology of scientific research programmes. In the period dominated by the MP&R Lakatos's notion of methodology encompassed a rational, non-psychologistic heuristic concerned with finding rules for discovering solutions to problems. In this respect the references to Polya and Bernays in the Introduction to "Proofs and Refutations" cannot be misunderstood[5] and my above definitions do not contradict Lakatos's views. However, in the second period of his work dominated by the MSRP, Lakatos explicitly separated heuristics from methodology. In that period he argued that a methodology only consists of rules for the appraisal of ready articulated theories independent of the considerations of the scientists involved in the development, referring heuristics to the domain of psychology and sociology.[6] The MSRP appraises research programmes, but does not give advice to an individual scientist about which of two rival programmes he should work in. Although on this point it may seem as if my definition of methodology is opposed to Lakatos's, it is not. The appraisal of a development, in which one particular research programme came to dominate its rivals as being rational, implies a judgement that the mathematical behaviour of the mathematical community as a whole was rational. Indirectly, then, the choices of opinion leaders, editors of scientific journals, research foundations etc. are appraised with respect to their rationality. According to Lakatos, editors of scientific journals should refuse papers written by scientists adhering to degenerating programmes.[7] So Lakatos's MSRP does not give any firm heuristic advice to individual scientists, but it definitely yields recommendations for a rational scientific community on the way it should act. Remember that at the macro level, in my terminology the word "act" often no longer refers to an act by an individual but concerns the preferences of the entire mathematical community.

---

[5] [Lakatos,1976], p.3

[6] [Lakatos,1978a], p.103; Cf. also [Lakatos,1971], p.174

[7] [Lakatos,1978a], p. 117

I.2 An example of a rational construction

1.2.1 Pythagoras' theorem

I shall now first give a simple example of a rational construction, which I owe essentially to Polya's work ([Polya,1954]). The starting point is Pythagoras' theorem. In Figure 1.1 squares have been extended on the three sides of a rectangular triangle.

In that situation I apply the following methodological rule:

Rule 1 When a problem P has been solved, it makes sense to look for an *analogous* problem P' and to try and solve P' by means of the same or an *analogous* method.

Assuming that Pythagoras' theorem has been proved, I apply the rule as follows. I consider a rectangular triangle with equilateral triangles extended on each of the sides (Figure 1.2). It is then easy to prove that the areas of the triangles on the sides with lengths a,b,c are respectively $(a^2/4)\sqrt{3}$, $(b^2/4)\sqrt{3}$, $(c^2/4)\sqrt{3}$, which also means that between the areas of equilateral triangles on the sides of a rectangular triangle, the relation of Pythagoras holds.

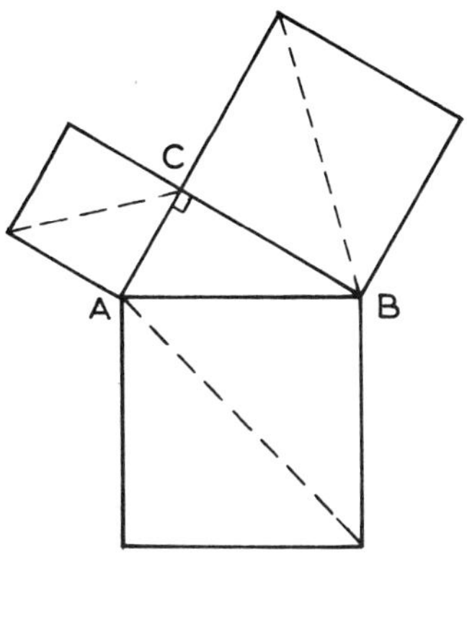

Figure. 1.1

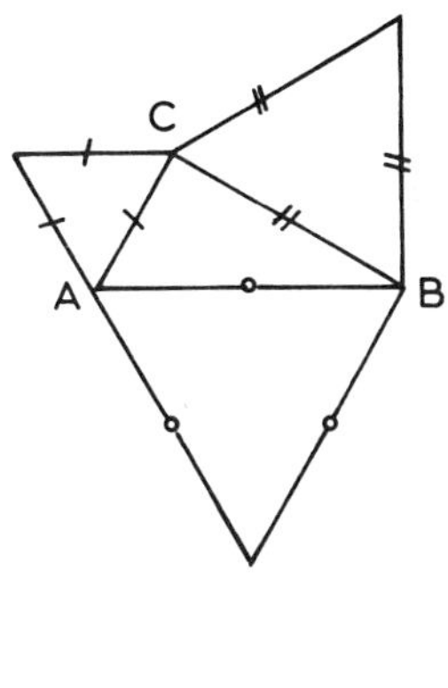

Figure 1.2

I now apply another methodological rule:

Rule 2 When two different but analogous situations appear to have non-obvious elements in common, then it makes sense to make the analogy *explicit* by means of an *appropriate generalization.*

The situations of Figures 1.1 and 1.2 are analogous and in both cases the relation of Pythagoras holds. Application of this rule may yield the conjecture that for all n, the areas of regular n-gons extended on the sides of an arbitrary rectangular triangle satisfy the relation of Pythagoras. Yet removing, for example, halves of the squares in Figure 1.1 shows that one should not perhaps restrict oneself to regular polygons. Rule 2 may then yield the conjecture : If we extend similar polygons on the sides of a rectangular triangle, then the areas of these polygons satisfy the relation of Pythagoras. The conjecture can be proved easily by means of the theorem which states that the ratio of the areas of two similar polygons equals the ratio of the squares of two corresponding sides.

The proof of our conjecture requires the validity of the relation of Pythagoras in one particular case. If one assumes Pythagoras' theorem, the requirement is, of course, satisfied. One can, however, ask oneself whether there exists a special case in which the truth of the relation is particularly obvious. Such a case would give us an alternative proof of Pythagoras' theorem and at the same time an alternative proof of its generalization.

Such a case exists. In the rectangular triangle ABC of Figure 1.3, I have drawn the perpendicular CD.

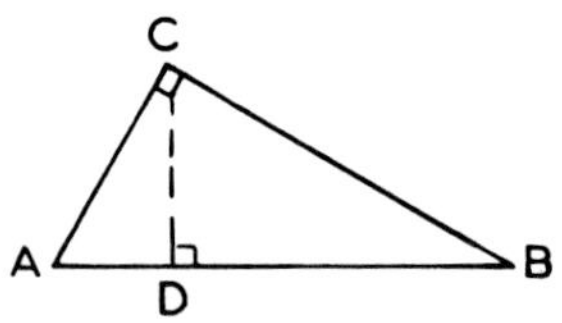

Figure 1.3

The triangles CBD, ACD and ABC are similar; they are extended inwards on respectively CB, AC and AB and, moreover, they obviously satisfy the relation of Pythagoras. [We could of course generalize Pythagoras' theorem even further towards curvi-linear similar figures (Application of Rules 1 and 2)].

The last part of this simple rational construction illustrates a third rule:

Rule 3 It makes sense to study *special cases* of a general statement or problem.

The simple rational construction given above is based upon three methodological rules which are of tremendous importance in mathematics. For example, Bourbaki, while emphasizing that the axiomatic method is also a heuristic method, writes:

"Where the superficial observer sees only two, or several, quite distinct theories, lending one another 'unexpected support' [...] through the intervention of a mathematician of genius, the axiomatic method teaches us to look for the deep lying reasons for such a discovery, to find the common ideas of these theories, buried under the accumulation of details properly belonging to each of them, to bring these ideas forward and to put them in their proper light" ([Bourbaki,1950], p.223).

What Bourbaki says here is that unexpected support implies unexpected analogies which should be made explicit by means of an appropriate generalization. Bourbaki urges the application of Rule 2. Of course, the analogous situations involved are now whole theories and no longer theorems which are more or less isolated. Moreover, since in Bourbaki's view all theories are axiomatic theories, the problem boils down to finding the axioms of the more general theory.

Rules 1 through 3 also appear in the work of Hilbert. In the lecture in 1900 in which he presented his famous list of unsolved problems in mathematics, Hilbert said:

"If we do not succeed in solving a mathematical problem, the reason frequently consists in our failure to recognize the more general standpoint from which the single problem before us appears only as a single link in a chain of problems" ([Hilbert,1900], p.6).

Clearly one way to reach such a standpoint runs via analogous problems, i.e. via the application of Rules 1 and 2.

In the same lecture Hilbert said:

"In dealing with mathematical problems, specialization plays, as I believe, a still more important part than generalization [...]. All depends, then, on finding out these easier problems, and on solving them by means of devices as perfect as possible and of concepts capable of generalization. This rule is one of the most important levers for overcoming mathematical difficulties and it seems to me that it is used almost always, though perhaps unconsciously" ([Hilbert,1900], p.6).

Here Hilbert urges the application of Rule 3. The Rules 1, 2 and 3 are examples of rules Polya dealt with in his work on plausible reasoning. Polya also gives a figure similar to Figure 1.4 ([Polya,1954], p.16).

Rules 1, 2 and 3 boil down to the following. The position of a mathematician always corresponds with a vertex of one or more triangles like the one in Figure 1.4. The rules actually say that it is always rational to move from that vertex to one of the other vertices.

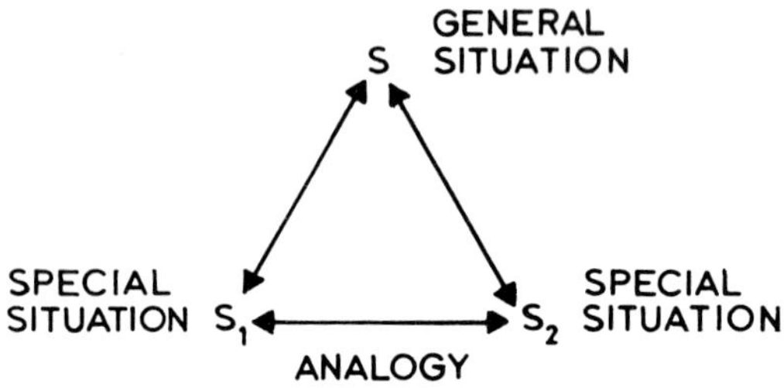

Figure 1.4

I.2.2 Some further remarks on Rules 1, 2 and 3

Rules 1, 2 and 3 do not precisely say what should be done. For example, it is possible to begin with Figure 1.1 and to pass by means of Rule 1 to an *arbitrary triangle* with squares extended on the sides, and to study the relation of the areas of the three squares. This can lead to the cosine-rule, which is a different generalization of Pythagoras' theorem. One obvious problem is that the notion of analogy is not at all clear. Of course it makes sense to say, for example, that two situations are analogous if and only if they possess a common "structure". And as soon as such a common structure can be made explicit, it is clear what the analogy amounts to in the case under consideration. The problem, however, is that from the point of view of methodology, analogies which have not been made explicit are the most interesting. From my point of view such implicit analogies play a central role in what Bourbaki calls a "special intuition",

> "which is not the popular sense-intuition, but rather a kind of direct divination (ahead of all reasoning) of the normal behaviour, which he (i.e. the mathematician - T.K.) seems to have the right to expect of mathematical beings, with whom a long acquaintance has made him as familiar as with the beings of the real world" ([Bourbaki,1950], p.227).

In my view this "special intuition" or tacit knowledge that professional mathematicians possess consists in the capability of recognizing all sorts of inexplicit analogies. These analogies determine their expectations concerning the objects they study.

One can distinguish between rules for generating ideas and rules for evaluating goals and intermediate results. Yet the two kinds of rules are closely related. I described Rule 1 more or less as a rule that may help to produce a new idea. However, compare Rule 1 with the following rule given by Polya (A and B are statements):

Premises:	A analogous to B
	B true
Conclusion:	A more credible

([Polya,1968], p.10)

This rule is closely related to Rule 1. Clearly, if theorem B has been proved, it makes sense to try to prove an analogous theorem A, because there is an indication that A may be true. I also described Rules 2 and 3 as rules that may help to generate ideas. However, compare Rule 3 with the following rule, also given by Polya:

Premisses:	A implies B
	B true
Conclusion:	A more credible

([Polya,1968], p.19)

This rule implies that if one cannot prove a conjecture A, it makes sense to turn to a special case B of A. If B turns out to be false, A is also false. If B appears to be true, this strengthens our faith in A.

Salmon has distinguished three kinds of plausibility criteria: formal criteria, material criteria and pragmatic criteria ([Salmon,1967], pp.125-126). Pragmatic criteria, although they are applied by mathematicians, are of less interest to us in this context. An example is the following: a conjecture becomes more credible if a great mathematician, without giving a proof, states that it can be proved. The rule that states that a conjecture A becomes more credible when it appears to imply a true statement B is an example of a formal criterion (because A and B are deductively related). The rule stating that a conjecture A becomes more credible because it appears to be analogous to a true statement B is an example of a material criterion (because A and B are not deductively related). It is interesting that Polya has shown that this rule concerning analogy can be derived in a certain way from the rule concerning implication. The derivation runs as follows. The assumption is that the analogy of two statements A and B is based upon the existence of a more general statement H implying both A and B. Then from "A is analogous to B" follows: "H implies B". If B appears to be true, then according to the above implication, rule H becomes more credible. Since we assume that H also implies A, then clearly A also becomes more credible ([Polya,1968], p.28).

I.3 Lakatos's reconstruction of the history of Euler's formula for polyhedra

I.3.1 The methodology of proofs and refutations

In this chapter I will discuss two rational reconstructions from the period in Lakatos's work which was dominated by the *method of proofs and refutations*. The two reconstructions first appeared in Lakatos's 1961 Cambridge Ph. D. thesis. The first reconstruction concerns the history of Euler's formula for polyhedra V-E+F=2 (V is the number of vertices of a polyhedron, E represents the number of edges and F is the number of faces). It was published under the title "Proofs and Refutations" ([Lakatos,1963-64]). The second reconstruction concerns the history of the notion of uniform convergence. It appeared posthumously in 1976 under the title "Another Case-Study in the Method of Proofs and Refutations" ([Lakatos,1976], pp. 127-141).

In "Proofs and Refutations", Lakatos gives "a sort of *rationally reconstructed or 'distilled' history*" ([Lakatos,1976], p.5 - The emphasis is Lakatos's) of the actual historical developments concerning Euler's formula for polyhedra V - E + F = 2. Regarding the aim of the reconstruction, Lakatos wrote:

> "Its modest aim is to elaborate the point that informal, quasi-empirical, mathematics does not grow through a monotonous increase of the number of indubitably established theorems but through the incessant improvement of guesses by speculation and criticism, by the logic of proofs and refutations" ([Lakatos,1976], p.5).

One notices in the quotation both the fallibility thesis ("incessant improvement of guesses") and the rationality thesis ("the *logic* of proofs and refutations"). Lakatos's paper deals with *informal, quasi-empirical mathematics*. What type of mathematics is this? Apparently Lakatos distinguished formal mathematics from informal mathematics. The introduction to "Proofs and Refutations" suggests that formal mathematics is what a suitably programmed Turing-machine can do. This would mean that informal mathematics in fact encompasses nearly all of mathematics, since only a very small part of mathematics consists in giving formal derivations within a formal system. This interpretation is supported by Lakatos's posthumously published paper: "A Renaissance of Empiricism in the Recent Philosophy of Mathematics?", completed in 1967 ([Lakatos,1978b], pp.24-42).[8] In this paper Lakatos defends the view that mathematics is a *quasi-empirical science* and about the methodology of such a science he remarks:

> "It starts with problems followed by daring solutions, then by severe tests, refutations. The vehicle of progress is bold speculations, criticism, controversy between rival theories, problemshifts. Attention is always focussed on the obscure borders. The slogans are growth and permanent revolution, not foundations and accumulation of eternal truths" ([Lakatos,1978b], pp.29-30).

This view is, of course, opposed to most of the traditional views concerning the nature of mathematics. Lakatos's defence of his view deserves a careful analysis. Popper had argued that the theories of natural science are necessarily fallible, because no number of true test-statements can imply the truth of an explanatory universal theory. This simple logical argument cannot be used for mathematics. In natural science a theory must explain the experimental data, but its starting point may in principle be completely arbitrary, and in the development of natural science there need not be any theoretical continuity, not even with respect to the same empirical domain. In a mathematical theory we do not have empirical statements that are to be explained. The starting point of a mathematical theory concerning a particular domain will usually consist of statements that are primarily considered to be true because of what they mean, and not because of what they logically imply. Because he has no logical arguments, Lakatos's major tool used for showing the fallibility of mathematical theories is the rational (re-)construction of mathematical developments. Lakatos's rational constructions are meant to show how mathematical theories are proposed and subsequently refuted. In other words the reconstructions are meant to show that the development of mathematics, when it is described truthfully on the basis of a falsificationist methodology without "justificationist" prejudice, exhibits the characteristics of the development of a fallible science. Lakatos is very critical of philosophers who recognize the fallibility of mathematics, but who appeal to analogies which he takes to be wrong, because the emphasis is on confirmation. These philosophers insufficiently investigate the possibility of refutations in mathematics. Lakatos writes:

---

[8]In Chapter II, I study this paper more thoroughly.

> "Such *soft fallibilism* divorces fallibilism from criticism and shows how deeply ingrained the Euclidean tradition is in mathematical philosophy" ([Lakatos,1978b], p.42).

In this context Lakatos quotes Bernays as saying

> "To a philosopher there can be nothing which is absolutely self-evident [...] But in practice there are, of course, many things which can be called self-evident [...] each method of research presupposes certain results as self-evident" (Ibidem),

and he objects to the second sentence. Lakatos's views apparently imply that *nothing* in mathematics is self-evident. Self-evidence in mathematics is an illusion. All we can say is that certain results have so far resisted all attempts to refute them.

I.3.2 A summary of Lakatos's reconstruction

"Proofs and Refutations" is extremely rich in content. It is moreover written in the dialogue form which makes its interpretation difficult. Yet in the dialogue a central position is occupied by a sequence of mathematical acts that are in accordance with the following global falsificationist methodological principle: *The vehicle of progress is daring solutions, bold speculations, bold generalizations on one hand, and severe tests, and refutations on the other* (compare with Lakatos's above characterization of a quasi-empirical science). Conjectures and also proofs are in Lakatos's view guesses, speculations. A proof is not a deduction from evidently true axioms, but a decomposition of a conjecture into subconjectures. Consequently proofs must also be subjected to severe tests.

The sequence of acts in "Proofs and Refutations" corresponds in some way to an actual historical development. When it is viewed in this way we are dealing with a rational reconstruction in my sense of the word. The reconstruction shows us quite a dramatic succession of versions of Euler's theorem for polyhedra, $V - E + F = 2$ (V is the number of vertices of a polyhedron, E represents the number of edges and F is the number of faces), each version being temporarily the best until it is refuted and has to be replaced by its successor. Each version includes a more or less clear definition of the notion of polyhedron. The succession of theorems is developed by an imaginary group of pupils with their teacher.

In order to be able to discuss the rational reconstruction, I shall first give a sketch of it. Of course the sketch is only a faint extract of the real dialogue. Slightly simplified, a major part of "Proofs and Refutations" can be depicted as in Figure 1.5. Here T_1, $T_{2'}$, $T_{2''}$, $T_{2'''} = T_2$, T_3, T_4, etc. are the different versions of the theorem that are being developed..The letters M_1, M_2, M_3 denote different methods of dealing with a refutation of a theory and of finding the next version.

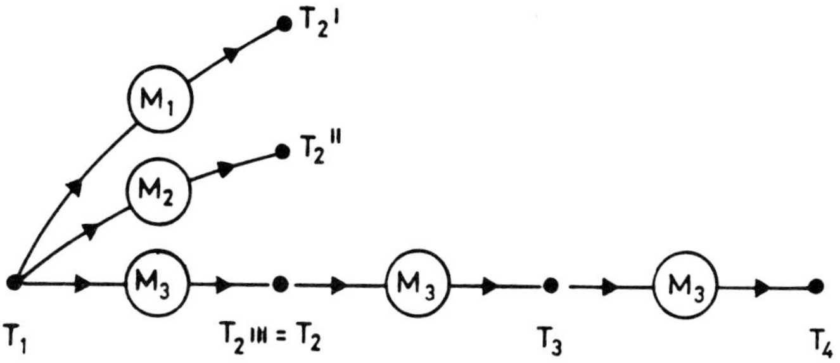

Figure 1.5

Theorem T_1 : For an arbitrary polyhedron V - E + F = 2 holds.
This theorem, given by the teacher, is the starting point. The teacher gives the following proof.

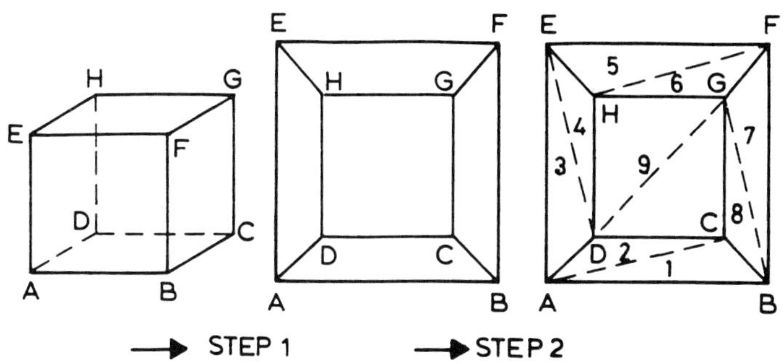

Figure 1.6 Face ABFE is cut out. The triangles are removed for example in the order 1, 2, 3, etc.

Proof: *Step 1*. One imagines the polyhedron to possess a surface of thin rubber. One face is cut out and the remaining surface is stretched out flat.
Step 2. By drawing diagonals in the polygons of the resulting network it is completely triangulated. The sum V - E + F, which was diminished by 1 in Step 1, does not change during this process of triangulation.
Step 3. The triangles are now removed one by one. During this process of removal, V - E + F does not change. The final result is *one* triangle, for which, of course, V - E + F = 1. This implies that for the original polyhedron V - E + F = 2. Q.E.D.
In "Proofs and Refutations" the three steps, or the three lemmas on which the proof is based, are illustrated by means of the example of a cube (Figure 1.6) Certainly from a modern point of view there is a bold generalization involved here: the three steps can be executed with the required result for a cube, but what about an arbitrary polyhedron?
Theorem T_1 is refuted by pupil Alpha by means of counterexamples, namely a pair of nested cubes (or a "hollow cube") and the "picture frame" (Figure 1.7). For the hollow cube we have: V - E + F = 4 and for the picture frame we have V - E + F = 0.

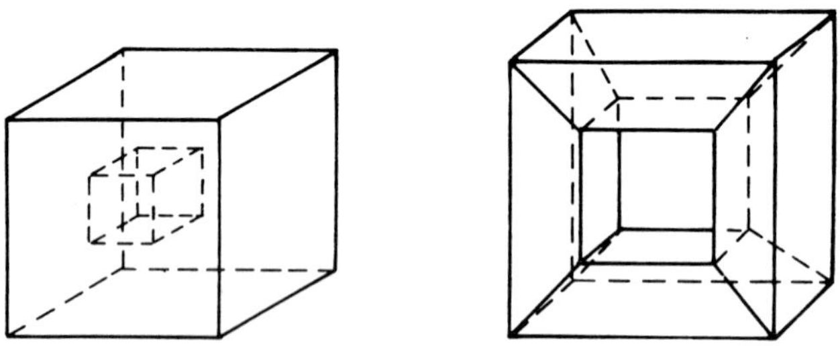

Figure 1.7 The "hollow cube" and the "picture frame"
([Lakatos,1976], p.13 and p.19)

A possible reaction is given by pupil Delta, who remarks:

"In the case of a genuine polyhedron, through any arbitrary point in space there will be at least one plane, whose cross-section with the polyhedron will consist of one single polygon" ([Lakatos,1976], p.21).

With this reaction Delta represents *the method of monster-barring*(method M_1 in my notation). By means of a suitable adjustment of the definition of a polyhedron, the counter-examples turn out to be monsters that do not belong, and were never meant to belong, to the set of objects the theorem refers to. The resulting version of the theorem, which only differs from Theorem T_1 in the sense that by means of Delta's proposal objects like the hollow cube and the picture frame are no longer considered to be genuine polyhedra, is in my notation T_2'.

The pupil Beta represents another possible reaction to counter-examples, namely *the method of exception-barring* (M_2 in my notation). With this method the definition of a polyhedron is not amended, but a condition excluding exceptional cases becomes part of the theorem. Application of this method yields:

Theorem T_2'': For all polyhedra that have no cavities (like the pair of nested cubes) and tunnels (like the picture-frame), V - E + F = 2 ([Lakatos,1976], p.26).

The teacher who objects to the ad hoc character of the methods M_1 and M_2 defends *the method of lemma-incorporation* (method M_3 in my notation). This method states that in the case of a counter-example that is both global (i.e. it refutes the theorem) and local (i.e. it refutes one of the steps (lemmas) in the proof), the domain of the theorem should be reduced to the domain on which the "guilty" lemma is still valid ([Lakatos,1976], p.34).

In the case of the hollow cube and the picture frame, the guilty lemma is the first lemma in the original proof of theorem T_1. Application of M_3 yields:

Definition 1: A *simple polyhedron* is a polyhedron, which is stretchable in the sense of step 1 of the proof of T_1.

Theorem T_2''' (= *Theorem* T_2): For all simple polyhedra the value of V-E+F is 2 ([Lakatos,1976], p.34).

Alpha refutes T_2 by means of another counter-example (Figure 1.8) where obviously V-E+F=3.

This time the guilty lemma in the proof is the lemma of step 2 of the proof. Application of M_3 yields:

Definition 2: A face of a polyhedron is by definition simply-connected if it falls into two pieces after drawing a diagonal.

Theorem T_3: For a simple polyhedron, with all its faces simply-connected, the value of V - E + F is 2 ([Lakatos,1976], p.36).

The development in the dialogue continues because the participants are not satisfied with the result thus far. A good theory should encompass more than simple polyhedra with simply-connected faces. It should, moreover, encompass on the one hand those polyhedra with an Euler characteristic (i.e. the value of V-E+F) different from 2.

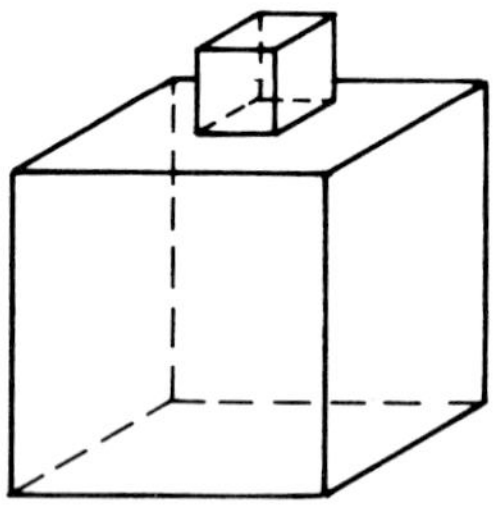

Figure 1.8

But it should also encompass polyhedra which are not simple or which possess faces that are not simply-connected and still have Euler characteristic 2.
The polyhedra that are not encompassed by the theory are even called *heuristic counter-examples* by Lakatos ([Lakatos,1946], p.83). In this way, by means of the introduction of such "heuristic refutation", he succeeds in depicting the discovery (or invention) of examples that are not *logical* counter-examples, but which nevertheless stimulate the development of further theory as a result of the application of his global methodological rule. They bring pupil Zeta to the development of a series of new versions of the theorem: T_4, T_5, etc. ([Lakatos,1976], pp.70-81). Zeta proceeds as follows. The formula for polygons V = E is his starting point. For one polygon we have: V-E+F=1. If we attach to a polygon another polygon, then the value of V-E+F does not change (Figure 1.9). We can do this in the way of Figure 1.9 (i), but also in the way of Figure 1.9 (ii). If we attach further polygons to the network obtained in this way, V-E+F does not change. We obtain a *closed* system of polygons by covering the *open* system of polygons obtained in this way by one, all-encompassing polygon whose border coincides with the outer border of the open system. If we blow up such a closed system of polygons like a balloon, we obtain a polyhedron. For such polyhedra we apparently have: V-

E+F=2. Zeta phrases the theorem that he has obtained in this way by "deductive guessing"[9] as follows:

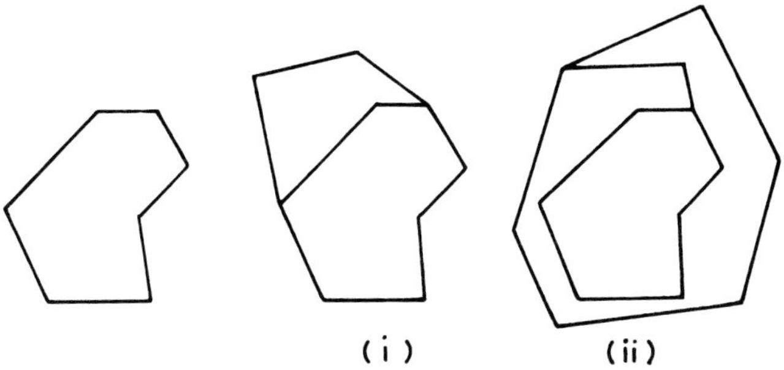

Figure 1.9

Definition 3: Polyhedra that can be obtained by means of the construction described above are called *closed normal polyhedra.*[10]
Theorem T_4 : For all closed normal polyhedra V-E+F=2. ([Lakatos,1976], p.76).
The same idea that led Zeta to construct the above networks of polygons brings him to "paste" polyhedra together. (Apparently he applies here Rule 1 from Section I.2.1). One of his results is the following:
Definition 4: A closed normal polyhedron is called *mono-spheroid* and a *(n+1)-spheroid* polyhedron is obtained by pasting a mono-spheroid polyhedron P_1 to a n-spheroid polyhedron P_n in such a way that two faces of P_1 fit exactly onto two corresponding faces of P_n. ([Lakatos,1976], pp.77-78).
Theorem T_5: For all n-spheroid polyhedra V-E+F = 2-2(n-1).
In this way Zeta succeeds in including picture frames etc. in his theory.

[9]The method is called deductive guessing because i) Zeta deduces new results from known results and ii) the "direction" of the deduction is based on guesses.
[10]A polyhedron is called normal by Lakatos if it can be obtained by means of Zeta's construction.

I.3.3 The real history

Lakatos's rational reconstruction, in particular the first part, is quite dramatic. It creates a sense of insecurity and raises the question as to how certain mathematical results are. That was, of course, Lakatos's intention: in his view mathematics is fallible and mathematical theories are essentially guesses. Mathematics develops by the methodology of proofs and refutations. In Section I.1.2, I pointed out that there are three ways to defend a particular methodology by means of rational constructions. One can show that the application of the methods involved leads to results, one can aim at recognition of the mathematical behaviour that is displayed, and finally, one can show that actual mathematicians (in the past) acted in accordance with the methodology. Lakatos never stated that "Proofs and Refutations" contained a factual rational reconstruction of an historical development in the sense in which I use those two words. Yet statements such as:

> "The dialogue form should reflect the dialectic of the story; it is meant to contain a sort of *rationally reconstructed or 'distilled' history. The real history will chime in the footnotes, most of which are to be taken, therefore, as an organic part of the essay*" ([Lakatos, 1976], p. 5, The emphasis is Lakatos's)

suggest a very intimate relationship between dialogue and actual history. Lakatos apparently thought that the dialogue captures an essential element of the actual historical development of Euler's theorem. The question is: If this is so, what is this essence?

What does the real history of Euler's formula look like? I shall give an outline in the next few pages. Then in I.3.4, I shall try to answer the question as to what this means for the interpretation of "Proofs and Refutations".

(1) Euler

In 1750 it occurred to Euler to try to determine the general properties of *solids bounded by plane faces,* i.e. polyhedra. He thought that it was obvious that general theorems could be found, analogous to the theorem that in a polygon the number of edges equals the number of angles. Having found the formula F+V=E+2 by trial and error, and having shown the correctness of the formula for a number of special polyhedra in a first paper ([Euler,1758a]), he gave a proof of the theorem in a second paper ([Euler, 1758b]). Euler consciously exploited the analogy with the theorem which states that in a polygon the number of edges equals the number of vertices. That theorem can be proved by cutting off triangles from the polygon until one triangle remains. Since the number of edges and the number of angles are diminished by the same number during the process of cutting off triangles, the theorem follows from the fact that it holds for a triangle. Euler proves the formula F+V=E+2 for polyhedra by cutting off triangular pyramids from an arbitrary polyhedron until one last triangular pyramid is left. The idea is as follows. Let O be a vertex and A, B, C, D, E, F the successive neighbouring vertices of O. We can then draw those edges of the (spatial) polygon ABCDEF, that are not edges of the original polyhedron, and also the diagonals of that polygon, for example all diagonals from C to the other vertices. (Figure 1.10).

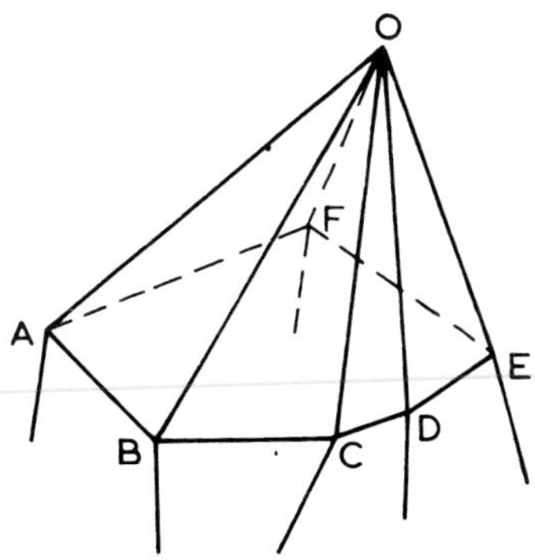

Figure 1.10

Euler cuts the polyhedron along the triangles ABC and AOC and removes the triangular pyramid OABC. Then he cuts along the triangles CAF and COF and he removes the triangular pyramid OCAF. During this process of removing all triangular pyramids with summit O the value of F+V-E does not change.[11] The process can be repeated with another vertex until finally one triangular pyramid is left for which F+V-E = 2.

[11]One should, of course, distinguish different cases. The counting procedure, for example, depends on whether or not an edge of the spatial polygon ABCDEF is also an edge of the original polyhedron. Although Euler's proof is not rigorous from a modern point of view, nineteenth century mathematicians were in general quite happy with it (Cf. [Brückner,1900], p. 59).

(2) Legendre

The second mathematician to deal with the formula was Legendre. In his *Élements de géometrie* ([Legendre,1794]), while treating spherical geometry, Legendre gave the following formula for spherical polygons :

$$\text{area} = \text{sum angles} - E.180^{o} + 360^{o}$$

where E is the number of edges of the polygon. Legendre applied the formula to polyhedra and gave a proof of the relation F+V=E+2 which was different from Euler's. He restricted himself explicitly to *convex* polyhedra. He wrote:

> "N.B. All polyhedra that we will consider are polyhedra with extending angles or *convex* polyhedra. That is what we call a polyhedron whose surface cannot be met in more than two points by a straight line".[12]

Considering a convex polyhedron located inside a sphere, Legendre projected it from a centre within the polyhedron onto the surface of the sphere. Using the formula for the area of a spherical polygon, the sum of the areas of all spherical polygons that the image consists of is then

$$V.360^{o} - 2E.180^{o} + F.360^{o}.$$

On the other hand, the sum of all the areas equals 720°, being the area of the sphere. The formula then follows immediately.

The examples in [Euler, 1758a] are all convex polyhedra. It is therefore reasonable to suppose that Euler only thought of convex solids bound by plane faces when he wrote [Euler,1758a] and [Euler,1758b] and that he did not imagine other polyhedra[13], whereas Legendre realized that non-convex counter-examples to his proof, and perhaps even to Euler's formula, might very well exist. Legendre then withdrew safely to convex polyhedra.

(3) L'Huilier

Presumably the many editions of Legendre's work made Euler's theorem well-known. Around 1810 Poinsot, L'Huillier, Cauchy and Gergonne dealt with the theorem in one way or another. Poinsot remarked on the one hand that Legendre's proof also applies to certain non-convex polyhedra (re-entrant solid angles are not a problem provided the projection does not cause an

---

[12]"N. B. Tous les polyèdres que nous considérerons sont des polyèdres à angles saillants ou polyèdres *convexes*. Nous appelons ainsi ceux dont la surface ne peut être rencontrée par une ligne droite en plus de deux points" ([Legendre, 1794], p.168).

[13]Federico wrote: "the context of the papers shows plainly that he was dealing with convex polyhedra; the etymology and the usage of terms in Euler's day would also indicate convex polyhedra" ([Federico,1982], p. 70).

overlap) and on the other hand pointed he out that not all polyhedra are Eulerian ([Poinsot, 1810]).[14] L'Huilier concentrated in particular on the exceptions. I will first consider his work. L'Huilier wrote in 1811:

> "I chiefly propose to show that Euler's theorem suffers from numerous exceptions and that it is only generally true for polyhedra which do not have re-entrant parts".[15]

First he gives in his paper a proof of Euler's formula for convex polyhedra which is different from the proofs of Euler and Legendre. Euler had written that he had vainly tried to prove the theorem by decomposing an arbitrary polyhedron into a set of pyramids with a common centre.[16] L'Huilier used that idea as follows:

- the formula holds for pyramids
- pyramids can be put together in different ways so that the formula holds for the resulting polyhedron
- an arbitrary convex polyhedron can be built out of pyramids in such a way that during the building process the formula continues to hold.

After having proved Euler's formula for convex polyhedra in this way[17], L'Huilier went on to what he considered to be the most important part of his paper - an enumeration of the exceptions that he had found. It is probable that these exceptions were in a sense generated by the proof[18], because L'Huilier

---

[14][Federico,1982], p. 70

[15]"Je me propose principalement de montrer que le théorème d'Euler souffre des exceptions nombreuses, et qu'il n'est vrai, d'une manière générale, que pour les polyèdres qui n'ont point de parties rentrentes" ([L'Huilier,1811]).

[16]Euler wrote: "This method of decomposing an arbitrary solid body into pyramids does not contribute enough for our present purpose" (hic modus solidum quodcunque in pyramides resolvendi ad praesens institutum parum confert" ([Euler,1758b], pp. 95-96).

[17]How did L'Huilier succeed in doing something rather simple where Euler failed? I do not know. It is possible that Euler would have considered L'Huilier's proof as being insufficiently rigorous and lacking in clarity. It is also possible that the occurrence of non-convex polyhedra in the course of L'Huilier's building procedure stopped Euler from taking that direction seriously.

[18]Lakatos seems to have missed this possibility. He categorically states that L'Huilier was led to the first kind of exception by certain crystals ([Lakatos, 1976], p.13, footnote 1). Lakatos refers to [Gergonne,1813], p.188 (Lakatos's text says p. 194. This is an error). But the reference does not justify such a categorical statement. Gergonne quotes L'Huilier as saying that he has seen examples of the first kind of exceptions in a crystallographical collection. That is all and L'Huilier's original paper says nothing more ([L'Huilier,1811], pp. 290-291).

first remarks that his proof is based upon the possibility of joining polyhedra together in such a way that two faces fit precisely onto each other and then adds:

> "Because of this supposition the proof is rigorous. I go on to the examination of the consequences of the supposition that the faces whose planes are joined do not coincide with each other".[19]

He then proceeds to give three types of exceptions:

(i) polyhedra obtained by pasting n Eulerian polyhedra onto one and the same Eulerian polyhedron in such a way that n ring-shaped faces are created ([L'Huilier,1811], pp. 286-289),.

(ii) polyhedra obtained by drilling holes through an Eulerian polyhedron ([L'Huilier, 1811], pp. 295-296),

(iii) polyhedra with n interior holes ([L'Huilier,1811], pp.300-301).

L'Huilier also gave formulae for these exceptional cases. For example, in case (i) we have the formula F+V=E+2+n and in case (iii) we have F+V=E+2(1+n).

So far the history is less dramatic than Lakatos's reconstruction. In the footnotes of "Proofs and Refutations", Lakatos does his best to interpret the history in accordance with his rational reconstruction.

When Poincaré remarks that "It is known that Euler proved that V-E+F=2 for *convex* polyhedra", this is for Lakatos a *falsification* of history, since "Euler of course stated his theorem for all polyhedra" ([Lakatos,1976], p.46, footnote 3). Lakatos suggests that Legendre only restricted himself to convex polyhedra "after having stumbled on exceptions never stated" ([Lakatos,1976], p.28, footnote 3).

As for L'Huilier, Lakatos only read Gergonne's abstract of L'Huilier's paper ([Gergonne,1813]) and he writes "L'Huilier and Gergonne seem to have been sure that L'Huilier had enumerated all the exceptions" ([Lakatos,1976], p.26, footnote 2). With respect to L'Huilier, however, there is no evidence for such a statement. Moreover, as we will see below, the attitude of Gergonne with respect to exceptions to a general statement was different from our modern attitude.

---

[19] "Conformément à cette supposition la démonstration est rigoureuse. Je passe à l'examen des conséquences de la supposition que les faces, dont les plans sont appliquées l'un à l'autre, ne coincident pas entre elles" ([L'Huilier,1811], p.287).

(4) Cauchy

Apparently unaware of L'Huilier's work in 1811, Cauchy generalized Euler's theorem by proving that if one decomposes a polyhedron, by choosing new vertices in the interior, into as many other polyhedra as one wants, one has

$$F+V=E+P+1$$

Here F, V, E and P represent respectively the total number of faces, vertices, edges and the total number of (non-overlapping) polyhedra in the resulting figure ([Cauchy, 1813a], p.77). For P=1 we have the original formula of Euler, which Cauchy proves directly in two different ways. Cauchy's first proof is the proof given in Lakatos's dialogue in the beginning by the teacher. Cauchy's second proof also shows up in the dialogue. It is the proof by pupil Zeta that we reproduced above ([Lakatos,1976], pp. 70-71).[20]

Cauchy gives two proofs of the generalized formula. His second proof is based upon the idea that the system of polyhedra can be successively reunited starting from an arbitrarily chosen polyhedron out of the system. For our purposes his first proof is more interesting because it actually yields another proof of the formula in its original form. The proof-idea is the following. In a polyhedron consisting of P polyhedra (N.B. P=1 corresponds to the original formula) Cauchy triangulates all faces. Then by connecting in all polyhedra one vertex with all other vertices of that polyhedron, Cauchy splits up the polyhedra into triangular pyramids, which are then removed successively beginning from the outside. This should be executed "in such a way that at the end only one [triangular pyramid - T.K.] remains, beginning with those that have triangles situated on the external surface of the given polyhedron, and only then removing those of which one or more faces will have been uncovered by previous eliminations".[21] In the course of this process the value of F+V remains equal to the value of E+P+1. The theorem then follows from the fact that it is true for a triangular pyramid.

Clearly the proof-ideas of the proofs of the generalization are based on analogy with the proofs that Cauchy gave of the original formula.

Concerning what he actually did prove with his proofs of the formula V-E+F=2, Cauchy expresses himself ambiguously. On the one hand he formulated Euler's theorem as follows: "In an *arbitrary polyhedron*, the sum of the

---

[20]Here Lakatos gave no historical reference.

[21]"de manière à n'en laisser subsister à la fin qu'une seule, en commençant par celles qui ont des triangles situés sur la surface extérieure du polyèdre donné, et n'enlevant dans la suite que celles dont une ou plusieurs faces auront été découvertes par des suppressions antérieures" ([Cauchy,1813a], p.83).

number of faces and the number of vertices surpasses the number of edges with two units" (Emphasis is mine - T.K.).[22] And although he was undoubtedly aware of the existence of non-convex polyhedra, he wrote about "the convex surface of a polyhedron" ([Cauchy, 1813a], p.81) as if in his view all polyhedra have a convex surface.

Lakatos noticed Cauchy's careless terminology and he explained it as follows: "This is the reason for Cauchy's casual use of 'convex': it was a failure to realize that concave polyhedra might give counter-examples, not a conscious effort to *eliminate* these counter-examples" ([Lakatos, 1976], p.87, footnote 1). Lakatos writes elsewhere about Cauchy's "inability to imagine a polyhedron that is not homeomorphic with the sphere" ([Lakatos,1976], p.12, footnote 1). There is, however, another possible explanation. Let us consider Gergonne's remarkable reaction to L'Huilier's paper. Gergonne gives an abstract of L'Huilier's paper in [Gergonne,1813], but adds various comments and some of his own contributions (e.g. another proof of the Euler-formula which I shall not be discussing). Gergonne noticed that L'Huilier's methods to build a *convex* polyhedron from pyramids "en passant" yield many *non-convex* polyhedra for which the Euler-formula holds. For example, L'Huilier proved that if two polyhedra for which the formula holds are stuck together along a pair of congruent faces, the formula also holds for the resulting polyhedron. This means that convex polyhedra can be used as building blocks to build many non-convex polyhedra for which the formula holds. L'Huilier had restricted Euler's theorem explicitly to convex polyhedra. Gergonne however thought that more could be said on the the basis of L'Huilier's results. In his paper he pointed out that the theorem which states that in an arbitrary non-convex *polygon* the sum of the angles equals (the number of edges - 2).180°, can be proved by decomposing the polygon into convex polygons for which the property is obvious. And he continued "by applying an analogous argument, one easily succeeds in convincing oneself that *Euler's Theorem is generally true for convex or non-convex polyhedra, save exceptions that will be discussed below*".[23] After sketching L'Huilier's proof of the theorem he discussed the three kinds of exceptions given by L'Huilier and wrote that they

[22]"Dans un polyèdre quelconque, la somme faite du nombre des faces et de celui des sommets, surpasse de deux unités le nombre des arêtes" ([Cauchy, 1813b], p.94).

[23]"en appliquant [...] un raisonnement analogue[...] on parviendra aisément à se convaincre que le *Théorème d'Euler* est vrai généralement, pour les polyèdres convexes ou non convexes, sauf les exceptions dont il sera parlé ci-après ([Gergonne,1813], p.177).

"appear to be the only ones Euler's theorem could be subject to".[24] Actually Gergonne's paper contains a nice illustration of what, for him, a general proof did prove. His proof of the theorem that in an arbitrary *polygon* the sum of the angles equals (the number of edges - 2).180° is preceded by the sentence: "Here is how one can obtain a proof which is *general and completely independent of the nature of the polygon*".[25] Immediately after this general proof, allegedly completely independent of the nature of the polygon, Gergonne tells us that theorem and proof do not hold for a "*ring-shaped polygon or polygonal crown*; in other words, for the plane space enclosed by two polygons of which one is inside the other".[26] I suspect that if someone had come up with another exception to Euler's theorem, Gergonne would have simply added it to his list of exceptions. From my point of view Cauchy's careless terminology and also Gergonne's reasoning is related to certain characteristic features of, say, 18th century mathematics, which only began to disappear in the second half of the 19th century. As Engelsman has pointed out ([Engelsman,1984], pp.12-13), 18th century mathematicians were quite happy to live with paradoxes. They had a global attitude towards mathematics, i.e. they were often satisfied with a proof of "global validity" and were not bothered by a few exceptions.

(5) Von Staudt

In the first half of the 19th century several other mathematicians dealt with Euler's formula. Steiner and Von Staudt, for example, gave different proofs ([Steiner,1826], [Staudt, 1847]) and, independently of L'Huilier, Hessel came up with exceptions.[27]

The proof given by Von Staudt in his *Geometrie der Lage* ([Staudt, 1847]) is remarkable because it is strictly topological. Von Staudt considers solids bounded by plane faces. He states that in such solids each edge separates two

---

[24]"paraissent être les seules auxquelles le théorème d'Euler puisse être sujet" ([Gergonne, 1813], p.188).

[25]"Voici comment on en peut obtenir une *démonstration générale, et tout à fait indépendente de la nature du polygone*" ([Gergonne, 1813], p.174).

[26]"*polygone annulaire ou couronne polygonale*, c'est à dire, à l'espace plan compris entre deux polygones décrits l'un dans l'autre" ([Gergonne, 1813], p.176).

[27]In 1832 Hessel, independently of L'Huilier, published the three kinds of exceptions ([Hessel,1832]). However, Hessel's paper also contains exceptions that L'Huilier had not given: e.g. two polyhedra with precisely one edge in common, or two polyhedra with one vertex in common.

faces and in each vertex three or more faces meet.[28] He phrases and proves the theorem as follows:

Theorem[29]: If each vertex of a polyhedron can be connected with any other vertex by an edge, or a line consisting of edges, and if moreover each closed line consisting of edges and not intersecting itself divides the surface of the solid into two parts, then the number E of vertices plus the number F of faces equals the number K of edges plus two.

Proof[30]: (i)If the solid has E vertices, then E-1 edges (of which the first edge connects two vertices, the second edge one of those two with a third, the third edge one of the previous edges with a fourth, etc.) are enough to be able to pass from one vertex to any other vertex. (ii) Since there is no closed line in such a system of edges, and since each of the other (free) edges forms a closed line with two or more edges of the system, the other edges are sufficient but also necessary to be able to pass, by crossing only these edges, from each of the F faces of the solid to each of the other faces. (iii) From which one can draw the conclusion that the number of other edges is F-1, the number of all edges is E+F-2 and consequently E+F=K+2.

The first part of the proof (i) actually consists of the theorem that a connected

---

[28]"Ein Körper, welcher von ebenen Flächen eingeschlossen ist, wird ebenflächig und nach der Anzahl n seiner Flächen, welche sämmtlich geradlinige Figuren sind, ein n Flach benannt. In jeder Kante eines Polyeders (ebenflächigen Körpers) stoßen zwei seiner Flächen, in jedem Eckpunkte aber drei oder mehrere Flächen (eine Raumecke bildend) und eben so viele Kanten zusammen" ([Staudt, 1847], p.19).

[29]"Wenn jeder Eckpunt eines Polyeders mit jedem andern durch eine Kante oder eine aus Kanten zusammengesetzte Linie verbunden werden kann und seine Oberfläche durch jede aus Kanten zusammengesetzte geschlossenen Linie, welche nicht öfter als einmal durch einen und denselben Punkt geht, in zwei Theile getheilt wird, so ist die Anzahl E der Eckpunkte mehr der Anzahl F der Flächen gleich der Anzahl K der Kanten mehr zwei" ([Staudt,1847], p.20).

[30]"Wenn nämlich der Körper E Eckpunkte hat, so sind E-1 Kanten, von welchen die erste zwei Eckpunkte unter sich, die zweite einen derselben mit einem dritten, die dritte einen der vorigen mit einem vierten u.s.w. verbindet, hinreichend um von jedem Eckpunkte auf jeden andern übergehen zu können. Da nun in einem solchen Systeme von Kanten keine geschlossene Linie enthalten ist, jede der übrigen (noch freien) Kanten aber mit zwei oder mehrern Kanten des Systems eine geschlossene Linie bildet, so sind die übrigen Kanten hinreichend aber auch alle erforderlich, um durch sie von jeder der F Flächen des Körpers auf jede andere übergegehen zu können, woraus man schließen kann, daß die Anzahl der übrigen Kanten F-1, mithin die Anzahl aller Kanten E+F-2 und demnach E+F=K+2 sey" ([Staudt,1847]. pp. 20-21).

graph contains a spanning tree. During the construction of the spanning tree the existence of the next connecting edge is guaranteed by the fact that the graph is connected. The argument of (ii) and (iii) runs as follows. An edge, which is still free after the construction of the spanning tree, connects two vertices that are also connected by a path of the spanning tree (given the construction of that tree). This path, together with the edge, obviously forms a closed line, which implies that all free edges are sufficient and necessary in order to be able to move from one face to any other face via free edges. Then the conclusion that the number of remaining edges is F-1 can be drawn. After all, by removing free edges, we can systematically unite the faces of the polyhedron: by the removal of one free edge we unite two faces, and by the removal of a free edge that belongs to the boundary of the two united faces, we obtain three united faces, etc. After the removal of all free edges we apparently have united all faces. Because there are F faces, this takes F-1 steps. So there are F-1 free edges.[31]

Lakatos seems to have missed von Staudt's proof in his historical search, since he does not mention it. It is quite possible that Staudt was acquainted with the exceptions to Euler's theorem via Hessel's or via L'Huilier's work, and that he hit upon the two conditions in his theorem after careful consideration of those exceptions. In that case one could think of such conditions and of the whole proof as being in a sense *counter-example generated.*

(6) Riemann

In his inaugural dissertation entitled *Grundlagen für eine allgemeine Theorie der Funktionen einer veränderlichen complexen Größe* ([Riemann, 1851]), Riemann introduced the notions "simply-connected", "multiply-connected" and "degree of connectivity". I will only indicate here the context in which the notions appeared. If $f(z)$ is a holomorphic function on the domain A in the complex plane bounded by the contour b of A, we have in the case of Figure.1.11 - for a closed contour C inside A the theorem

$$\int_C f(z)=0$$

This theorem is usually attributed to Cauchy. However, the theorem ceases to hold if the contour encloses a singularity, a point where the function is not holomorphic. In such cases we can isolate the circularity with a small circle b_1

[31]Mulder gives an similar analysis of Von Staudt's proof in a paper on the history of the notion of tree ([Mulder, 1988]).

(Figure 1.11). We then get the domain A_1, which is the area between the contours b and b_1. However, in order to exclude contours that enclose the singularity, it is necessary that we also cut A_1 by means of a cross-cut connecting b and b_1 (the dotted line in Figure 1.11). The theorem is then applicable because by the cross-cut domain A_1 has become "simply-connected". In this context Riemann gives the following definition: "A connected surface is called simply-connected if it falls into two pieces by any cross-cut; otherwise it is termed multiply-connected".[32]
In the situation of domain A_2 (Figure 1.11), where b_1 and b_2 enclose singularities, we need two cuts before the theorem can be applied.

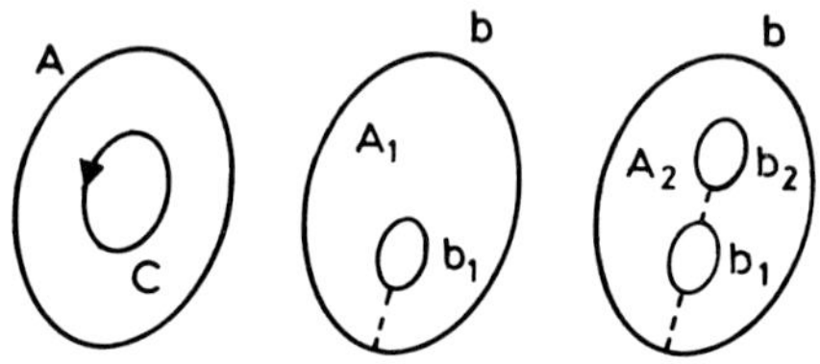

Figure 1.11

The orders of connectivity of the domains A_1 and A_2 are respectively 1 and 2 because 1 and 2 cuts are necessary before the domain becomes simply-connected. I shall not further discuss Riemann's notion of "order of connectivity" here. In Chapter V, I will briefly return to Riemann's work on the theory of functions of a complex variable.

(7) Listing, Jordan, Becker
In the 1860s, three mathematicians independently applied the notions introduced by Riemann in the theory of polyhedra: they were Listing, Jordan and Becker. The theorem that is called T_3 in my sketch of Lakatos's reconstruction - "For a simple polyhedron, with all its faces simply-connected, V-

[32]"Eine zusammenhängende Fläche heißt, wenn sie durch jeden Querschnitt in Stücke zerfällt, eine einfach zusammenhängende, andernfalls eine mehrfach zusammenhängende" ([Riemann, 1851], p.9).

E+F=2" - occurs in their work. Their work also encompasses L'Huilier's and Hessel's exceptions. These three mathematicians in fact succeeded in solving the problem which Becker in 1862 had expressed as follows:

"But also the far better known theorem of Euler occurs in most text-books in a way that gives little satisfaction. In particular, one does not learn anywhere *for which polyhedra it holds, and for which it does not hold*".[33]

I shall not discuss the work of Listing, Jordan and Becker in detail. I shall restrict myself to one remark about the man to whom the priority for the application of the notion of connectivity to polyhedra belongs - Listing. His theory is very general and deals with arbitrary *spatial complexes*, as he calls them, i.e. "arbitrary aggregates of points, lines (straight or curved) and surfaces (plane or curved)".[34] Listing's generalization of Euler's theorem encompasses all earlier results that we discussed. The notion of order of connectivity of constituents is of crucial importance.

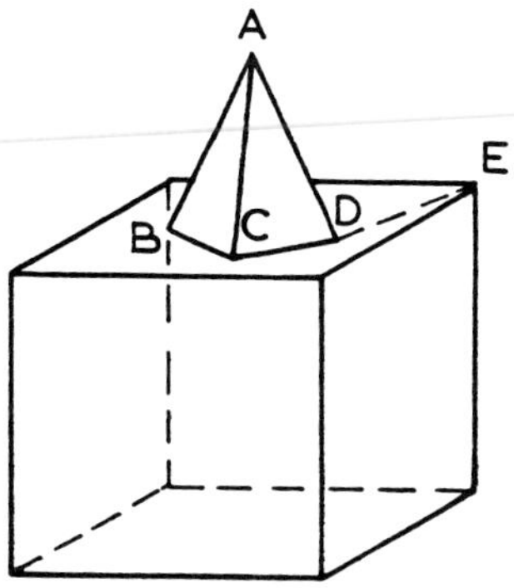

Figure 1.12

In this context also the work of Matthiesen and De Jonquières deserves to be mentioned. In 1863 Matthiesen showed that L'Huilier's exceptions could be turned easily into decent examples by also taking into account certain "hidden faces and edges" ([Matthiesen, 1863], p.450).

33"Aber auch der viel bekanntere *Euler*'sche Satz tritt in den meisten Lehrbüchern in einer Weise auf, die wenig befriedigt. Namentlich erhält man nirgend bestimmten Aufschluss, *für welche Polyeder er gilt, für welche nicht*" ([Becker, 1862], p.346).

34"beliebige Aggregate von Punkten, Linien (gerade oder krumm) und Flächen (eben oder gekrümmt)" ([Listing,1862], p.3).

In the case of a polyhedron like the one of Figure 1.12 consisting of a pyramid ABCD on top of a cube, where Euler's formula does not hold because of the presence of a ringshaped face, one necessarily must assume according to Matthiesen the existence of an edge DE which turns the ringshaped face into a "true polygon" ([Matthiesen, 1863], p.449). In 1890 De Jonquières came up with similar ideas ([Jonquières,1890]). Although these reactions are interesting from a methodological point of view (and as a result they play a rather important role in Lakatos's dialogue, ([Lakatos, 1976], pp.38-39), they were not taken very seriously by the mathematical community, undoubtedly because of their ad hoc character. This is apparent from the fact that surveys such as [Brückner,1900] and [Steinitz,1916] hardly pay any attention to them.

I.3.4 Evaluation in terms of the rationality thesis

There is some resemblance between Lakatos's reconstruction concerning the formula of Euler for polyhedra and the real history. In the reconstruction, and also in the real history, the introduction of the notion of connectivity is the crucial step forward in the developments. Yet the rational construction deviates considerably from the chronological order in which things actually happened. For example, although Euler published a proof of the conjecture in 1758, the dialogue starts with a proof given by Cauchy in 1813. And although monster-barrer de Jonquières only published a paper on Euler's formula in 1890, his work occurs immediately after Cauchy's in the dialogue. Legendre's 1794 proof, which obviously preceded Cauchy's work, is given somewhere in the middle of the dialogue, and the second of Cauchy's 1813 proofs only appears at the end of the dialogue. In other words, as far as chronology is concerned, there is not much in common between dialogue and real history. There is no doubt that "Proofs and Refutations" contains a highly counterfactual rational reconstruction.[35]

What then is the status of Lakatos's rational reconstruction? The deviations from the chronology are numerous. But there is more. Lakatos compresses a development, which lasted for over a hundred years, into a dialogue that could take place, in principle, in a few days. This leads to an interpretation of the primary sources which is anachronistic in the sense that, from a historio-

---

[35]A statement like "On trouvera une histoire des démonstrations de la formule d'Euler (accompagnés de considerations méthodologiques et philosophiques) dans Imre Lakatos, Proofs and Refutations [...]" ([Dieudonné, 1978], Vol. 2, p.266) points at a serious misunderstanding.

graphical point of view, the context from which the historical material is drawn is taken insufficiently into account. I have previously given several instances where this happens, but the main problem in this respect is that the account in "Proofs and Refutations" represents and tries to combine two contradicting methodological attitudes. It is to a certain extent based upon the critical methodology that came to dominate mathematics in the course of the 19th century. It is also based upon a less critical methodology that quickly presents generalizations and speculations as theorems in a way which is alien to the critical methodology. The dramatic effect is caused by the combination of those two contradictory elements. It is possible to rewrite the dialogue and to replace all bold generalizations by conjectures in the sense in which this word is usually now used in mathematics. The dramatic effect is then immediately gone and we are left with an essay in heuristics. This is actually the most fruitful way to read Lakatos's reconstruction of the history of Euler's theorem. One should forget about the actual history. "Proofs and Refutations" must be seen as a rational construction that illustrates the rationality of mathematics on the micro level. As a rational construction it is convincing mainly because it shows recognizable mathematical behaviour, but it is considerably less convincing in its relation with history.

Undoubtedly in mathematics also the pattern "theory - criticism - new theory - criticism - new theory etc." is a major pattern of progress. And undoubtedly on the micro level a theorem and its proof are repeatedly daring and the criticism repeatedly consists in logical refutation. "Proofs and Refutations" illustrates this very well. However, if on the macro level we consider the long range, the criticism does not usually consist of logical refutation. As far as long-range developments are concerned, "Proofs and Refutations" is less convincing.

In his posthumously published paper "The deductivist approach versus the heuristic approach"[36] Lakatos writes:

> "But mathematical activity produces mathematics. Mathematics, this product of human activity, 'alienates itself' from the human activity which has been producing it. It becomes a living, growing organism, that *acquires a certain autonomy* from the activity which has produced it; it develops its own autonomous laws of growth, its own dialectic. The genuine creative mathematician is just a personification, an incarnation of these laws which

---

[36]Published in [Lakatos,1976] (pp. 142-154). The text is based on a part of chapter 3 of Lakatos 1961 Cambridge Ph. D. thesis.

can only realize themselves in human action. Their incarnation, however, is rarely perfect. The activity of human mathematicians, as it appears in history, is only a fumbling realization of the wonderful dialectic of mathematical ideas. But any mathematician, if he has talent, spark, genius, communicates with, feels the sweep of, and obeys this dialectic of ideas. Now heuristic is concerned with the autonomous dialectic of mathematics and not with its history, though it can study its subject only through the study of history and through the rational reconstruction of history" ([Lakatos,1976], p. 146).

Lakatos clearly distinguishes a rational reconstruction of a series of mathematical acts from the real history and he would definitely have argued that the counter-factuality of the reconstruction in "Proofs and Refutations" merely reflects the difference between the "autonomous dialectic of mathematics" and its "fumbling realization" in history. Unfortunately, as far as I know, in the period of his work dominated by the MP&R, Lakatos never came up with a criterion to distinguish in real history the facts reflecting the "autonomous dialectic" from the facts reflecting the "fumbling" character of its realization. His "History of science and its rational reconstructions" of 1971 is dedicated entirely to the relationship between notions of rationality and actual history, but it concerns primarily the methodology of scientific research programmes and does not refer to the MP&R. Its core idea is, however, that the success of a methodology is proportional to its success in explaining the rationality of the history of science.[37] Application of this criterion to the MP&R would mean that one could argue as follows. If the rationality of mathematics were restricted to deductive logic, most of the history of mathematics would be irrational. "Proofs and Refutations" reveals rational patterns of mathematical behaviour all of which occurred or could have occurred in one way or another in history and thus, according to Lakatos's criterion of 1971, the MP&R is a better theory of rationality. However, in order to draw this conclusion one does not need to refer to Lakatos's later publications on the relation of history and its rational reconstructions. In my view, as an attempt to show the rationality of mathematics, "Proofs and Refutations" is mainly convincing because it shows recognizable mathematical behaviour and not so much because it reflects the real history of Euler's formula for polyhedra. When taken as a rational construction showing recognizable mathematical

---

[37]The criterion already occurs in Lakatos "On Popperian Historiography" ([Lakatos, 1978b], pp. 201-210) which was presumably written in 1966 or 1967.

behaviour, "Proofs and Refutations" supports the rationality thesis. We then read "Proofs and Refutations" as an essay in methodology, i.e. not as a rational re-construction but only as a rational construction which in different ways illustrates the methodological behaviour of mathematicians.
This conclusion in no way diminishes the value of "Proofs and Refutation". The dialogue wonderfully shows the importance of falsification (in a broad sense of the word) for the growth of mathematical knowledge and it shows many different possible rational reactions to logical falsification or "heuristic" falsification. In this respect Lakatos beyond any doubt made his point: the dialogue reveals a rich situational logic for the working mathematician which is not recognized by formalist philosophy.[38]

I.3.5 Strong fallibility or weak fallibility?

Lakatos's work on the philosophy of mathematics is primarily methodological and the emphasis is very much on the rationality thesis. The fallibility thesis is only defended indirectly. Questions concerning the "degree of fallibility" of mathematical knowledge are not explicitly taken into consideration by Lakatos. Yet such questions are important. If mathematical knowledge is fallible, how fallible is it? In this context I propose to distinguish a *strong fallibility thesis* from a *weak fallibility thesis* with respect to mathematics. These theses concern mathematical theories, and in my definition a mathematical theory consists of at least an (implicit) definition of a domain of investigation, and at least one theorem and a proof of that theorem. The distinction should serve to create clarity with respect to the degree of fallibility of mathematical theories.

The strong fallibility thesis consists in the view that, in spite of the fact that mathematics differs from natural science, at least at first sight, it is fallible to an extent comparable to the fallibility of natural science in Popper's view. *Strong fallibility implies that in principle a whole theory can be refuted. Strong fallibility implies that the impressive continuity in the development of mathematics and the accumulation of "truths" that seems to take place in mathematics could be accidental.*

The weak fallibility thesis is far more moderate. According to this thesis mathematical theories are considered to be fallible but they are not considered to be refutable in their totality. The bulk of the theorems of an older theory is always taken over by newer theories concerning the same subject matter.

---

[38][Lakatos,1976], p.4

Older theories are repeatedly refuted in the sense that from a new perspective, old definitions appear to be one-sided; originally unintended interpretations of the definitions yield counter-examples to the older theories. Nevertheless the resulting changes can usually be characterized as "a refinement", "a qualification" etc., and hardly as "a replacement of a refuted theory by a different theory". *From the point of view of the refined theory the older theory looks imperfect and one-sided, but never totally wrong. Weak fallibility implies that the continuity of mathematics and the accumulation of truths in mathematics are not accidental but are due to the nature of mathematics.*

In his "A Renaissance of Empiricism in the Recent Philosophy of Mathematics" ([Lakatos,1978b], pp. 24-42), which I will discuss more extensively in Chapter II, Lakatos writes about "a radical assimilation of mathematics to science" ([Lakatos,1978b], p.24), where science is considered from a Popperian perspective. When he wrote this, Lakatos may only have thought in methodological terms. However, the phrase suggests that mathematical theories might be fallible to an extent comparable to theories in natural science. Also Lakatos's characterization of a quasi-empirical science in terms of "bold speculations" and "daring solutions" followed by refutations suggests more than weak fallibility.

Let us consider both Lakatos's rational construction and the real history of Euler's formula from this point of view. The reconstruction in "Proofs and Refutations" clearly at most supports a weak fallibility thesis. What the process described by Lakatos shows is 'refinement' or 'qualification' of previous notions and theorems rather than strong fallibility, i.e. the complete rejection of results accepted earlier. If we contrast the concepts of polyhedron and of the faces of a polyhedron in theory T_1 at the beginning of the rational construction with their successors in the course of the construction, we see that the notions change. For example, in T_4, distinctions are made by means of the notion of connectivity which are not made in T_1. Here we have different forms of a process of conceptual change that occurs frequently in mathematics, which I will call *conceptual refinement by dissociation.*[39] Let C be a concept and let C be considered to be applicable to a particular set of situations Ext(C), the "extension" of C. There are at least two ways in which concept C can be refined by dissociation.

---

[39] The notion of "dissociation" is also used by Lautman with approximately the same meaning ([Lautman,1937a], p.293). Cf. also[Fine,1978].

I) Refinement by dissociation type I: New concepts C_1, C_2 ...C_n are introduced such that $Ext(C) = Ext(C_1) \cup Ext(C_2) .... \cup Ext(C_n)$. This occurs when one discovers, while studying a group of objects, that there are different types. For example, while studying triangles, one discovers that there are three kinds of triangles - obtuse, right and acute triangles. Or, when studying singularities of a function of a complex variable, one discovers that essential singularities have to be distinguished from non-essential singularities.
II) Refinement by dissociation type II: This type of dissociation has the following form. In a situation where one has concept C, new concepts C_1, C_2,..C_n are being introduced such that $Ext(C) = Ext(C_1) \cap Ext(C_2) .... \cap Ext(C_n)$. Obviously this type of refinement by dissociation is closely related to generalization - a notion becomes the conjunction of more general notions. Some examples of conceptual developments that exhibit this pattern are the following. Prior to the 19th century the geometrical notions "unbounded" and "infinite" were synonymous. Let C denote these synonymous notions. In the course of the 19th century C_1 = "unbounded" was distinguished from C_2 = "infinite". Or consider the fact that one can define the notion of absolute value in terms of the notion of order on an ordered field:

$$|x| = x \text{ iff } x>0 \text{ or } x=0, \text{ or else } |x|=-x,$$

and subsequently the notion of a closed field by means of the notion of absolute value. In these definitions closed fields are necessarily ordered. It is possible to dissociate the notions of order and absolute value and to define the notion of absolute value for a field that is not ordered. The absolute value then becomes a mapping from the field into an ordered field satisfying some axioms.[40]

Refinement by dissociation may occur in situations where all notions involved are precisely defined. In Lakatos's rational construction, however, counter-examples reveal that the notion polyhedron has *possible unintended interpretations* and that the border of its extension is obscure. This means that the process of refinement by dissociation is at the same time a process of defining certain notions more precisely.

What happens in the rational construction from a weak fallibilist point of view is the following. Since the possible interpretations that are revealed by the counter-examples are unintended, we have refutations only in a formal sense of the word. All methodological reactions to the counter-examples are forms of refinement by dissociation of type I, aimed at separating the Eulerian

---

[40] I owe this example to Lautman ([Lautman,1937a]).

polyhedra from the others. Monster-barring and exception-barring are, however, methodologically inferior reactions because they are obviously ad hoc. The method of proof-analysis followed by lemma incorporation is an attempt to clarify the situation concerning the obscure border. It is not an ad-hoc reaction: it is an attempt to understand why the counter examples are counter examples. By means of the right dissociations, the proof of theorem T_1 is saved in a non ad-hoc way.

The real history of Euler's theorem also supports at most a weak fallibility thesis. Consider for example the following sketch of the history of Euler's theorem during the second half of the 18th and the first half of the 19th century given by Steinitz:

> "At first interest concentrated only on *Euler*'s theorem which people attempted to prove in a simple and rigorous way. It could not remain unnoticed that the theorem does not hold completely generally. Later on attempts were made in two directions. On the one hand, the point was to phrase the conditions for validity in a precise way; and on the other hand, the problem is to replace the formula by a more general one without adding special conditions. With respect to the first point a restriction to convex polyhedra was unnecessarily severe. However, it was the easiest way out as long as the notion of *connectivity* was not understood clearly. The condition of convexity is explicitly given in many of the later proofs, as with Cauchy".[41]

This sketch, which at most supports the weak fallibility thesis, is much more factual than Lakatos's rational reconstruction in "Proofs and Refutations". With reference to this passage Lakatos writes:

> "After saving Cauchy by claiming that he 'could not possibly miss' polyhedra which are not simple and that therefore he 'categorically' restricted the

---

[41] "Zunächst konzentrierte sich das Interesse lediglich auf den *Euler*schen Satz (3) [i.e. the formula V-E+F=2 - T.K.], den man sich bemühte, mit einfachen und strengeren Beweisen zu versehen. Denn es konnte nicht unbemerkt bleiben, daß der Satz nicht in voller Allgemeinheit gilt. Hiernach gehen die Bestrebungen in zwei Richtungen. Einmal handelt es sich darum, die Bedingungen für die Gültigkeit genau zu formulieren; zweitens bei Vermeidung besonderer Voraussetzungen die Formel (3) durch eine allgemeinere zu erzetzen. Was den ersten Punkt betrifft, so bedeutete es zwar eine unnötig weitgehende Beschränkung, wenn man nur konvexe Polyeder in Betracht zog. Es war das aber der einfachste Ausweg, solange der Begriff des *Flächenzusammenhanges* nicht klar erfaßt war. Die Voraussetzung der Konvexität wird denn auch in vielen der späteren Beweise, wie bei A. Cauchy, ausdrücklich angegeben" ([Steinitz, 1916], p.20).

> theorem to the domain of convex polyhedra, the respectful historian now has to explain why Cauchy's borderline was 'unnecessarily' narrow. Why did he ignore non-convex Eulerian polyhedra? Steinitz' explanation is this: the correct formulation of the Euler - formula is in terms of connectivity of surfaces. Since in Cauchy's period this concept was not yet 'clearly' grasped, the 'simplest way out' was to assume convexity [...]. So Steinitz explains away a mistake that Cauchy never made" ([Lakatos, 1976], p.87, footnote 1).

Lakatos interprets the whole passage as referring to Cauchy, while Steinitz writes about the mathematical community and only mentions Cauchy in his last sentence as an example of a mathematician who restricted himself to convex polyhedra. In Lakatos's view, Steinitz is one of those "respectful historians [who] because they do not understand that concepts grow, create for themselves a maze of problems" ([Lakatos, 1976], p.87, footnote 1). Lakatos's reading of Steinitz is amazing and one suspects that Lakatos might have been willing to defend the strong fallibility of mathematics. After all, if concepts grow, they could change over time beyond recognition. The view that concepts are grasped more clearly over a period of time does not exclude weak fallibility, but it is incompatible with the strong fallibility thesis. It has a Platonist ring and excludes radical changes in the development of concepts.

The final section of "Proofs and Refutations" is interesting in this respect. It does not primarily concern the reconstruction of history but is devoted to a discussion about concept formation. Lakatos introduces the term *concept-stretching* to denote what happens when unintended interpretations of a notion show up: an old concept disappears and is replaced by a new concept of the same name with a greater extension. An important theme in the discussion is whether the process of concept-stretching ever comes an end. Kappa argues that there are no "inelastic, exact, concepts". According to him, as long as propositions mean anything the process of refutation by concept-stretching can continue forever. The participants in the discussion do not really succeed in refuting Kappa's views. Lakatos, however, remarked in a footnote as follows:

> "Nineteenth-century mathematical criticism stretched more and more concepts, and shifted the meaning-load of more and more terms onto the *logical form* of the propositions and onto the meaning of the few (as yet) unstretched terms. In the 1930s this process seemed to slow down and the demarcation line between unstretchable ('logical') terms and stretchable ('descriptive') terms seemed to become stable. A list, containing a small

number of logical terms, came to be widely agreed upon, so that a general definition of logical truth became possible [...]" ([Lakatos,1976], pp. 103-104)

Indeed, established modern mathematical theories can usually be considered as logical theories and the mathematical notions involved are defined by systems of axioms.[42] Moreover, through their abstract nature the notions involved are often so general in comparison with their nineteenth century predecessors of the same name that one could use the word "concept-stretching" for the transition involved. Yet, in spite of the fact that the new abstract definitions allow new and originally unintended interpretations, the actual history shows remarkably few refutations of earlier theorems and remarkably few situations where the concept-stretching took place along the lines of Lakatos's reconstruction of the history of Euler's theorem. It is much more appropriate to characterize the transition involved as a process of abstraction whereby older notions were replaced by more general, more abstract notions, and the earlier results were taken over without logical refutation by the newer theories.

I.4 Uniform convergence: a second Lakatosian reconstruction on the basis of the methodology of proofs and refutations

I.4.1 A survey of the relevant events

I now turn to Lakatos's second rational reconstruction on the basis of the methodology of proofs and refutations that Lakatos gave. It concerns the history of the notion of uniform convergence and appeared posthumously in 1976 under the title "Another Case-Study in the Method of Proofs and Refutations" ([Lakatos,1976], pp. 127-141). I will also in this case compare the reconstruction with the actual history. I shall turn things around here and give first a chronological survey of the main events and then Lakatos's reconstruction.[43]

(i) In 1787 L'Huilier wrote: "if a variable quantity at all stages enjoys a certain property, its limit will enjoy the same property" (quoted by Boyer in [Boyer,1959], p.256). He expressed here *Leibniz' law of continuity*, "the rules

---

[42]Cf. footnote 2

[43]When discussing Lakatos's reconstruction I restrict myself to the events that are important. I do not intend to give a complete survey of the history of the notion of uniform convergence. For more details I refer to [Spalt,1981].

of the finite also succeed in the infinite" ([Leibniz,1971], p.93)), a notion that still persisted in the 19th century.

(ii) In 1807 Fourier presented a treatise on heat diffusion to the Institut de France. The treatise contains important mathematical results. He came up with Fourier series with a discontinuous limit function (in the modern sense of the word "discontinuous"). He writes, for example:

"One concludes from the preceding that the function of u expressed by

$$\cos u - \frac{1}{3}\cos 3u + \frac{1}{5}\cos 5u - \text{etc.}$$

is such (1) that the second member has the value $\pi/4$ whatever the arc u is, provided that it is included between $-\pi/2$ and $\pi/2$; (2) that the second member is zero when the arc u equals $\pi/2$; (3) that the second member has the value $-\pi/4$ whenever the arc u has an arbitrary value included between $\pi/2$ and $3\pi/2$".[44]

(iii) In 1821 Cauchy in his *Cours d'analyse* presented the following theorem (concerning a series (I): $u_0 + u_1 + u_2 + ...$):

"Theorem I - Whenever the different terms of the series (I) are functions of one and the same variable x, continuous with respect to this variable in the neighbourhood of a particular value for which the series is convergent, the sum s of the series is also a continuous function of x in the neighbourhood of that particular value".[45]

It is obvious that this theorem, which is completely in accordance with Leibniz' principle of continuity, is contradicted by Fourier's results, since in the example above we have a convergent series of continuous fonctions converging to a discontinuous limit.

(iv) In 1826, in a famous treatise on the binomial series, Abel gave this theorem:

"Theorem V. Let $v_0 + v_1\,\delta + v_2\,\delta^2 + ...$ be a convergent sequence, in which $v_0, v_1, v_2, ...$ are continuous functions of one and the same variable quantity

---

[44] "On conclut de ce qui précède que la fonction de u exprimée par $\cos u - \frac{1}{3}\cos 3u + \frac{1}{5}\cos 5u$ - etc. est telle (1) que le second membre a pour valeur $\pi/4$, quel que soit l'arc u, pourvu qu'il soit compris entre $-(\pi/2)$ et $\pi/2$; (2) que ce second membre est nul lorsque l'arc u est égal à $\pi/2$; (3) que ce second membre a pour valeur $-(\pi/4)$ toutes les fois que l'arc u a une valeur quelconque comprise entre $\pi/2$ et $3\pi/2$"([Grattan-Guinness, 1972], pp. 157-158).

[45] "Théorème I - Lorsque les différents termes de la série (I) sont des fonctions d'une même variable x, continues par rapport à cette variable dans le voisinage d'une valeur particulière pour laquelle la série est convergente, la somme s de la série est aussi, dans le voisinage de cette valeur particulière, fonction continue de x" ([Cauchy, 1821], p.120).

x between the bounds x = a und x = b, then the sequence $f(x) = v_0 + v_1\alpha + v_2\alpha^2 + ...$ with $\alpha < \delta$, is convergent and a continuous function of x between the same bounds".[46]

In a footnote he remarked that the theorem proved by Cauchy in 1821 has numerous exceptions. For example, $\sin\phi - \frac{1}{2}\sin 2\phi + \frac{1}{3}\sin 3\phi$ - etc. is discontinuous for all values $\phi=(2m+1)\pi$ when m possesses an integer value (Op. cit. , p. 9).

(v) In 1847 Seidel wrote that the contradiction between Fourier series with discontinuous limit functions on the one hand, and Cauchy's 1821 theorem on the other hand, had brought him to an analysis of Cauchy's proof. Seidel considered a convergent series of continuous functions.

$$s(x) = u_0(x) + u_1(x) + ... u_n(x) + r_n(x) = s_n(x) + r_n(x).$$

Then

$$s(x+\varepsilon) - s(x) = s_n(x+\varepsilon) - s_n(x) + r_n(x+\varepsilon) - r_n(x)$$

Seidel carefully considered the expression $r_n(x)$ on the interval $[x,x+\mu]$ and the fact that for all x on the interval, $r_n(x)$ can be made arbitrarily small by making n large enough. Undoubtedly because of the existence of Fourier series with discontinuous limit functions, Seidel realized that the size of n required to bring the value of $r_n(x)$ below a certain level depends on the value of x. In his paper Seidel distinguishes two cases (ρ is an arbitrarily chosen positive number) :

(I) there exists a smallest natural number N such that $|r_n(x)| < \rho$ on $[x,x+\eta]$ for $n \geq N$

(II) there does not exist a smallest natural number N such that $|r_n(x)| < \rho$ on $[x,x+\eta]$ for $n \geq N$.

With respect to (I) Seidel wrote: "In this first case one obtains indeed *Cauchy*'s theorem" ([Seidel,1847], p.40). For the case (II) he introduced the notion "arbitrarily slow convergence" ("beliebig langsam convergieren"). Seidel's conclusion is that in case of a discontinuity in the limit function, there is always arbitrarily slow convergence, although the converse need not be true ([Seidel, 1847], p.44).

---

[46]"Lehrsatz V. Es sei $v_0 + v_1\delta + v_2\delta^2 + ...$eine convergente Reihe, in welcher $v_0, v_1, v_2, ...$ continuierliche Functionen einer und derselben veränderlichen Grösse x sind zwischen den Grenzen x = a und x = b, so ist die Reihe $f(x) = v_0 + v_1\alpha + v_2\alpha^2 + ...$wo $\alpha<\delta$, convergent und eine stetige Function von x zwischen denselben Grenzen" ([Abel,1826], p.8).

I.4.2 Lakatos's reconstruction

The rational reconstruction of these events which Lakatos gave in 1961 runs as follows.

(i) First there is the primitive conjecture that the limit function of a convergent series of continuous functions is a continuous function.

(ii) In 1821 Cauchy gave a proof of that conjecture.

(iii) Fourier's counter examples refuted the conjecture.

(iv) In 1826 Abel reacted, applying the *method of exception-barring* he gave a proof for a restricted class of functions.

(v) In 1847 Seidel discovered the *method of proof-analysis* (also called the method of lemma-incorporation): an analysis of Cauchy's proof revealed a hidden lemma and yielded the proof-generated concept of uniform convergence.

There are obvious questions concerning the reconstruction:

1. Why did Cauchy in 1821 prove a theorem that contradicted Fourier's well-known 1807 results?
2. Why did Cauchy prove the theorem? After all, no one before him felt the need to prove the theorem.
3. Why was the hidden lemma discovered only in 1847 and not earlier?

According to Lakatos, Cauchy knew Fourier's results but did not believe that they were correct. Yet Fourier's results challenged the conjecture, and so Cauchy decided to prove it. In this way Lakatos answered the first two questions ([Lakatos,1976], p.130). As for the third question according to Lakatos the generation before Seidel applied exception-barring because within the prevailing Euclidean tradition with its infallibilist philosophy of mathematics, the idea that a proof can be at the same time respectable and fallible did not fit ([Lakatos,1976], pp.138-140). In Lakatos's view the forties of the last century show "a new fallibilist mood" which made Seidel's discovery possible (Ibid., p.139, footnote 1).

These are interesting hypotheses. Yet Lakatos's reconstruction is far from satisfactory. According to Lakatos, the footnote in Abel's 1826 paper, in which Abel remarks that Cauchy's 1821 theorem has exceptions, is the key to a proper understanding of that paper. Lakatos wrote: "Abel states the problem, which I claim to be the basic background problem of this celebrated paper on the binomial series, only in a footnote" ([Lakatos,1976], p.133). The established view is entirely different. For example, in 1895 Wangerin wrote on Abel's paper:

"Yet it is of great historical interest, because it contains first of all an investigation of the binomial series for complex values of the exponent and a determination of its sum in instances where it converges (Cauchy in his *Cours d'analyse* did consider complex values of the base, but only real values of the exponent), and secondly because it was an example of rigorous argumentation at a time when the treatment of infinite series often lacked the necessary rigor. The general theorems about series contained in the first section are also important".[47]

Is Wangerin wrong? Why did Abel hide the basic background problem of his paper in a footnote? Lakatos does not answer this question and, unfortunately, Lakatos's interpretation is plagued by other problems. If one reads Seidel's 1847 paper, one notices that Seidel was primarily interested in explaining why the limit of a convergent sequence of continuous functions can be discontinuous. The only theorem that Seidel explicitly phrases in his paper runs as follows.

"If one has a convergent series representing a discontinuous function of a quantity x, whose individual members are continuous functions, then it must be possible to indicate values of x in the immediate neighbourhood of the spot where the function is discontinuous, for which the sequence converges *arbitrarily slowly*".[48]

The analysis of Cauchy's proof was only a means to reach that result. It was not Seidel's goal to give an improved version of Cauchy's theorem. Lakatos does not explain why the most important aspect in his reconstruction has a secondary position in Seidel's paper. If Lakatos is correct, it is also strange that Seidel appears to be completely unaware of the fact that he made an important methodological discovery. Also at this point Lakatos's interpretation is not supported by sufficient arguments. The same holds for the hypothesis

---

[47]"Doch hat sie ein großes historisches Interesse, weil in ihr zuerst die binomische Reihe für complexe Werthe des Exponenten untersucht und für diese, falls sie convergirt, summiert ist (Cauchy hat in seinem "Cours d'analyse" zwar complexe Werthe der Grundzahl, aber nur reelle des Exponenten betrachtet), sodann weil sie in einer Zeit wo man es bei Behandlung unendlicher Reihen oft an der nötigen Strenge fehlen ließ, ein Muster exacter Beweisführung war. Wichtig sind auch die im ersten Abschnitt enthaltenen allgemeinen Sätze über Reihen" ([Wangerin, 1895], p. 41).

[48]"Hat man eine convergirende Reihe, welche eine discontinuierliche Function einer Größe x darstellt, von der ihre einzelnen Glieder continuierliche Functionen sind, so muß man in der unmittelbaren Umgebung der Stelle, wo die Function springt, Werthe von x angeben können, für welche die Reihe *beliebig langsam* convergirt" ([Seidel, 1847], p.37).

that Seidel was a representative of a new fallibilist mood, which appeared in the forties of the 19th century.

I.4.3 A different interpretation of the events

The fact that Cauchy proves in the *Cours d'analyse* that the limit of a convergent series of continuous functions is continuous can be explained very well from the immediate context of the theorem.

Elsewhere in the book Cauchy proves the theorem that the binomial series

$$\psi(\mu) = 1 + \frac{\mu}{1} x + \frac{\mu(\mu-1)}{1.2} x^2 + \text{etc.}$$

equals $(1+x)^\mu$ for real μ and $|x|<1$ ([Cauchy, 1821], pp.146-147).

Let us briefly consider the proof. Cauchy proves first that $\psi(\mu)$ converges for $|x|<1$. Because $\psi(\mu)$ is clearly a convergent series of polynomials in μ, Cauchy can use the theorem on the limit of a convergent sequence of continuous functions to prove that $\psi(\mu)$ is a continuous function in μ. Then it follows immediately that $\psi(\mu)$ and $(1+x)^\mu$ are equal because they are both continuous solutions of the functional equation $\psi(\mu).\psi(\mu')=\psi(\mu+\mu')$ for which $\psi(0)=1$. Elsewhere in the *Cours d'analyse* Cauchy shows that that particular functional equation has only one continuous solution (if one value of the function is given).

It is very reasonable to suppose that Cauchy proved the theorem on the limit of a convergent sequence of continuous functions because he needed it in his treatment of the binomial series in the way sketched above. Moreover, after having given precise definitions of the notions "continuous function", and "convergent series", the theorem could no longer be assumed as an evident truth. As for Fourier's counter-examples, they must have belonged in Cauchy's thinking to a different context which did not relate to the material which Cauchy included in his lectures for students, out of which grew the *Cours d'analyse*.

Abel's 1826 paper is primarily a continuation of Cauchy' work on the binomial series. Cauchy only dealt with real values of the exponent. Abel studied the case of complex values. When Abel became aware of the contradiction between Cauchy's 1821 theorem and certain Fourier series, an essential building block in the theory broke down. Being primarily interested in the binomial series and not in general theorems on arbitrary series of continuous functions, he proved, ad hoc, what he needed. Seidel, on the contrary, was not interested in the binomial series, but in the discontinuities in the limit function of certain Fourier series.

My conclusion is that Lakatos's rational reconstruction is based upon mathematical acts which belong to two different threads of development. On the one hand we have a thread which concerns the binomial series. First Cauchy treated the binomial series in the *Cours d'analyse* in the context of his new foundation of analysis. Because he needed the continuity of the limit function of a series of polynomials, he tried to prove it and realized in doing so that he could prove a more general theorem. There he made a "mistake". In 1826 Abel, when he tried to generalize Cauchy's results, became aware of that mistake and ad hoc provided another proof good enough for his purposes. On the other hand we have a thread which goes from Fourier's work on Fourier series towards Seidel. Seidel explained how it is possible that the limit function of certain Fourier series is discontinuous.

Mathematicians of the 18th century and the first half of the 19th century were primarily interested in what were, from a later point of view, particular cases (binomial series, Fourier series, etc.) and were less interested in results valid for larger classes of functions (e.g. Cauchy's 1821 theorem). No one was very much interested in theorems on functions in general. For everybody involved, the theorem on the limit function of a convergent sequence of continuous functions played a secondary role. When Lakatos called Abel an exception-barrer, he interpreted Abel's work from the point of view of mathematicians from the second half of the 19th century. He forgot that Abel's primary goal was to get results concerning the binomial series. In Lakatosian terms Abel actually practised proof-analysis: Cauchy's theorem acted as a refuted lemma in a theory concerning the binomial series, which he replaced by a sound lemma.

I.4.4 Conclusion

The events concerning the appearance of the notion of uniform convergence confirm the conclusion of section I.3. They are much more in accordance with a weak fallibility thesis than with a strong fallibility thesis. Obviously Cauchy's theorem was untenable because of Fourier's results, but the resulting reactions show more of a conceptual refinement, over the course of time, than a dramatic abandonment of older theories. Cauchy made a mistake[49] in the sense that the theorem that he proved turned out to be less generally valid than he supposed, but the theorem was not totally abandoned. Its domain of validity was restricted.

---

[49] I will return to that mistake in Chapter III.

Lakatos's reconstruction of the history of the notion of uniform convergence is rather counter-factual. Yet, if taken as a rational construction, independent of the real history, namely as a rational construction, just as Lakatos's rational construction based on the history of Euler's formula, it illustrates heuristic patterns in the development of mathematics that had not been noticed before, at least not so explicitly.[50] In this sense of a fictitious story pieced together out of realistic heuristic steps, the construction also supports the rationality thesis.

[50]In a private conversation Pólya said to Feferman that in his (Polya's) view Lakatos's method of proofs and refutations consists of the "alternating procedure" going back to Pólya and Szegö in 1925. That procedure boils down to working alternately on proofs and counterexamples as follows. If we cannot prove a clearly stated assertion A, we try to prove instead a weaker proposition (which we have a greater chance to prove). And, if we cannot disprove the assertion, we try to disprove a stronger assertion (which we have a greater chance to disprove) ([Feferman, 1978], p.325). This, obviously, was written from an infallibilist position.

Chapter II The Methodology of Scientific Research Programmes

II.1 Mathematics as a quasi-empirical science

In the first chapter I critically discussed two Lakatosian rational constructions of developments in mathematics which belong to that period in Lakatos's work in which the emphasis was on the methodology of proofs and refutations. We have seen there that these rational constructions reveal important heuristic patterns and in that sense they definitely support the rationality thesis. We have also seen that the reconstructions at most support a weak fallibility thesis. The present chapter deals with Lakatos's methodology of scientific research programmes (MSRP) and its applicability to mathematics.

The MSRP is usually and rightly considered as Lakatos's rationalist answer to Kuhn's views on the development of science. Yet I will argue that when Lakatos developed that methodology, problems generated by his work in the philosophy of mathematics. I will show that in principle the MSRP offered a solution to the problem.

At the International Colloquium on the Philosophy of Science in London,1965, Lakatos replied to the empiricist views on mathematics expressed in a paper by Lászlò Kalmár.[1] Kalmár had argued i) that the research in the foundations of mathematics had failed to show that mathematics is a firmly founded pure deductive science, ii) that the axioms of any interesting branch of mathematics were originally abstracted, directly or indirectly, from empirical facts and that the rules of inference had shown their validity not a priori, but in the course of our actual thinking practise, and iii) that the consistency of most of our formal systems is an empirical fact ([Kalmár,1967], p.192). According to Kalmár, intuitive evidence is ultimately a product of practical testing in experience. Lakatos very much agreed and he expanded his original remarks into a longer paper which he completed in 1967. The paper, entitled "A Renaissance of Empiricism in the Recent Philosophy of Mathematics?" (from now on to be called "A Renaissance"), was at the time withheld from publication by Lakatos and only published posthumously ([Lakatos,1978b], pp.24-42). It is clear that Lakatos must have been dissatisfied with the paper. We will see that there were good reasons for dissatisfaction.

---

[1] [Lakatos,1967]

In his reply to Kalmar, Lakatos distinguished two kinds of theories - *Euclidean theories* and *quasi-empirical theories*. He defined a Euclidean theory as follows: *truth* flows characteristically from *the top* (the axioms) to *the bottom* and inundates *the whole system*. Quasi-empirical theories he defined as theories in which *the crucial truth-flow* is *the upward re-transmission of falsity*, from a set of *basic statements*" to *the axioms*.

The two kinds of theories correspond to two different methodologies. In the case of a Euclidean ideal the basic methodological rule is

> "to search for self-evident or at least plausible axioms - Euclidean [...] methodology is puritanic, antispeculative".[2]

However, if one aims at a quasi-empirical ideal, the basic methodological rule is

> "to search for bold, imaginative hypotheses with high explanatory and heuristic power; indeed, it advocates an uninhibitedly speculative proliferation of alternative hypotheses to be pruned by severe criticism" ([Lakatos,1967],p. 202).[3]

Euclidean and quasi-empirical theories also develop in a different way. In the extended version of the paper Lakatos wrote:

> "The development of Euclidean theory consists of three stages: first the naive pre-scientific stage of trial and error which constitutes the prehistory of the subject; this is followed by the foundational period which reorganizes the discipline, trims the obscure borders, establishes the deductive structure of the safe kernel; all that is left is the solution of problems inside the system, mainly constructing proofs or disproofs of interesting conjectures [...].
>
> The development of a quasi-empirical theory is very different. It starts with problems followed by daring solutions, then by severe tests, refutations. The vehicle of progress is bold speculations, criticism, controversy between rival theories, problemshifts. Attention is always focussed on the

---

[2]In the elaborated version of the paper this becomes "to search for self-evident axioms" ([Lakatos,1978b], p.29). Why did Lakatos drop the "or at least plausible"? Presumably because he writes about the Euclidean ideal which, unlike Euclidean practice, is not satisfied with plausibility.

[3]Here Lakatos refers to a paper which does not exist in print for an elaboration of these ideas: "Empiricism in the Contemporary Philosophy of Mathematics", British Journal for the Philosophy of Science, 18, 1967. I suppose that Lakatos originally intended to publish his extended remarks under that title.

> obscure borders. The slogans are growth and permanent revolution, not foundations and accumulation of eternal truths.
>
> The main pattern of Euclidean criticism is suspicion: Do the proofs really prove? Are the methods used too strong and therefore fallible? The main pattern of quasi-empirical criticism is proliferation of theories and refutation" ([Lakatos, 1978b], pp. 29-30).

According to Lakatos, mathematics is quasi-empirical. His argument in the elaborated version of the paper is organized as follows. First of all he quotes a considerable number of mathematicians whose words imply that the developments concerning the foundations of mathematics in the beginning of the century had damaged mathematics' appearance of "absolute certainty". For example, he quotes Russell as saying: "The splendid certainty which I always hoped to find in mathematics was lost in a bewildering maze" ([Lakatos,1978b], p.28). After having defined the notions of quasi-empirical theory and Euclidean theory, Lakatos turns to the background of those remarkable developments and gives a brief account of the failure of the Frege - Russell approach and of the Hilbert approach to the foundational problems. The Frege - Russell approach aimed at reducing all of mathematics to indubitably true logical axioms, at turning mathematics into a "unified classical Euclidean theory". The Hilbert approach was an attempt to found all of mathematics in an indubitably true Euclidean *meta-theory*. However, the logicists, and set-theorists in general, had to accept the fact that at least some of the axioms they had to introduce possessed a hypothetical character. The formalist programme in its turn was "refuted" by Gödel's theorem. Understandably, because he wants to show that mathematics is a quasi-empirical science, Lakatos then concentrates on the problem of the flow of falsity in mathematical theories from the bottom to the top, in other words on the problem of the potential falsifiers in mathematics. Apart from potential logical falsifiers, i.e. statements of the form "P&¬P"([Lakatos, 1978b],p.36), Lakatos discusses "heuristic" falsifiers. For example, if within a formalization of an informal theory a theorem can be derived formally which contradicts a theorem from the informal theory, then this informal theorem falsifies the formal theory. If the informal proof can be formalized in the formal theory, then we have a logical falsification. If it cannot, we have a heuristic falsification. We are forced to choose: either we drop one of the axioms of the formal theory or we drop one of the arguments in the informal proof.

II.2 Some problems

Lakatos has difficulties to point out precisely what are the potential falsifiers in mathematics. He calls contradictions potential (logical) falsifiers. Contradictions, however, falsify a theory in a way that is different from the way in which basic statements falsify scientific theories. In one sense a contradiction transmits falsity "upwards" into the system. Yet one can also argue that it transmits truth "upwards" into the system because an inconsistent theory can be considered as a theory whose statements are all at the same time false and true.[4] Considering contradictions as potential falsifiers is not very satisfactory. Moreover, what precisely are heuristic falsifiers? Lakatos admits that the concept of a heuristic falsifier is "slippery":

> "A heuristic falsifier after all is a falsifier only on a Pickwickian sense: it does not falsify the hypothesis, it only suggests a falsification - and suggestions can be ignored. It is only a rival hypothesis.[...] *The crucial role of heuristic refutations is to shift problems to more important ones*, to stimulate the development of theoretical frameworks with more content. One can show of most classical refutations in the history of science and mathematics that they are heuristic falsifications. The battle between rival mathematical theories is most frequently decided also by their relative explanatory power"([Lakatos,1978b], p. 40).

With the sentence concerning the crucial role of heuristic falsifiers, Lakatos obviously shifts the problem. Within the Popperian view of science the 'basic statements' or 'potential falsifiers' are singular spatio-temporal statements. Lakatos defined a quasi-empirical theory as a theory in which the *crucial* flow of truth values is the flow from the basic statements to the top. However, being unable to locate the mathematical analogon of the singular spatio-temporal statements, Lakatos switches to a different perspective in which theories are compared on the basis of their explanatory power.

Another problem Lakatos runs into is the following. He writes:

> "The history of quasi-empirical theories is a history of daring speculations and dramatic refutations. But new theories and spectacular refutations (whether logical or heuristic) do not happen every day in the life of quasi-empirical theories, whether scientific or mathematical. There are long *stagnating periods* when a single theory dominates the scene without having rivals or acknowledged refutations. Such periods make many forget about the criticizability of the basic assumptions. Theories, which look

---

[4]If we define truth and falsity in terms of provability.

counterintuitive or even perverted when first proposed, assume authority. Strange methodological delusions spread: some imagine that the axioms themselves start glittering in the light of Euclidean certainty, others imagine that the deductive channels of elementary logic have the power to retransmit truth (or probability) 'inductively' from the basic statements to the extant axioms" ([Lakatos,1978b], p.41).

Clearly mathematics in particular suffers from these "long, stagnating periods", and from the "strange methodological delusion" that the established results are certain. Moreover, it is not so easy to find "spectacular refutations" in the history of mathematics. Lakatos's view in 1967 of mathematics as a quasi-empirical science offers no explanation of this phenomenon. Yet the phenomenon is highly remarkable if mathematics is indeed a quasi-empirical science in the sense that the crucial flow of truth values is from the basic statements, whatever they are, to the top.

A related problem concerns the way in which mathematicians ought to react to the refutation of a theory. Lakatos's methodology of a quasi-empirical theory implies that the proper reaction would be the development of a daring new speculative theory. In principle, the old theory could be thrown away in its totality. However, all this seems very much opposed to mathematical practice.

In "A Renaissance" Lakatos wanted to show that mathematics is fallible in a way analogous to the fallibility of natural science in Popper's philosophy, but he did not succeed. A quotation such as "Whether the system of logic which results from our postulates is adequate for the development of mathematics, and whether it is wholly free from contradiction, are questions which we cannot now answer except by conjecture" (Church quoted by Lakatos ([Lakatos,1978b], p.31) shows that Church admitted a conjectural element in logic and mathematics. The quotation points at the possibility of inconsistency. And it is possible, although it is rather artificial, to describe the effect of the discovery of an inconsistency as a flow of falsity from the bottom (i.e. the contradiction) to the top (i.e. the axioms) whereby falsity inundates the whole system. Lakatos reads this quotation and all the others that admit a conjectural element in mathematics as statements supporting his hypothesis that mathematics is quasi-empirical in the sense that the "crucial", or "characteristic" or "important" (these are all Lakatos's words) flow of truth-values goes from the bottom to the top. Yet the quotation, and most of the others that Lakatos gives in order to illustrate the quasi-empiricist views of many mathematicians, are much more in accordance with a view of mathematics according to which the characteristic flow of truth-values is nevertheless from the top to the bottom (although some flow the other way around is not excluded).

II.3 The Methodology of Scientific Research Programmes

"A Renaissance" shows very well that the form of Popperian falsificationism which was on Lakatos's mind in the period of his work dominated by the MP&R, corresponds only to a certain extent to the actual development of mathematics. I will now show that at least in principle the MSRP offered a more adequate notion of rationality for mathematics.

The notion of "research programme" appears for the first time in Lakatos's published work in the same year 1967, in which he completed "A Renaissance" in his "Changes in the Problem of Inductive Logic" ([Lakatos,1968], pp.315-417)[5]. In that paper Lakatos discusses Carnap's *research programme* (Lakatos's term) in epistemology. Lakatos wrote:

> "A successful research programme bustles with activity. There are dozens of puzzles to be solved and technical questions to be answered; even if some of these - inevitably - are the programme's own creation. But this self-propelling force of the programme may carry away the research workers and cause them to forget about the problem background. They tend not to ask anymore to what degree they have solved the original problem, to what degree they gave up basic positions in order to cope with internal technical difficulties. Although they may travel away from the original problem with enormous speed, they do not notice it. Problem-shifts of this kind may invest research programmes with a remarkable tenacity in digesting and surviving almost any criticism. Now problem-shifts are regular bed-fellows of problem-solving and especially of research programmes. One frequently solves very different problems from those which one has set out to solve. One may solve a more interesting problem than the original one. In such cases one may talk about a 'progressive problem-shift'.[6] But one may solve some problems less interesting than the original one; indeed, in extreme cases, one may end up

---

[5]The preface of the book is dated September,1967.

[6]Lakatos's notion of "problem-shift" is similar to Brouwer's notion of "jump from goal to means"([Brouwer,1905], p. 20). In a chapter (called "The Fall by the Intellect") dedicated to the negative role of our intellect, Brouwer describes how our intellect easily shifts its attention from an original goal to the problem of the means by which to reach that goal. If such a shift takes place repeatedly we may finish solving problems that correspond to goals that are completely opposed to our original goal.

with solving (or trying to solve) no other problems but those which one has oneself created while trying to solve the original problem. In such cases we may talk about a 'degenerating problem-shift'" ([Lakatos,1968], pp.316-317).

The quotation shows how Lakatos became fascinated by the phenomenon of the "self-propelling force" that carries scientists in a particular direction in spite of the fact that the theories which they develop are strictly speaking falsified by the data that they set out to explain. Soon Lakatos would use the term "positive heuristic" to indicate this force. The earliest published extensive discussion of the MSRP is in "Criticism and the Methodology of Scientific Research Programmes" ([Lakatos,1968])[7], presumably written in 1967. According to Lakatos, great scientific achievements are not characterized by an isolated hypothesis, but rather by a *research programme.* A research programme manifests itself in the form of a series of successively developed scientific theories, characterized by a certain *continuity* which connects the members of the series.

I will describe the MSRP by means of an example also given by Lakatos - Newton's gravitational research programme. In this research programme, four conjectures play a central role: the three laws of mechanics and the law of gravitation. These four conjectures return in all theories produced by the programme; they are irrefutable by the methodological decision of those working in the programme. The three laws of mechanics and the law of gravitation form what Lakatos calls the *hard core* of the programme. Part of the execution of the programme took place, according to Lakatos, roughly as follows. Newton first worked out his programme for a point-like sun and one single point-like planet. Newton then turned to more planets, consciously forgetting about interplanetary forces. After that he worked out the situation when sun and planets are no longer point masses but mass-*balls*. In the course of time he studied spinning balls. He included interplanetary forces and finally also considered bulging instead of spherical planets. The rule or the rules that yield such a chain of ever more complicated models simulating reality are part of the *positive heuristic* of the research programme. The guiding idea in this case is formulated by Lakatos as a "metaphysical" (his word) principle: "the planets are essentially gravitating spinning tops of roughly spherical

[7]The following account is primarily based upon "Falsification and the Methodology of Scientific Research Programmes" ([Lakatos,1978a],pp.8-101; written in 1968-69 and first published in 1970).

shape" ([Lakatos,1978a], p.51). While constructing this chain of models, the Newtonian scientist ignores to a large extent the fact that the available data strictly speaking falsify the models. Yet the goal of the construction of the chain of models simulating reality is, of course, the explanation of the available data and the prediction of new facts. In the case of Newton's gravitational programme, the positive heuristic also encompassed guidelines concerning the way in which to modify and use a vast set of auxiliary hypotheses. "For instance, if a planet does not move exactly as it should, the Newtonian scientist checks his conjectures concerning atmospheric refraction, concerning propagation of light in magnetic storm, and hundreds of other conjectures which are all part of the programme. He may even invent a hitherto unknown planet and calculate its position, mass and velocity in order to explain the anomaly" (Lakatos,1978a], p.4). Such auxiliary hypotheses may be rejected or changed during the course of the execution of the programme. New ones may also be invented. This set of auxiliary hypotheses is called by Lakatos the *protective belt* of the hard core, because it protects the hard core from refutation: falsifications of a prediction are always taken as a refutation of some auxiliary hypothesis. The heuristic of the Newtonian programme also included a complex mathematical apparatus - the calculus, etc.
Why should one consider Newton's programme as a great scientific success? Lakatos's answer is simple: the programme repeatedly led to the discovery of hitherto unknown facts. Halley for example predicted, on the basis of Newton's theory, the return in seventy-two years of the comet that bears his name now. He calculated very precisely when and where in the sky it would be seen again. Such predictions constitute in Lakatos's terminology *theoretical progress* of a research programme. Halley's comet returned as predicted seventy-two years later. Such a corroboration of a prediction constitutes *empirical progress* of a programme in Lakatos's terminology. Next to theoretical and empirical progress, Lakatos also has a notion of *heuristic progress*: the successive modifications of the protective belt must be in the spirit of the heuristic of the programme (E.g. [Lakatos, 1978a], p.179). In Lakatos's terminology, Newton's gravitational programme was a great scientific achievement because for a long time it consistently showed theoretical progress, heuristic progress and intermittingly empirical progress.

II.4 The applicability of the MSRP to mathematics

Lakatos thought that the MSRP was also valid for mathematics. And indeed, without detailed rational reconstructions, there is no doubt that the MSRP in principle solves the problems which are so clear in "A Renaissance". The periods of stagnation in the development of a science - when there are no new theories and spectacular refutations and when, moreover, axioms that perhaps once looked counter-intuitive start "glittering in the light of Euclidean certainty" - are no longer mysterious from the point of view of the MSRP. They represent periods in which certain successful research programmes dominate. The certainty of the axioms is no longer a "methodological delusion", but a "methodological decision" of the proponents of the programme. The MSRP also solves Lakatos's difficulties in pointing out potential falsifiers in mathematics in the sense that falsification in the sense of logical refutation plays a subordinate role in the MSRP.

As regards the development of Lakatos's work with respect to mathematics, one could argue as follows. In "Proofs and Refutations", and in particular in "A Renaissance", Lakatos attempts to apply to the development of mathematics the Popperian scheme: bold speculations and severe refutations succeed each other. However, on one hand mathematicians always try to give theories a firm foundation and these can only be considered as speculative to a limited extent, while on the other hand logical refutation of an established theory hardly ever occurs. For mathematics the MSRP offered in these respects a far more promising alternative.

Although Lakatos believed that the MSRP is applicable to mathematics, he unfortunately never went further than a few hints. Referring to "Proofs and Refutations" Lakatos wrote in a footnote:

> "Unfortunately in 1963-64 I had not yet made a clear terminological distinction between theories and research programmes, and this impaired my exposition of a research programme in informal, quasi-empirical mathematics." ([Lakatos,1978a], p. 52)

It seems that in restrospect Lakatos viewed "Proofs and Refutations" as the description of the development of a research programme. This is understandable. I emphasized above certain discontinuities in the development of Lakatos's work and I argued that the MSRP succeeded the MP&R because it solved problems the MP&R could not answer. These discontinuities certainly exist. However, in the period dominated by the MSRP Lakatos in many respects further developed ideas that are already present in some form in his earlier work. When Lakatos tried to apply the Popperian model to mathematics in "Proofs and Refutations", given the nature of mathematics, he necessarily had to modify it. In an interesting paper De Vries places much emphasis on the continuity in Lakatos's work ([De Vries,1981]). He argues that the MSRP not only solves some of the problems that the MP&R did not, but that it was in many ways also positively implied by Lakatos's earlier work. Indeed, "Proofs and Refutations" is concerned with a series of theories and certain parts of it possess a clear heuristic unity. For example the consistent application of the method of lemma-incorporation by the teacher yields theorems T_2 and T_3, and Zeta's "deductive guessing" is also an example of someone following a positive heuristic ([Lakatos,1978a], p.50, footnote).[8] The method of monster-barring or the method of exception-barring are both quite suitable for defending a hard-core. The methods discussed in "Proofs and Refutations" all return in one way or another in Lakatos's later work. Yet it is not at all clear exactly how Lakatos would have applied the conceptual apparatus of the MSRP to the research programme that we find in "Proofs and Refutations". Is there a positive heuristic characterizing the whole programme? What would Lakatos have identified in this case as the hard core of the programme?

There exists quite a gap between the conceptual apparatus of the MSRP and mathematics. Nearly everything that Lakatos wrote on the MSRP explicitly concerns the natural sciences. However, in 1973 Lakatos briefly returned to the subject of "Proofs and Refutations" ([Lakatos,1978a], 93-103). At that time he distinguished the following pattern in the developments concerning Euler's formula for polyhedra. Firstly *the stage of naive trial and error*, lasting from Euclid to Descartes. Secondly *the stage characterized by proof procedures* during which the conjecture was subjected to sophisticated attempts at refutation. These proof procedures then led to the third stage when they generated a rich *research programme*: hidden lemmas, after having been articulated, turned into the hard core of the research programme of algebraic topology. The derivative of that hard core consists of the axioms of algebraic topology ([Lakatos,1978a], pp. 96-97). In the same paper Lakatos gave a second example of a research programme in mathematics. He suggested that the Greeks inherited from the Babylonians and Egyptians *a body of naive mathematical conjectures* which they subjected to *test- and proof-thought-experiments*. After hundreds of proof-procedures, certain lemmas kept popping up and were finally turned into *the hard core of a research programme* (an 'axiomatic system') by Euclid ([Lakatos,1978a], pp.99-100).

[8]In Zeta's positive heuristic analogy appears to play the major role: he builds polygons from edges, polygonal systems from polygons, polyhedra from polygonal systems and, finally, systems of polyhedra from polyhedra.

In these two examples the hard cores (or the derivatives of the hard cores) of the research programmes are systems of axioms. Unfortunately Lakatos gives no characterization, either of the positive heuristics of the research programmes, or of the protective belt surrounding the hard core.

When one considers Lakatos's definition of the methodology of scientific research programmes, it is clear that its application to mathematics is not a trivial matter. The application of the MSRP to mathematics would require the identification in mathematics of research programmes by means of notions like hard core, protective belt and positive heuristic. It would also require that in mathematics we distinguish theoretical, heuristic and empirical progress. And finally it would require examples of competing research programmes in mathematics.

There are many examples of axiom systems that have been, or are, considered to be true of certain intended models. One could call them the hard cores of corresponding research programmes. However, as soon as one might want to apply the notions of *protective belt* (consisting, in Lakatos's definition, of statements of a clearly hypothetical nature) *positive heuristic* (in the sense of a rule listing chains of ever more complicated models simulating reality) and *theoretical versus empirical progress* to mathematics, one is faced with the problem that mathematics and science are quite different. The point is that in mathematics there is no clear analogon of the "empirical data" in science. True, working mathematicians often have a more or less clear goal in mind which they try to reach. Sometimes it is a conjecture that a mathematician tries to prove. Sometimes the statement which needs to be proved is no conjecture but has already been proved differently. Yet it is not obvious in what sense one can say in such situations that what those mathematicians are doing is "explaining data". The fact of the matter is that there is no major difference between, say, a general conjecture and a very special case of it, comparable to the difference between Newton's simplest models and the empirical facts. Another problem is the following. In Lakatos's rational reconstructions on the basis of his MSRP there are always rival research programmes concerning a particular aspect of empirical reality. However, there seems to be considerably less competition in mathematics than in science, at least at first sight.

In summary, Lakatos has given only a few examples of research programmes in mathematics. He has discussed none of them thoroughly, restricting himself instead to a few remarks in passing. His remarks suggest that the application of the MSRP to mathematics will require modifications of that methodology.

II.5 The fallibility thesis and the rationality thesis

In the following chapters I will first put the emphasis on the problem of the rationality of the development of mathematics and on possibilities for applying the MSRP to mathematics. I will investigate the validity of the following picture of the rationality of mathematics.

On the micro level MP&R is the overall mathematical methodology: conjectures, attempts at proofs, refutations, and critical proof-analysis are all part of the methodology of mathematics. Unexpected refutations leading to large-scale overthrow of theories, however, are rather rare, particularly when the mathematicians involved do not prematurely consider plausible conjectures as theorems. On the micro level theorems and proofs are fallible in the sense that they can be replaced by better ones along the lines of the MP&R.

On the macro level the situation is different. There, the developments can no longer be described in terms of the MP&R. The development of mathematics shows that, in practice, coherent clusters of considerable numbers of theorems become established. After an initial phase, when the patterns of "Proofs and Refutations" may still play an important role and counter examples reveal unintended interpretations and repeated conceptual refinement takes place then such clusters are no longer subject to logical refutation, but form the core of a theory or a set of theories that is being developed further. Such developing theories or sets of theories can be described as "research programmes". On the level of those "research programmes" one could also speak of a form of refutation. The refutation of a "research programme", however, is not a logical refutation but an expression of the fact that the programme has lost the competition with another programme. It is not logical refutations but the degree of progress of the rival programmes which determines the development ([Lakatos,1978a], p.50). In other words, at the macro level, research programmes are only 'fallible' in the sense that they can be replaced.

Only in Chapter X will I draw some more definite conclusions on the fallibility of mathematics. In the next chapter I will discuss a posthumously published paper by Lakatos, which, although written before the notion of research programme was introduced, in fact contains a reconstruction of two competing research programmes in mathematics. This is the main reason for examining that paper. As we have seen, although Lakatos believed that the MSRP is applicable to mathematics, he mainly applied it to natural science.

There is another reason for dealing with the paper. Suppose programme A is highly incompatible with programme B, and suppose that B replaces A completely, the knowledge produced by A would then have to be considered as "refuted" by B as a guess, as a conjecture, which was proved "wrong" by programme B. Such an example in mathematics would be a good argument to take a strong fallibility thesis with respect to mathematics into serious consideration. The paper that I will examine in the next chapter contains a reconstruction showing two competing research programmes in mathematics that are very much opposed. Unfortunately, as we shall see, the reconstruction is highly problematic.

Chapter III "Cauchy and the continuum"

III.1 Introduction

In 1966, before he introduced the concept of research programme, Lakatos delivered a paper at the International Logic Colloquium in Hannover, called "Cauchy and the continuum: the significance of non-standard analysis for the history and the philosophy of mathematics" ([Lakatos,1978b], pp.43-60).[1] As we have seen in section I.4, in 1961 Lakatos still construed Cauchy's theorem on the continuity of the limit function of a convergent series of continuous functions in accordance with the views of most historiographers of mathematics: Fourier's counter-examples refuted the theorem and eventually a hidden lemma concerning the uniformity of the convergence was discovered in the proof. In "Cauchy and the continuum", Lakatos returned to the subject and gave a radically different interpretation of Cauchy's "mistake". Robinson had pointed out that by means of non-standard analysis, certain results of Cauchy and other mathematicians, that in one way or another use infinitesimals, could be understood differently from the received interpretation.[2] After reading Robinson's work, Lakatos concluded that we can distinguish two competing theories of the continuum in the history of the calculus. On the one hand, according to Lakatos in 1966, there is in the history of calculus a theory based on the standard continuum, with Weierstraß as a prominent representative; while on the other hand, we have a theory based on a richer continuum, which contains not only the standard real numbers but also infinitesimals and infinitely large numbers; in Lakatos's view Leibniz was an early representative of this theory. According to Lakatos, in "Cauchy and the continuum" Cauchy also worked in the tradition of Leibniz, and in order to show this, he interprets Cauchy's work in a way that differs radically from the received view.

"Cauchy and the continuum" had been accepted for publication, but Lakatos withheld it and it appeared only posthumously. Because Lakatos wrote very little on mathematical research programmes, the paper should be taken seriously whenever one discusses the applicability of Lakatos MSRP' to mathematics. According to the editor of the paper - Cleave - there are no indications that Lakatos contemplated alterations of the main points of his

---

[1] Below I will call the paper "Cauchy and the continuum".

[2] [Robinson, 1966], pp. 260 - 282

arguments.[3] If Lakatos had written the paper a few years later, he would undoubtedly have interpreted the developments as an example of two competing research programmes in mathematics. In this chapter I will first briefly describe Lakatos's paper. Then I will concentrate on the question of which interpretation of Cauchy's analysis is to be preferred - the received interpretation, or an interpretation based upon the assumption that Cauchy worked with a non-standard continuum. I will also briefly discuss some work of Cleave and Laugwitz, who under the influence of Robinson and Lakatos attempted to give non-standard interpretations of Cauchy's mathematics.

III.2 Lakatos's views on Cauchy's mathematics

Lakatos concentrates on Cauchy's proof of the theorem in the *Cours d'analyse*, which states, as we have seen in Chapter I, that on an interval the limit function of a convergent series of continuous functions is continuous. Historians generally assume that Cauchy made the mistake of not distinguishing between convergence and uniform convergence. Lakatos in "Cauchy and the continuum" had a different opinion. He writes:

> "Cauchy made absolutely no mistake, he only proved a completely different theorem, about transfinite sequences of functions which Cauchy-converge on the Leibniz-continuum" ([Lakatos,1978b] , p.50).

Lakatos interpreted Cauchy's proof as follows:

> "Let $\lim s_n(x)=s(x)$, where $s_n(x)$ are continuous. Then in order to prove that $s(x)$ is continuous at some x_1, we have to show that $s(x_1+\alpha)-s(x_1)$ is infinitesimal for all infinitesimal α [...]. Now
>
> $$|s(x_1+\alpha)-s(x_1)| = |s_n(x_1+\alpha)-s_n(x_1) + r_n(x_1+\alpha)-r_n(x_1)| \leq |s_n(x_1+\alpha)-s_n(x_1)|+|r_n(x_1+\alpha)|+|r_n(x_1)|$$
>
> where r_n are the remainders. Cauchy thought the left hand side was infinitesimal for all infinitesimal α since $|s_n(x_1+\alpha)-s_n(x_1)|$ is infinitesimal for all n because of Cauchy's definition of *continuity*; and $|r_n(x_1+\alpha)|$ and $|r_n(x_1)|$ are again infinitesimal for all infinitesimally large n because of Cauchy's definition of the *limit*: a_n->0 if a_n is infinitesimal for infinitely large n" ([Lakatos,1978b], p.49).

And Lakatos added:

> "Of course this argument implies that $s_n(x)$ should be defined and continuous and converge not only at standard Weierstrassian points but at *every* point of the 'denser' Cauchy continuum, and that the sequence $s_n(x)$

[3]([Lakatos,1978b] , p.43, footnote by J.P. Cleave)

> should be defined for infinitely large indices n and represent continuous functions at such indices. Cauchy's 'limit' and 'continuity' are only defined for 'transfinite' sequences of functions defined on his overdense continuum. For such sequences of functions Cauchy's theorem is indeed true; and Fourier-Abel counterexamples[4] are not continuous either at infinitely large indices or at non-Weierstrassian points of functions with finite indices. Alternatively, these recalcitrant sequences may not S-converge in the entire monad of points of discontinuity, and it is very likely that Cauchy may have suspected something like this in 1821" ([Lakatos,1978b], pp.49-50).

According to Lakatos, Cauchy's theorem concerns sequences of functions $\{s_n(x)\}$defined upon a non-standard continuum, i.e. a continuum consisting not only of standard real numbers but also of non-standard real numbers. In such a non-standard continuum the number zero is surrounded by a "cloud" of infinitely small numbers α - i.e. numbers unequal to zero - with an absolute value smaller than any positive standard real. Because the normal rules of calculation hold for standard and non-standard numbers, every standard real a is surrounded by such a "cloud" of non-standard reals $a+\alpha$. In the non-standard continuum there are also infinitely large numbers, and among them infinitely large natural numbers.

Lakatos argues that Cauchy's theorem concerns functions $s_n(x)$ that are defined not only for standard natural numbers n, but also for infinitely large natural numbers; and not only for standard real numbers x, but also for non-standard real numbers. For such functions Cauchy proved the following: If for all n (including the infinite ones) $s_n(x)$ is continuous (for all real numbers involved, including the non-standard ones) and the sequence$\{s_n(x)\}$ converges (again for all real numbers involved, including the non-standard ones) to a limit function s(x), then s(x) is also continuous.

The received view is that in 1821 Cauchy made a mistake which was discovered by Abel in 1826 and corrected satisfactorily only after the notion of uniform convergence had been introduced. It is an interesting fact that Cauchy in 1853 published a paper in which he returned to his 1821 "mistake". The received view is that Cauchy corrected the mistake in that paper: he introduced uniform convergence and gave a correct proof of the theorem. In 1966 Lakatos also interpreted Cauchy's 1853 work in a totally different way:

---

[4]See Chapter I, section I.4.1

> "In a paper read to the Academy in March 1853, Cauchy stubbornly restated his theorem, and claimed that the recalcitrant sequences do *not* converge everywhere, and in particular not in the infinitely small neighbourhood of the points of discontinuity" ([Lakatos,1978b] , p.47).

According to Lakatos, Cauchy repeated the theorem in 1853 and showed that the alleged counter-examples simply do not satisfy the conditions of the theorem: there is no convergence where it is necessary. In Lakatos's 1966 view, the fact that Cauchy's work has been misunderstood is a result of the fact that in the course of the 19th century the rival theory - based upon a standard continuum without infinitely small and infinitely large elements - began to dominate the scene. Why did this happen? Lakatos rejects Robinson's answer that the adherents of the infinitesimal theory failed to develop a *consistent* non-Archimedean system. In his view inconsistency was not the problem. Lakatos:

> "The downfall of Leibnizian theory was then not due to the fact that it was inconsistent, but that it was capable only of limited growth. It was the heuristic potential of growth - and explanatory power - of Weierstrass's theory that brought about the downfall of infinitesimals. The crucial hidden lemmas which emerged under the pressure of the criticism of its proofs were not independently testable: this discouraged the advocates of infinitesimals [...] and made them finally disappear for a century from the history of mathematics" ([Lakatos,1978b] , p.54).

I read this as follows: the "Weierstrassian" criticism of Cauchy's proofs should have been countered by making explicit the implicit assumptions concerning the non-standard part of the continuum that were present in Cauchy's work. The advocates of the infinitesimal theory did not succeed in this. The crucial point here seems to be that in Lakatos's view, the Leibnizian theory of the continuum did not possess a positive heuristic leading to a satisfactory *definition* of non-standard numbers, which made the construction of examples and counter-examples very difficult. The great advantage of the Weierstrassian theory was that counter-examples, easily specifiable non-uniformly convergent sequences, could be defined easily.

"Cauchy and the continuum" shows that Lakatos's thinking in 1966 was on the way towards the methodology of scientific research programmes. In this respect the following quotation is remarkable. Lakatos writes about "justificationist" historiography:

> "A very widespread device to try to maximize continuity is to demarcate in mathematical theories a *hard formal core* which is uncontested and indubitable, eternal; and a *'metaphysical' interpretation* of the formalism which is controversial, 'soft', changing" ([Lakatos,1978b], p.58).

Remember that this was written in 1966, a year before the MSRP shows up in Lakatos's published works. The notions of "an indubitable hard core" and "a soft metaphysical belt" may have been born here, in a mathematical context. Obviously, the notions that are used by Lakatos to describe the Leibnizian theory do not yet have their later meaning. The hard core of the Leibnizian calculus is not yet a set of principles that distinguished that theory from Weierstraß' theory, but rather the bulk of formal results which was taken over without problems by the conquering Weierstrassian calculus. It is remarkable that in this context the identification of a hard core is a device that is challenged by Lakatos, because it creates an illusion of continuity where there is, in Lakatos's view, discontinuity. Also the soft belt of that hard core does not yet have a (positive) heuristic function; it consisted of the metaphysical justifications of the use of infinitesimals which the Weierstrassians rejected.

III.3. Interpreting Cauchy

III.3.1 Cauchy's 1821 text

With respect to Cauchy's theorem we have an interesting situation in the historiography of mathematics. On one hand we have the traditional interpretation - that Cauchy works with a standard continuum and that in the proof of the theorem in 1821 he made a mistake because he did not distinguish between point-wise convergence and uniform convergence. In the traditional view Cauchy corrected his mistake in 1853. On the other hand we have non-standard interpretations - that Cauchy worked with a non-standard continuum, and his 1821 and 1853 proofs of the theorem should be interpreted accordingly. Several authors have attempted to give a non-standard interpretation. From a historiographical point of view the matter is important because the traditional view is that Cauchy was a precursor of Weierstraß, representing a break with the eighteenth century mathematics. If the advocates of a non-standard interpretation are right, that view has to be corrected. However, if Lakatos's reconstruction is correct the situation is also interesting from a philosophical point of view. We might then have a nice example of two competing research programmes in mathematics, and we could discuss the consequences for Lakatos's fallibility and rationality thesis. I will argue below that unfortunately Lakatos's reconstruction is counter-factual.

Let us have a look at Cauchy's text. The text of the theorem is:

> "Theorem I - Whenever the different terms of the series ($u_1 + u_2 + ...u_n +$ ..etc. -T.K]) are functions of one and the same variable x, continuous with respect to this variable in the neighbourhood of a particular value for which the series is convergent, the sum s of the series is also a continuous function of x in the neighbourhood of that particular value".[5]

The proof of the theorem given by Cauchy in 1821 runs as follows:

($s_n = u_1 + u_2 + ...u_{n-1}$ and r_n is the difference between the limit s and s_n.)

> "That done, let us consider what the increments which these three functions receive [s_n, r_n and the limit function s (= $s_n + r_n$) - T.K.] will be if one lets x grow with an infinitely small quantity α. The increment of s_n, for all possible values of x, will be an infinitely small quantity; and the increase of r_n will become imperceptible together with r_n, if one assigns to n a very large value. Consequently the increment of the function s can only be an infinitely small value".[6]

Let us consider the context of the theorem. The second chapter of Cauchy's book is dedicated entirely to the principal properties of infinitely small and infinitely large quantities. It is obvious that in Cauchy's mathematics infinitely small and infinitely large quantities occur. The question is, however, what their status is. Cauchy writes:

> "One can say that a variable quantity becomes *infinitely large* if its numerical value grows indefinitely in such a way that it converges to the limit ∞".[7]

---

[5][Cauchy, 1821], p.120 (Cf. Chapter I, footnote 45).

[6]"Cela posé, considérons les accroissements que reçoivent ces trois fonctions [s_n, r_n and the limit function s = $s_n + r_n$ - T.K.], lorsqu'on fait croître x d'une quantité infiniment petite α. L'accroissement de s_n sera, pour toutes les valeurs possibles de n, une quantité infiniment petite; et celui de r_n deviendra insensible en même temps que r_n, si l'on attribue à n une valeur très considérable. Par suite, l'accroissement de la fonction s ne pourra être qu'une quantité infiniment petite" ([Cauchy,1821] , p.120).

[7]"On dit qu'une quantité variable devient *infiniment grande*, lorsque sa valeur numérique croît indéfiniment de manière à converger vers la limite ∞" ([Cauchy,1821], p.38; the emphasis is Cauchy's).

And about infinitely small quantities he writes[8]:

> "One can say that a variable quantity becomes *infinitely small* if its numerical value decreases indefinitely in such a way that it converges to the limit 0".

The usual interpretation of these definitions, which I will call a *standard interpretation*, is that Cauchy worked with a standard continuum. In this view the two quotations are saying that an infinitely large quantity is defined by Cauchy as a variable converging to the limit ∞ and an infinitely small quantity is defined by him as a variable converging to the limit zero. Those definitions do not then imply an extension of the object of investigation. I will contrast this standard interpretation with the following non-standard interpretation of the definitions, defended, for example, by Laugwitz: Cauchy worked with a non-standard continuum also containing infinitely large and small elements, and the quoted sentences show that Cauchy found it useful to *represent* the non-standard elements by means of variables converging to a limit ([Laugwitz, 1987]).

So we are dealing with two opposing views:

i) Cauchy investigated functions defined on a standard continuum and he considered infinitesimals only as a useful notion definable in terms of the standard continuum by means of variables converging to a limit,

ii) Cauchy investigated functions defined on a non-standard continuum and he considered infinitesimals "as mathematical entities of their own specific rank, which can be considered 'directly', though it may sometimes be useful to think of them as *represented* by sequences" ([Laugwitz, 1987], p.263).

We need some more quotations. Here is how Cauchy defines the continuity of a function on an interval:

> "If, starting from a value of x between those two bounds, one assigns to the variable x an infinitely small increment α, the increment of the function itself will be the difference $f(x+\alpha)-f(x)$ that will depend at the same time on the new variable α and the value of x. Having written that, the function f(x), between the two bounds assigned to the variable x, will be a continuous function of this variable if for each value of x between those bounds, the numerical value of the difference $f(x+\alpha)-f(x)$ decreases indefinitely with the value of α. In other words, *the function f(x) will*

---

[8]"On dit qu'une quantité variable devient *infiniment petite*, lorsque sa valeur numérique décroît indéfiniment de manière à converger vers la limite zéro" ([Cauchy,1821], p.37; the emphasis is Cauchy's).

remain continuous with respect to x between the given bounds, if, between those bounds, an infinitely small increment of the variable always produces an infinitely small increment of the function itself".[9]

And on the convergence of series Cauchy wrote:

"By virtue of the principles that have been established above, in order that the series [...] $u_0, u_1, u_2, \ldots, u_n, u_{n+1}, \ldots$

converges, it is necessary and sufficient that growing values of x make the sum

$$s_n = u_0+u_1+u_2+u_3+\ldots+u_{n+1}$$

converge indefinitely to a fixed limit s; in other words, it is necessary and sufficient that, for infinitely large values of the number n, the sums

$$s_n, s_{n+1}, s_{n+2}, \ldots$$

differ in infinitely small quantities from s, and consequently from each other".[10]

In a standard interpretation of these two quotations the emphasis is in both cases on the first part, and the second part introduced by "in other words" ("en d'autres termes") is considered to be simply a paraphrase of the first part. A non-standard interpretation, on the contrary, puts the emphasis in both quotations on the second part, and therefore the first part in both cases concerns only one possible way to apply the second part.

---

[9]"Si, en partant d'une valeur de x comprise entre ces limites, on attribue à la variable x une accroissement infiniment petit α, la fonction elle-même recevra pour accroissement la différence $f(x+\alpha)-f(x)$, qui dépendra en même temps de la nouvelle variable α et de la valeur de x. Cela posé, la fonction f(x) sera, entre les deux limites assignées à la variable x, fonction *continue* de cette variable, si, pour chaque valeur de x intermédiaire entre ces limites, la valeur numérique de la différence $f(x+\alpha)-f(x)$ décroît indéfiniment avec celle de α. En d'autre termes, *la fonction f(x) restera continu par rapport à x entre les limites données, si, entre ces limites, un accroissement infiniment petit de la variable produit toujours un accroissement infiniment petit de la fonction elle-même*" ([Cauchy,1821], p.43).

[10]"D'après les principes ci-dessus établis, pour que la série [...] u_0, u_1, $u_2, \ldots, u_n$, $u_{n+1}, \ldots$soit convergente, il est nécessaire et il suffit que des valeurs croissantes de n fassent converger indéfiniment la somme $s_n = u_0+u_1+u_2+u_3+\ldots+u_{n+1}$ vers une limite fixe s; en d'autres termes, il est nécessaire et il suffit que, pour des valeurs infiniment grandes du nombre n, les sommes $s_n, s_{n+1}, s_{n+2}, \ldots$diffèrent de la limite s, et par conséquent entre elles, de quantités infiniment petites" ([Cauchy,1821], p.115).

III.3.2 Lakatos's non-standard interpretation of the 1821 proof versus a standard interpretation

Let us now consider Lakatos's *non-standard interpretation* of Cauchy's 1821 proof. If one confines oneself strictly to the text of the proof, Lakatos's interpretation, quoted above, seems quite possible. After all, Cauchy phrased the proof almost entirely in terms of infinitesimals. Yet Lakatos's interpretation is problematic. In Lakatos's view Cauchy in fact considered a transfinite, convergent sequence of continuous functions $\{s_n(x)\}$. Cauchy did not consider the continuity of the limit function of such sequences obvious, because otherwise he would not have felt the need to prove the theorem. However, in Lakatos's interpretation in the proof Cauchy implicitly uses *the continuity of* s_n *for transfinite n*. Considering the low level of sophistication in dealing with the transfinite in Cauchy's work this amounts to begging the question. For example, when Cauchy elsewhere in the *Cours d'analyse*[11] shows that

$$1+\mu x+[\mu(\mu-1)/2!]x^2+ \text{ etc.} = (1+x)^\mu,$$

he first shows that the convergent series $1+\mu x+[\mu(\mu-1)/2!]x^2+$ etc. is a continuous function of μ by means of the theorem that we are discussing. The finite partial sums here are, apparently, polynomials continuous in μ. If Lakatos's interpretation of the theorem is correct, application of the theorem in this case requires that the *transfinite* partial sums of the series are continuous in μ! Cauchy neither proves nor mentions this, although it concerns an assumption which is just as hard to prove as the theorem itself. Lakatos saw no problem here. He just wrote that there is no reason why Cauchy could not think that $s_n(x_1+\alpha)-s_n(x_1)$ is infinitely small for infinitely large values of n.[12]

There is another problem. Why, in the following sentence - "[the increment] of r_n will become imperceptible together with r_n, if one assigns to n a very considerable value" - did Cauchy write 'imperceptible' instead of 'infinitely small' and 'very considerable' instead of 'infinitely large'? Those who defend Lakatos's non-standard interpretation have to assume that Cauchy was sometimes rather casual in his terminology.

Conclusion: If one restricts oneself to the text of the 1821 proof and the "definitions" of the major notions that accompany that proof, Lakatos's interpretation of the text is definitely a possible interpretation. Yet we must

---

[11] Cf. Chapter I, section I.4.3

[12] [Lakatos, 1978b], p.56

assume then that Cauchy was not always consistent concerning terminology and secondly, that the proof contains a serious flaw: it is practically circular.

Let us now first consider a *standard interpretation* of the text of the theorem in order to compare it with Lakatos's interpretation. A standard interpretation could run as follows. Cauchy repeatedly mentions infinitely small quantities in the proof. Yet wherever he does so, he simply echoes his definition of a continuous function in terms of infinitesimals. The sentence - "[the increment] of r_n will become imperceptible together with r_n, if one assigns to n a very considerable value" - can now be taken literally. Cauchy then actually says: It is possible to make n so large that the remainder r_n can be neglected (i.e. we may put r_n equal to 0) and then the increment of r_n can also be neglected. From the standard point of view one could say this: Terms like "become imperceptible" and "a very considerable value" are not technical terms in Cauchy's work. If Cauchy had meant respectively "infinitely small" and "infinitely large", he would have said so. In the proof we have:

$$s(x_1+\alpha)-s(x_1) = s_n(x_1+\alpha)-s_n(x_1)+r_n(x_1+\alpha)-r_n(x_1)$$

Cauchy makes n so large that r_n can be neglected. One then has

$$s(x_1+\alpha)-s(x_1) = s_n(x_1+\alpha)-s_n(x_1)$$

and, obviously, when α converges to zero, the right-hand side $|s_n(x_1+\alpha)-s_n(x_1)|$ converges to zero (because of the continuity of s_n) and so does the left-hand side.

If one confines oneself to the text of Cauchy's proof and the definitions surrounding it, this standard interpretation is definitely possible. Yet in this interpretation the proof is not very satisfactory either. The assumption that for large n the remainder r_n can be neglected, trivially implies the theorem and Cauchy's proof proves as little as a circular proof.

III.3.3 Cauchy's 1853 text

Let us now have a look at Cauchy's 1853 revision of the theorem involved. Perhaps that will help us to decide between the two interpretations. The standard interpretation of the 1853 paper is that Cauchy corrected his mistaken 1821 version of the theorem and added the condition of uniform convergence. The defenders of a non-standard interpretation argue that Cauchy repeated the same theorem.

In 1853 Cauchy deals with a convergent series of continuous functions on a certain interval:

(1) $s = u_1+u_2+u_3+ \ldots$

The reason for returning to the theorem is the existence of problematic series like

(2) $\sin x + \frac{\sin 2x}{2} + \frac{\sin 3x}{3} + ...$

Cauchy remarks:

"it is easy to see how one must modify the wording of the theorem so that it does not give rise to any further exception".[13]

This can be understood as meaning that Cauchy only wanted to modify the phrasing of the theorem and not its contents, but it can obviously also be understood as meaning that Cauchy wanted to re-phrase the theorem because he had discovered that it was wrong in its 1821 form.

By definition again $s=s_n+r_n$ with $s_n=u_1+u_2+ ... u_{n-1}$. The following is a free translation of Cauchy's argument[14]:

"Suppose that, by taking n large enough, for all values of x involved, we could make the expression $|u_n+u_{n+1}+...+u_{n'-1}|$ (for arbitrary n'), and consequently r_n, smaller than an arbitrarily small ε. Because $|s_n(x+\alpha)$-

---

[13]"il est facile de voir comment on doit modifier l'énoncé du théorème, pour qu'il n'y ait plus lieu à aucune exception" ([Cauchy, 1853], pp.31-32).

[14]"Si l'on nomme n' un nombre entier supérieur à n, le reste r_n ne sera autre chose que la limite vers laquelle convergera, pour des valeurs croissantes de n', la différence (3) $s_{n'}-s_n = u_n+u_{n+1}+...+u_{n'-1}$ Concevons, maintenant, qu'en attribuant à n une valeur suffissamment grande on puisse rendre, pour toutes les valeurs de x comprises entre les limites données, le module de l'expression (3) (quel que soit n'), et, par suite, le module de r_n, inférieurs à un nombre ε aussi petit que l'on voudra. Comme un accroissement attribué à x pourra encore être supposé assez rapproché de zéro pour que l'accroissement correspondant de s_n offre un module inférieure à un nombre aussi petit que l'on voudra, il est clair qu'il suffira d'attribuer au nombre n une valeur infiniment grande, et à l'accroissement de x une valeur infiniment petite, pour démontrer, entre les limites données, la continuité de la fonction $s = s_n+r_n$. Mais cette démonstration suppose évidemment que l'expression (3) remplit la condition ci-dessus énoncée, c'est-à-dire que cette expression devient infiniment petite pour une valeur infiniment grande attribuée au nombre entier n. D'ailleurs, si cette condition est remplie, la série (1) sera évidemment convergente. En conséquence, on peut énoncer le théorème suivant:

Théorème I. - *Si les différents termes de la série*

(1) $u_0, u_1, u_2, ..., u_n, u_{n+1}, ...$

sont des fonctions de la variable réelle x, continues, par rapport à cette variable, entre des limites données; si, d'ailleurs, la somme

(3) $u_n + u_{n+1} + ...+ u_{n'-1}$

devient toujours infiniment petite pour des valeurs infiniment grandes des nombres entiers n et n' > n, la série (1) sera convergente, et la somme s de la série (1) sera, entre les limites données, fonction continue de la variable x" ([Cauchy, 1853], pp.32-33).

$s_n(x)|$ can be made as small as one wants by taking the increment α small enough, it is clear that it suffices to attribute to n an infinitely large value, and to the increment α of x an infinitely small value, in order to demonstrate on the given interval the continuity of the function

$$s = s_n + r_n.$$

But this demonstration obviously supposes that the expression $u_n+u_{n+1}+\ldots+u_{n'-1}$ satisfies the condition expressed above, i. e. the expression becomes infinitely small for an infinitely large value attributed to the number n. Moreover, if that condition is fulfilled, the series $s = u_1+u_2+u_3+\ldots$ will obviously be convergent. Consequently one can formulate the following theorem:

Theorem I - If the different terms of the series

(1) $u_0, u_1, u_2, \ldots, u_n, u_{n+1}, \ldots$

are continuous functions of the real variable x on the given interval; if, moreover, the sum

(3) $u_n+u_{n+1}+\ldots+u_{n'-1}$

becomes always infinitely small for infinitely large values of the integers n and $n'>n$, the series (1) will be convergent and the sum s of the series (1), between the given limits, will be a continuous function of the variable x".

III.3.4 A standard interpretation of the 1853 text

Let us first consider a possible *standard interpretation* of this text. When Cauchy says: "But this demonstration obviously supposes that the expression $u_n+u_{n+1}+\ldots+u_{n'-1}$ satisfies the condition, expressed above, that it becomes infinitely small for an infinitely large value attributed to the number n", which above-mentioned condition does he mean? Only one of the statements concerned is clearly phrased as a condition, viz. the one expressed in this sentence:

(A) Suppose that by taking n large enough, for all values of x involved, we could make[15] the expression $|u_n+u_{n+1}+\ldots+u_{n'-1}|$ (for arbitrary n'), and consequently r_n, as small as we want.

Cauchy considers this condition to be equivalent with

(B) The expression $u_n+u_{n+1}+\ldots+u_{n'-1}$ becomes infinitely small for an infinitely large value attributed to the number n.

[15] I am grateful to Stanko Lasić for verifying that in 18th/19th century French, the expression "Concevons qu'on puisse rendre" clearly introduces a condition and not a statement which the reader is supposed to accept unconditionally.

In the text of the theorem he once more expresses the condition as follows:
(C) The sum $u_n+u_{n+1}+...+u_{n'-1}$ becomes always infinitely small for infinitely large values of the integers n and n'>n.
Cauchy expresses one and the same condition in three different ways. When one considers the rest of Cauchy's paper it becomes clear what he means. There he considered one of the counter-examples to the 1821 theorem:

$$(2) \qquad \sin x + \frac{\sin 2x}{2} + \frac{\sin 3x}{3} + \ldots$$

and calculated in that case the value of the remainder r_n:

"If one substitutes the series (2) in the series (1), the expression (3), reduced to the sum

$$(4) \quad \frac{\sin (n+1)x}{n+1} + \frac{\sin (n+2)x}{n+2} + \ldots + \frac{\sin n'x}{n'}$$

will vanish for x=0; but for those values of x which are very close to zero, for example for x=1/n, n being a very large number, it could differ considerably from zero; and if, when assigning to n a very large value, one simply puts x=1/n and n'=∞, the sum (4) - or the remainder r_n of series (2) which amounts to the same thing - will noticeably reduce itself to the integral

$$\int_1^\infty \frac{\sin x}{x}\, dx = \frac{\pi}{2} - 1 + \frac{1}{1.2.3}\frac{1}{3} - \frac{1}{1.2.3.4.5}\frac{1}{5} + \ldots = 0.6244\ldots\text{"}^{16}$$

16"Si à la série (1) on substitue la série (2), l'expression (3), réduite à la somme

$$(4) \quad \frac{\sin(n+1)x}{n+1} + \frac{\sin(n+2)x}{n+2} + \ldots + \frac{\sin n'x}{n'}$$

s'évanouira pour x=0; mais pour des valeurs de x très voisines de zéro, par example pour x=1/n, n étant un très grand nombre, elle pourra différer notablement de zéro; et si, en attribuant à n une très grande valeur, on pose seulement x = 1/n , mais encore n'=∞, la somme (4), ou, ce qui revient au même, le reste r_n de la série (2) se réduira sensiblement à l'intégrale

$$\int_1^\infty \frac{\sin x}{x}\, dx = \frac{\pi}{2} - 1 + \frac{1}{1.2.3}\frac{1}{3} - \frac{1}{1.2.3.4.5}\frac{1}{5} + \ldots = 0{,}6244\ldots\text{"}$$ ([Cauchy,1853], p.33).

What happens is this. Cauchy concentrates on

$$r_n(x) = \sum_{i=1}^{\infty} \frac{\sin (n+i)x}{n+i}$$

for very large n, and calculates an approximation the of r_n for x=1/n. If n is large, this x is close to zero. The value of the remainder is then

$$r_n = \sum_{i=1}^{\infty} \frac{\sin (1+\frac{i}{n})}{1+\frac{i}{n}} \frac{1}{n}$$

which is approximately equal to the integral

$$\int_1^{\infty} \frac{\sin x}{x} dx = \int_0^{\infty} \frac{\sin x}{x} dx - \int_0^{1} \frac{\sin x}{x} dx$$

The first integral on the right hand side equals $\pi/2$ and the second integral on the right hand side is presumably determined by Cauchy after substitution of a power series for sin x. The example shows that in this case the sum $|u_n+u_{n+1}+...+u_{n'-1}|$ does not always become as small as we want if n is large enough and $n' \geq n$! No matter how large n is, we can always find an x very close to zero such that r_n equals approximately 0.6. From a standard point of view, the example shows that Cauchy was aware of the fact that $|u_n+u_{n+1}+ ...+u_{n'-1}|$ should "uniformly" vanish for growing n and n' ($n' \geq n$). In other words Cauchy's condition is uniform convergence.

The proof from a standard point of view can then be paraphrased as follows: Suppose we have uniform convergence. We can then make r_n, and consequently also its increments, as small as we want independently of α by choosing n large enough. Moreover, after having fixed (a large enough) n, we can make the increment of s_n as small as we want by taking α small enough. Because $s = s_n+r_n$, we can make the increment of s as small as we want by taking α small enough. Q.E.D. Before he re-phrases the theorem, Cauchy repeats the condition of uniform convergence and draws attention to the fact that it implies convergence, which means that the condition of convergence can be dropped when we require uniform convergence.

From a standard point of view Cauchy cannot resist expressing himself in terms of infinitely small and infinitely large quantities, although that language lacks clarity in situations like this. Yet his proof is acceptable from a standard point of view.

III.3.5 A non-standard interpretation of the 1853 text

Let us now consider a non-standard interpretation of the 1853 text. Because Lakatos did not thoroughly discuss Cauchy's 1853 proof, I will consider the non-standard interpretation given by Cleave in the footnotes to his edition of Lakatos's manuscript. From the point of view of Lakatos and Cleave, Cauchy in 1853 simply repeated his 1821 theorem in a slightly modified form. Cauchy's modified theorem does *not* then contain an extra condition. Cleave's view is that instead of merely referring to *convergent* sequences of functions, Cauchy in 1853 explicitly included his notion of a convergent sequence of functions in the text: "the sum $u_n+u_{n+1}+ ... +u_{n'-1}$ becomes always infinitely small for infinitely large values of the integers n and n'>n".

According to Cleave, Cauchy's notion of convergence implies uniform convergence: by taking n large enough for all x involved, r_n (and consequently its increment) can be made smaller than an arbitrary ε.[17] Moreover, the increment of s_n can be made as small as one wants by taking α small enough. The continuity of $s=s_n+r_n$ clearly then follows by taking n infinitely large and the increment α of x infinitely small. This last sentence echoes the 1821 proof. In this non-standard interpretation, Cauchy's 1853 proof differs from Cauchy's 1821 proof in Lakatos's interpretation in the sense that Cauchy attempted to be more explicit. Yet it is not a very clear proof.[18]

III.3.6 An argument in favour of the standard interpretation

If one compares the standard and the non-standard interpretations which I described above of Cauchy's 1821 and 1853 versions of the theorem without adding other arguments, it is not absolutely clear which interpretation is to be preferred. The reader may prefer one of the two interpretations, but he will have to admit that the other interpretation can only be rejected on the basis of further evidence. I will argue that a standard interpretation of the texts that we discussed should be preferred to a non-standard interpretation because a non-standard interpretation produces more problems than it solves. In other words, a standard interpretation along the lines given above is simpler. Lakatos wrote: "Cauchy's story is a fairly mysterious one, everywhere dense

---

[17] [Lakatos, 1978b], p.52, footnote by Cleave.

[18] Cleave gives a modern non-standard proof of the theorem that Cauchy's notion of convergence implies uniform convergence ([Lakatos, 1978b], pp. 50-51, footnote by Cleave). The argument is based too much upon modern notions for it to help us understand Cauchy.

with problems. Robinson's theory, however, gives us the crucial clue to their solution".[19] This was wishful thinking, as I will show now.

Standard interpretations of the 1821 and 1853 texts may differ as to what Cauchy precisely thought when he wrote down each line of his proofs; they agree as follows as to the meaning of the theorems and their proofs. In 1821 Cauchy made a mistake by not distinguishing between convergence and uniform convergence. In 1853 Cauchy corrected the mistake: he incorporated a hidden lemma - the requirement of uniform convergence. As far as I can see, there is essentially only one standard interpretation in this sense.[20] There are, however, several quite different non-standard interpretations. The first non-standard interpretation of Cauchy's theorem on a convergent sequence of continuous functions was given by Robinson. Lakatos was influenced by Robinson, but the Lakatos-Cleave interpretation given above is different from Robinson's. Lakatos-Cleave believed that Cauchy twice proved the same correct non-standard theorem concerning sequences of functions converging everywhere on a non-standard continuum. The second proof is then simply a more elaborate version of the correct 1821 proof. Robinson, however, assumed that Cauchy's original theorem only referred to sequences of functions converging in standard points. He interprets the theorem as a standard theorem and only the proof as being a non-standard proof. In Robinson's view there is a gap in the proof.[21] According to Robinson the gap was filled in 1853: the proof is still a non-standard proof, but the condition that Cauchy introduced amounts precisely to uniform convergence.[22]

A third, and again different non-standard interpretation, was given by Laugwitz. Laugwitz wrote about the 1821 proof:

> "I read the proof as follows. I assume that convergence is postulated for all values of x of the variable in the given interval, and moreover for all $x+\alpha$, α infinitesimal; the latter assumption is needed in the proof, but it is not

---

[19][Lakatos,1978b] p.47

[20]In 1984 Giusti gave a slightly different standard interpretation of Cauchy's "mistake" concerning the limit function of a convergent sequence of continuous functions ([Giusti,1984]). In Giusti's view Cauchy possessed no notion of point-wise convergence of a series of functions. Cauchy in 1821 proved a theorem for a uniform convergent sequence of continuous functions and all he did in 1853 was to write down the theorem and the proof somewhat more extensively. Also in this interpretation Cauchy in 1853 made explicit a lemma that was hidden in the 1821 proof.

[21][Robinson, 1966], p.272

[22][Robinson, 1966], p.273

obvious in the statement of the theorem. Let $s(x)=s_n(x)+r_n(x)$ and $s(x+\alpha)=s_n(x+\alpha)+r_n(x+\alpha)$, where $s_n = u_0+u_1+...+u_{n-1}$. Then s_n is, for finite n, a continuous function, and $s_n(x+\alpha)-s_n(x)$ is infinitesimal. According to the convergence both $r_n(x)$ and (!) $r_n(x+\alpha)$ will become 'imperceptible' for very large (but still finite) numbers n. In other words, for any given finite $\varepsilon>0$, $|r_n(x)|<\varepsilon$ and $|r_n(x+\alpha)|<\varepsilon$ for some finite n. Hence,

$$| s(x+\alpha)-s(x) | \leq | s_n(x+\alpha)-s_n(x) |+| r_n(x+\alpha) |+| r_n(x) | < 3\varepsilon$$

for each finite $\varepsilon > 0$, i.e., $s(x+\alpha)-s(x)$ is infinitesimal. It follows that s(x) is a continuous function" ([Laugwitz, 1987], p.264).

This interpretation leads to a correct non-standard proof. Laugwitz interprets the 1853 text similarly. However, Laugwitz comments that the text of the 1853 theorem is, "to put it mildly, at least misleading" ([Laugwitz,1987], p.265) and that "the proof is more involved than that in the Cours". In the 1853 paper Laugwitz seems mainly to appreciate Cauchy's discussion of the "counter-example" that is following the proof.[23] Of course, this raises the question as to why Cauchy's work degenerated and why he did not simply write down his 1821 proof a bit more extensively.

Lakatos remarked in his "Cauchy and the continuum", after having discussed Cauchy's 1821 and 1853 proofs: "So slowly everything seems to fall into place, and an exciting story of two rival theories of the calculus is revealed[...]".[24] The end of the sentence is: "at, however, a surprisingly low degree of articulation". It is precisely this low degree of articulation, that is bothersome. Whenever Cauchy explicitly wrote *about* infinitesimals in his foundational work, he did so in terms of variables converging to zero assuming standard numbers as values. The assumption that in Cauchy's reasoning infinitesimals are more than such sequences, and that the sequences are only an expedient to handle infinitesimals, ought to lead to simpler, more natural interpretations. In the case of Cauchy's theorem on convergent series of continuous functions this is not so. The choice for a non-standard

---

23 About the 1821 proof Laugwitz writes: "[...]if 'très considérable' meant infinitely large Cauchy would have said so[...]" ([Laugwitz,1987], p.264). In the case of the 1853 example, however, he has a different view, because he paraphrases Cauchy's "x = 1/n, n étant un très grand nombre" as follows: "Let x=1/n for n infinitely large" ([Laugwitz,1987], p.265). And he must do so, because Laugwitz' interpretation is based upon the assumption that Cauchy in 1821 and in 1853 postulates convergence for all x and all $x+\alpha$ with α infinitesimal. The 1853 counter-example must then be interpreted as referring to the infinitesimal neighbourhood of x = 0.

24 [Lakatos,1978b] p.53

interpretation only creates problems.[25] The "low degree of articulation" of Cauchy's non-standard notion of the continuum leaves us groping in the dark as to precisely what he intended to prove, and creates room for several different interpretations.[26]

It is interesting to compare in Cauchy's work the situation of the infinitesimals with the situation of the complex numbers. Cauchy defines complex numbers in terms of real numbers and everywhere in his foundational work he carefully distinguishes functions defined on a set of real numbers and functions defined on a set of complex numbers. Cauchy's definition of infinitesimals in terms of sequences of real numbers can be compared with his definition of complex numbers in terms of real numbers. If Cauchy had really wanted to consider functions defined on sets of infinitesimals, isn't it then highly improbable that he would not have explicitly said so?

In the course of history the status of infinitesimals has very often been problematic. Cauchy also seems to have his reservations, at least in his foundational work. A sentence such as:

---

[25]Fisher quoted Cauchy as follows: "When the successive numerical values of the same variable decrease indefinitely, in such a way as to fall below any given number, this variable becomes what one calls *an infinitely small* or an *infinitely small quantity*. A variable of this kind has zero for limit" ([Fisher,1978], p. 316) and he writes that it "*could* mean that when the numerical (i.e. absolute) values of a variable decrease in such a way as to be less than any positive number, then the variable takes on infinitesimal values" ([Fisher,1978] , p.316). He asks: "Can Cauchy's infinitely small quantities, these variables with limit zero, take on infinitesimal values?" ([Fisher,1978], p.318). Fischer's conclusion is that nowhere in the Cours d'Analyse does Cauchy explicitly exclude this possibility. One can say more. It is true that nowhere does Cauchy write: Infinitely small quantities, i.e. variables with limit zero, *cannot* take on infinitesimal values. But if the answer to Fisher's question is yes, then we have in Fischer's interpretation only two possibilities. *Either* there are two kinds of infinitesimals - the ones that are variables with limit zero and the ones that these variables can take on as values - *or* Cauchy's infinitesimals are variables that can take on themselves as values. It seems out of the question that Cauchy ever considered these two possibilities. There are, as far as I know, no contemporaries of Cauchy who pointed at such interpretations, which, had they been serious, would have been quite disturbing in a book meant for students (based upon lectures that Cauchy had already given) in which Cauchy, moreover, wanted to be as rigorous as possible.

[26]Of course, one should not restrict oneself to the theorem that I discussed in this chapter if one wants to compare the standard and non-standard interpretations of Cauchy's work. If elsewhere in Cauchy's work non-standard interpretations would be much simpler than non-standard interpretations, then that would be a good argument for taking seriously a non-standard interpretation of the theorem on the limit of a converging sequence of continuous

> "When speaking about the continuity of functions, I have not been able to avoid explaining the properties of the infinitely small quantities, properties that serve as basis of the infinitesimal calculus"[27]

does not suggest that the infinitesimals were for Cauchy an essential part of the continuum. It suggests that he would have preferred to have investigated the continuity of functions without using infinitesimals. As for matters of existence, Cauchy is very careful in his foundational work. If the advocates of a non-standard interpretation are right, Cauchy must have assumed that one can calculate with infinitesimals roughly in the way one calculates with reals. How should we then reconcile Cauchy's explicit rejection of the generality of algebra[28] with the fact that he pays relatively little attention to calculation with infinitesimals? The chapter on infinitesimals in the *Cours d'analyse* is mainly dedicated to ratios of infinitesimals. If in an arbitrary analytical expression, as far as Cauchy was concerned, variables really varied over a non-standard continuum, it is strange that he did not pay more attention to the rules of calculation with infinitesimals. The choice of a non-standard interpretation also creates problems in this respect.[29]

---

functions. Laugwitz discussed several other examples of Cauchy's reasoning from a non-standard point of view. In Chapter IX, footnote 28, I will briefly discuss one of those examples.

[27]"En parlant de la continuité des fonctions, je n'ai pu me dispenser de faire connaître les propriétés principales des quantités infiniment petits, propriétés qui servent de base au calcul infinitésimal" ([Cauchy, 1821], p.ij).

[28]Compare Chapter VIII, section VIII.4.4.

[29]Some of the difficulties involved are already clear in Lakatos's text. In section 5 of his paper Lakatos draws our attention to the fact that there is a big difference between Robinson's non-standard continuum and Cauchy's non-standard continuum. He admits so far having analysed Cauchy's work as relating to a *static* non-standard continuum. However, to Lakatos, Cauchy's 1853 paper reveals that Cauchy's continuum does not consist of a set of actual points but of *moving points*. Cauchy's continuum appears to be *dynamic*. In Lakatos view Cauchy in 1853 showed that the recalcitrant sequence

$$\sin x + \frac{\sin 2x}{2} + \frac{\sin 3x}{3} + .$$

does not converge at the moving point $x=1/n$ where $n \to \infty$. Lakatos writes: "The fact that the sequence does not converge at this moving point is in fact what later came to be known as the Gibbs-phenomenon and the corresponding condition - namely that $\Sigma f_n(x)$ is uniformly convergent in I if for all $\{x_n\}$ in I the corresponding remainders $r_n(x_n)$ tend to zero - can be shown to be equivalent to Weierstrassian uniform convergence" ([Lakatos,1978b], p.57). The striking thing is that Lakatos in this sentence is very close to

III.4 Concluding remarks

It was necessary to study "Cauchy and the continuum" for several reasons. Firstly, for the sake of completeness, it was inevitable that in the present study the paper - which is one of Lakatos's rare texts on mathematics related to the MSRP - should be taken into consideration. Moreover, as we have seen in section III.2, "Cauchy and the continuum" shows that in 1966, Lakatos's thinking was developing in the direction of the MSRP and that the appearance of the MSRP in Lakatos's work was more than a reaction to Kuhn's philosophy of natural science. The main reason for studying the paper, however, was that if the reconstruction had been factually correct we would have had a beautiful example of two competing research programmes in mathematics, and we could have attempted to draw conclusions from it. Unfortunately the reconstruction is very unsatisfactory. In the next chapter I will show that the above considerations concerning "Cauchy and the continuum" nicely illustrate certain circularity problems that historians or philosophers are sometimes faced with when they attempt to reconstruct mathematical developments. The central question is: How can we avoid our selection and interpretation of the historical material to be so biased that the resulting rational reconstruction no longer describes what happened but merely illustrates our subjective preferences. In the next chapter I will concentrate on those problems. I will show that they are serious, but that they can be solved, at least conditionally. A brief discussion of an attempt by Spalt to characterize the rationality of mathematics from a Lakatosian perspective will be my point of departure.

---

reducing the differences between Weierstrass and Cauchy in 1853 to mere terminological differences! In other words, Lakatos moves towards a standard interpretation.

Chapter IV A meta-methodological intermezzo

IV.1 Spalt's "Vom Mythos der mathematischen Vernunft"

An attempt to expand the ideas advanced in "Cauchy and the continuum" has been made by Spalt. His *Vom Mythos der mathematischen Vernunft* ("On the Myth of Mathematical Reason" [Spalt,1981]) contains the conversations of an imaginary group of students over a period of three days. The group is interested in the problem of the rationality of mathematics, and a large part of Spalt's book is dedicated to an attempt to define this rationality by means of an application of Lakatos's MSRP to the history of analysis. In the conversations on the second day, which fill more than two thirds of the book ([Spalt,1981], pp. 34-284), Spalt depicts the history of analysis as the history of the competition between two research programmes. On one hand we have what Spalt calls the *continuity programme* (das Kontinuitäts-programm), and on the other hand what he calls the *finitary programme* (das Finitärprogramm). The two programmes are characterized by means of their positive heuristics. The positive heuristic of the continuity programme is the following:

> "*One calculates explicitly with infinite numbers; everything that holds for all finite natural numbers also holds for the infinite numbers*".[1]

The rival programme, the finitary programme, has the positive heuristic:

> "*The form of the infinite (or a limit process) is the means to relate quantities of different natures to one another* - irrational numbers are limits of rational numbers; discontinuous functions are limits of continuous functions; the infinite serves as a bridge that connects the previously unconnected".[2]

The positive heuristic of the finitary programme led to the banning of infinite numbers from analysis and also to the use of $\varepsilon-\delta$ -methods. Mathematicians who allegedly worked in the continuity programme were, for example, Leibniz, Fourier, Cauchy and Riemann, while the finitary programme is represented, amongst others, by Newton, Stokes, Seidel and Weierstraß. In

---

[1]"*Es wird ausdrücklich mit unendlichen Zahlen gerechnet; dabei gilt für die unendlichgroßen Zahlen alles das, was für sämtliche endlichen natürlichen Zahlen gilt*" ([Spalt,1981], p.73).

[2]"*Die Form des Unendlichen (oder: ein Grenzprozess) ist das Hilfsmittel, um Größen unterschiedlicher Natur zueinander in Beziehung zu setzen* - irrationale Zahlen sind Grenzwerte rationaler Zahlen; unstetige Funktionen sind Grenzwerte stetiger Funktionen. Das Unendliche dient als Brückenschlag, der zuvor Unverbundenes miteinander verbindet" ([Spalt,1981], p.74, emphasis is Spalt's).

Spalt's reconstruction the finitary programme completely replaced the continuity programme in the course of the nineteenth century. So far Spalt's reconstruction is more or less in accordance with "Cauchy and the Continuum". However, the discussions then take a surprising turn. The victory of the finitary programme is connected with the rise of "Bourbakistic mathematics", and an attempt is made to show the *irrational character* of this development by means of the following criterion for progress in mathematics. If, within a programme, a new *example* appears, then we have *internal progress* for that programme. If the example is new from the point of view of all other programmes, then we have *over-all scientific progress* ([Spalt,1981], p.78). According to the imaginary character Peter, examples are for mathematics what facts are for natural science (Ibid.).[3] The conversations on the second day contain several nineteenth century results that allegedly show the superiority of the continuity programme over the finitary programme and contain an attack on Bourbakistic mathematics.[4] On the third day, the dialogue again takes a remarkable turn. The participants in the discussion discover that the authors of nineteenth century textbooks cannot usually be clearly identified as representatives of one programme or another. A third programme is discovered - a programme aiming at the reduction of analysis to algebra ("das Algebraisierungs-Programm"), whose first representative was Lagrange - and it seems that the authors of nineteenth century textbooks often represent several programmes at the same time. The conclusion is:

> "A comprehensive study of the original literature reveals the defectiveness of the methodology of scientific research programmes. The more one pen-

---

[3]Spalt does not attempt to define the notion "example". Yet it is problematic. A particular cyclic group is an example of a group, and a group is an example of an algebraic structure. Whatever we study in mathematics can be considered as an example as soon as a more general notion shows up.

[4]The character Anna says: "*wie kann dann die Mathematik dennoch zu diesem dogmatischen Lehrgebäude gerinnen*, das wir heute vor uns sehen?" ([Spalt,1981], p.170) and Peter says : "Die Wahrheit ist keineswegs erreicht, sondern die Unerschütterlichkeit des vorgestellten Wissenschafts-gebäudes beruht einfach auf einer *systematischen Ausblendung aller Zweifel und Kritik*. Die heute vorgestellte Mathematik ist ein blutleeres, saft- und kraftloses *Kunstprodukt* des modernen Mathematiker(kreise)s *Bourbaki* und der von ihm propagierten und verwendeten *axiomatischen Methode*; mit der früher gewachsenen und auch noch heute ständig wachsenden wirklichen Mathematik hat dieses Gespenst nichts gemein"(Op.cit., p.171).

etrates into the mathematics which was written down, the more the idea of the competing research programmes becomes blurred".[5]

The message is this. Mathematical rationality in the sense of Bourbaki is a myth because it is the rationality of deductive logic, and true mathematics develops in quite a different way. However, mathematical rationality in the sense of Lakatos's MSRP is also a myth. A unique mathematical rationality simply does not exist!

IV.2 The first circularity problem: concerning the identification of competing units

Although I have argued in Chapter III that Cauchy's theorem on the limit function of a converging series of continuous functions should be interpreted in the standard way, the discussion illustrates very well the *dilemma of confirmation*[6] that historians are sometimes faced with: in certain situations every argument in favour of an hypothesis A can be matched by a counter-argument in favour of an alternative hypothesis B. Staunch advocates of a non-standard interpretation of Cauchy's mathematics need not be convinced by my arguments in favour of a standard interpretation; and, vice versa, a convinced advocate of the standard interpretation will always find an ad hoc solution, should an impressive argument in favour of a non-standard interpretation come up. The case illustrates that the choice of a particular interpretation in the historiography of mathematics may in the final analysis depend on subjective preferences. Whenever such preferences are projected into the historical material, the resulting interpretation obviously should not be taken as an adequate description of the facts, because it suffers from a vicious circularity. The interpretation of Cauchy's theorem on the limit function of a converging series of continuous functions concerns the micro-level. Fortunately, there is among historians of mathematics in general considerable consensus on the facts regarding the micro level, and we need not worry too much about the circularity problem there. However, in the case of rational reconstructions of macro level developments that encompass many micro-level developments, there exists a circularity problem, similar to the one described above, which requires our special attention. Spalt's book illustrates

---

[5]"Ein breit angelegtes Studium von Originalliteratur offenbart die Mangelhaftigkeit der Methodologie der wissenschaftlichen Forschungsprogramme. Je eingehender sich der forschende Blick auf die niedergeschriebene Mathematik richtet, desto stärker verschwimmt die Idee der konkurrierenden Forschungsprogramme" ([Spalt,1981], p.285).

[6]The term is also used by Stegmüller ([Stegmüler, 1977], p. 7 and further).

this very well. Spalt defends the thesis that a unique mathematical rationality in the sense of Lakatos does not exist. The essence of Spalt's defence consists in the failure of one particular rational re-construction of the history of analysis by means of an adapted version of Lakatos's MSRP: the reconstruction, although very promising in the beginning, presented with much rhetoric and a wealth of facts, turns out to be counter-factual in the end. There is no need to discuss Spalt's book in detail. The dialogue form makes the book attractive on the one hand but creates difficulties on the other hand. The participants in the discussion express many quite divergent ideas. Moreover no consensus is reached in the end. Yet one of the problems that Spalt's attempt at a rational reconstruction of the history of analysis clearly suffers from is a lack of precision in the definition of the two research programmes. The continuity programme and the finitary programme do not possess clearly defined hard cores, neither do Spalt's characterizations of the positive heuristics clearly imply "preconceived plans" according to which "problems are being defined, auxiliary hypotheses are being constructed" etc. (cf. Lakatos's characterization of a positive heuristic). I argue that what happened is this. Spalt selected a complex of methods that were often accepted in proofs during the eighteenth century, and declared them to be the core of a research programme - *the continuity-programme*. The opposing programme, *the finitary programme* is, on the one hand, characterized by negation - the rejection of the "extrapolation" from the finite to the infinite (which characterizes the continuity-programme) - and, on the other hand, by the use of Weierstrassian ε-δ methods for dealing with limits (or their less precise 18th century predecessors). In the course of time, analysis underwent a rather slow process of rigorization: certain arguments which in the eighteenth century were acceptable in a proof, kept only some heuristic value or else disappeared altogether. Lagrange for example, in his attempt to found analysis, rejected infinitesimals, but treated infinite sums in many respects as finite sums. Spalt's reconstruction is an attempt to force the history of analysis into too simple a framework. The course of the dialogue shows the failure of the reconstruction; it does not show that mathematical rationality is a myth.

Criticism along these lines of Spalt's book, although correct, would be too easy. Apart from the precise way in which Spalt has organized and presented his case against the unity of mathematical reason, *Vom Mythos der mathematischen Vernunft* raises serious questions concerning global models of rationality such as Lakatos's MSRP. Even if Spalt's proof of the thesis is not satisfactory, the thesis that something like *the* mathematical rationality in the sense of Lakatos does not exist could still be true. Spalt's dialogue raises some important "meta-methodological" questions concerning the status of large units such as research programmes which encompass vast sections of mathematics. *How can we, when we attempt to rationally reconstruct developments in mathematics, protect ourselves against the danger of selecting from the overwhelming multitude of mathematical texts precisely those that support our preconceptions, and ignore or consider as less relevant those texts that are not in accordance with our model of rationality?* Isn't Spalt's dialogue very representative in the sense that whatever model of rationality we have, it will always be possible to corroborate it first by selecting the right facts, and to falsify it afterwards by selecting different facts? Isn't the whole enterprise doomed from the outset because of an insoluble dilemma of interpretation? Every argument in favor of an hypothesis A can be matched by a counter-argument in favor of an alternative hypothesis B.

Let me illustrate this by an example from Spalt's book. The problem concerns the identification of units such as research programmes, and assigning mathematicians to these units. Let us consider what Spalt writes about Fourier, a mathematician allegedly working in the continuum programme ([Spalt,1981], pp.132-143). Spalt considers a passage from Fourier's work ([Grattan-Guinness, 1972], pp.146-157)[7] in which Fourier solves the equation

(1) $1 = a.\cos u + b.\cos 3u + c.\cos 5u + \ldots etc.$

with unknowns a, b, c, etc. as follows. He differentiates infinitely often, and after substituting u = 0 he obtains a system of infinitely many linear equations in a, b, c,...etc. Fourier shows that if one breaks off the infinite sum (1) after n terms followed by differentiating n times and substituting u = 0, one obtains n equations in n unknowns which yield:

$$a = \frac{3^2}{3^2 - 1} \cdot \frac{5^2}{5^2 - 1} \cdot \frac{7^2}{7^2 - 1} \;\ldots\; \frac{(2n-1)^2}{(2n-1)^2 - 1}$$

By means of Wallis's infinite product

$$\frac{2}{\pi} = \frac{1.3.3.5.5.7\ldots}{2.2.4.4.6.6\ldots}$$

---

[7] Spalt refers to Fourier's *Théorie analytique de la chaleur* (1822). I am referring to an earlier version of the same passage from 1807, which can be found in [Grattan-Guinness,1972]. As regards my arguments the differences between the two editions are inessential. See also footnote 8 of this chapter.

Fourier concludes that the original system has the solution $a = 4/\pi$. Spalt sees here a clear application of the positive heuristic of the continuity-programme, which consists of the idea that everything which holds for all natural numbers holds also for infinitely large natural numbers. This makes sense: Fourier handles the infinite sum (1) as if it were finite. Moreover, in order to determine the solution of the infinite system of linear equations that he then obtains, Fourier assumes that the formula for a that applies to finite n also applies to $n = \infty$. Fourier, who obtains the roots b, c, d,...etc. of the infinite system in a similar way, finally arrives at the following result:

$$(2) \quad \frac{\pi}{4} = \cos u - \frac{1}{3}\cos 3u + \frac{1}{5}\cos 5u - \frac{1}{7}\cos 7u + \dots \text{ etc.}$$

Possibly because he is not sure that (2) holds for all values of u under consideration, Fourier verifies his answer in a different way. He considers the expression:

$$y = \cos x - \frac{1}{3}\cos 3x + \frac{1}{5}\cos 5x - \dots \frac{1}{2m-1}\cos(2m-1)x$$

about which he says:

"If this function of x and m were familiar, one would examine how it varies when m increases, and one would make sure that it continuously approaches the limit $\pi/4$, whatever the value of x would be. That means more briefly that if m becomes infinite, the function reduces to $\pi/4$."[8]

After some algebra Fourier gets[9]:

$$(3) \quad y = c - \frac{1}{2}\left\{\frac{\cos 2mx \,.\, \sec x}{2m} + \frac{\sin 2mx \,.\, \sec' x}{2^2 m^2} - \frac{\cos 2mx \,.\, \sec'' x}{2^3 m^3} + \dots \text{ etc.}\right\}$$

and he remarks:

[8]"Si on connaissait cette fonction de x et de m, on examinerait comment elle varie à mesure que m augmente, et l'on s'assurerait si elle approche continuellement de la limite $\pi/4$, quelle que soit la valeur de x. C'est-à-dire plus brièvement, qu'en fesant m infini la fonction doit se réduire à $\pi/4$" ([Grattan-Guinness, 1972], p.159). In the 1822 version of the text Fourier wrote about the expression involved "On considérera la valeur de cette suite comme une fonction de x et de m. On réduira la valeur de la fonction en une série ordonnée suivant les puissances négatives de m; et l'on reconnaîtra que cette valeur approche d'autant plus d'être constante et indépendente de x, que m est un plus grande nombre"([Fourier,1822], p.158).

[9]$\sec^{(n)}x$ is the n-th derivative of $1/\cos x$.

"One can see now that the more the number m increases, the more the value of y approaches the value of the constant c. That is why, when the number m is infinite, the quantity $\cos u - \frac{1}{3}\cos 3u + \frac{1}{5}\cos 5u - \ldots$ etc. has a constant value independent of x".[10]

Spalt sees here also an application of the positive heuristic of the continuity programme ([Spalt,1981], p.140). In Spalt's interpretation, Fourier substitutes $m = \infty$ in (3) and concludes that all terms vanish.

It is clear of course that Fourier assumes that certain properties of finite sums also hold for infinite sums. Yet, on the other hand, the two quotations show that, when Fourier uses the word "infini", he at least *also* thinks of a limit in the sense of the finitary programme, and one could even argue that when Fourier talks in the two statements quoted about the number m being infinite, this, is for him, a manner of speaking. In his reconstruction Spalt understandably concentrates on those aspects of Fourier's text that are in accordance with the assumption that Fourier worked in the continuity programme. It is precisely here that the problem becomes clear. How can we be sure that our selection and interpretation of the historical micro-level facts, that we use to identify programmes (e.g. to put persons into them), is not influenced to such extent by our biases in a rational reconstruction of a macro-level development that the whole enterprise of historically oriented methodology is bound to fail because of a vicious circularity? I call this the *first circularity problem*: How can we avoid vicious circularity when we identify large units like research programmes?

IV.3 The second circularity problem: concerning criteria of acceptance or rejection

When we want to corroborate a global notion of mathematical rationality, such as Lakatos's MSRP, we are faced at least twice with a circularity problem; first when we attempt to identify large units like research programmes. In this respect, we are confronted with a circularity problem concerning the *descriptive* part of the methodology. Moreover, we are faced with a circularity problem concerning the *normative* part of the methodology. I call this the *second circularity problem.* A considerable number of philosophers of

---

[10]"Il est visible maintenant que plus le nombre m augmente, plus la valeur de y approche de celle la constante c. C'est pourquoi lorsque le nombre m est infini la quantité cos x - (1/3)cos3x + (1/5)cos5x - ....etc. a une valeur constante et indépendante de x" ([Grattan-Guinness,1972], p.161).

science have written in particular about this second circularity problem. The first circularity problem regarding the identification of the competing units, for example research programmes, has not been their primary concern. This is understandable, because as far as rational reconstructions of macro-developments are concerned, the situation in natural science differs from the situation in mathematics. In natural science there are many well-known cases of competition: Newtonian mechanics versus Aristotelian mechanics, the theory of relativity versus Newtonian mechanics, etc. Furthermore, in the philosophy of science there exists an extensive literature on the rationality of natural science, and a considerable number of philosophers have offered different notions of rationality and, in particular, different criteria for theory rejection or acceptance. Lakatos's MSRP is only one example. Philosophers of science tend to concentrate on the question of what criteria for rejection or acceptance of theories and research programmes are preferable.

In order to clarify the issue further and to obtain an idea of a possible solution, it is useful to consider Laudan's characterization of the second circularity problem and the solution that he has come up with. Laudan characterizes the problem as follows:

> "If the writing of history of science presupposes a philosophy of science and if philosophy of science is then to be authenticated by its capacity to lay bare the rationality held to be implicit in the history of science, how can we avoid automatic self-authentication, since the history of science we write will presuppose the very philosophy which the written history will allegedly test?"[11]

In writing about natural science and concentrating on norms concerning theory rejection and theory acceptance, Laudan proposes the following solution. First of all he proposes to differentiate between what he calls HOS_1, which denotes the actual past of science and HOS_2, which denotes the writings of historians about that past. In Laudan's view the writings of philosophers who rationally reconstruct scientific developments are part of HOS_2. Although they have different aims, both historians and philosophers base their reconstructions of historical developments on certain norms. Historians do this less explicitly than do philosophers. Within HOS_1 there is a subset of cases of theory rejection or acceptance - about which most scientifically educated persons have strong normative intuitions - that is independent of the notion of scientific rationality which we want to test by means of rational reconstruc-

[11] [Laudan, 1977], p.157

tions. Laudan wants to use these "*preferred pre-analytic intuitions about scientific rationality*" (PI's for short) to test notions of scientific rationality. These preferred intuitions include, for example, the conviction that it was rational by, say, 1800 to reject Aristotelian mechanics and to accept Newtonian mechanics, and also the conviction that it was irrational after 1920 to believe that the chemical atom had no parts. Laudan states:

> "the degree of adequacy of any theory of scientific appraisal is proportional to how many of the PIs it can do justice to".[12]

We find the same solution of the (second) circularity problem also with Lakatos. Lakatos's 1971 paper "History of science and its rational reconstructions" ([Lakatos,1978a], pp.102-138) is devoted entirely to the relationship between notions of scientific rationality and the actual historical development. According to Lakatos, the rational reconstruction of the development of science on the basis of the MSRP requires that we identify competing research programmes in the history of science and that we describe their development in terms of theoretical and empirical progress. The reconstruction of the competition between two rival programmes also shows the slowly emerging victory of one of the two programmes over the other in accordance with the MSRP. In a rational reconstruction on the basis of the MSRP nothing occurs that is irrational in the light of the MSRP ([Lakatos,1978a], pp.118-120). The result is that many rational reconstructions will necessarily be counter-factual. Lakatos says: "even the greatest scientists make false steps and fail in their judgement" ([Lakatos,1978a], p.134). Thus in Lakatos's view, notions of scientific rationality will remain forever submerged in an ocean of anomalies. The only thing one can do is to diminish the number of such anomalies by devising new notions of rationality. Lakatos argues that

> "*[...] progress in the theory of scientific rationality is marked by discoveries of novel historical facts, by the reconstruction of a growing bulk of value-impregnated history as rational.* In other words, the theory of scientific rationality progresses if it constitutes a 'progressive' historiographical research programme" ([Lakatos, 1978a], pp.133-134, emphasis by Lakatos).

According to Lakatos, in comparison with its predecessors, the MSRP constituted progress in the theory of scientific rationality; he shows this at great length, unfortunately restricting himself entirely to natural science. Lakatos primarily concentrated on the problem of progress and the circularity prob-

---

[12][Laudan, 1977], p.161

lem does not seem to have troubled him very much. Yet in his proposal notions of rationality must be tested against the *accepted basic value judgements of the scientific élite.*[13] Because these are precisely Laudan's PI's, Laudan's solution of the (second) circularity problem is merely a more explicit version of Lakatos's views.

Let me illustrate Laudan's solution by means of a simple example.[14] Suppose that a philosopher or historian has a notion of rationality implying that crucial experiments are the only possible decisive grounds for abandoning a theory. This philosopher or historian will then tend to focus his account on crucial experiments and he will undoubtedly succeed in making some rational reconstructions that support that notion of rationality. Even if those reconstructions are factual, it is not clear as to how relevant these reconstructions are for the rationality of science as a whole since there are many cases in which scientists have not abandoned a theory in the face of apparently crucial negative experimental results.. Laudan solves this problem as follows: all notions of scientific rationality must be tested against a pre-determined (i.e. independent of a particular explicit notion of rationality) set of cases of theory rejection or acceptance. That particular set is determined by the strong "preferred pre-analytic intuitions about scientific rationality" of the scientific community.

IV.4 Objections to the solution

Several author's have criticized Laudan's solution of the circularity problem. I will briefly deal with them. Laudan sharply distinguishes between HOS_1 (the actual past of science) and HOS_2 (the writings about that past). "HOS_2" is value-impregnated and HOS_1 is not. The set of cases used to test notions of rationality, about which we have clear intuitions, exists in Laudan's proposal in HOS_1. Murphey and Hendrick have criticized Laudan on this point. They argue that HOS_1 and HOS_2 cannot be distinguished. Referring to philosophers of history such as Croce, Collingwood and Becker, Murphey and Hendrick argue:

> "The actual occurrences and events of the scientific past (HOS_1) are forever beyond our reach; there is only one kind of history, reconstructed history (HOS_2). Both Laudan's and Lakatos's attempt to ground [notions of rationality like Lakatos's MSRP's and others - T.K.] in the historical record is

[13]E. g. [Lakatos, 1978a], p.124

[14]The example is Laudan's ([Laudan,1974], p.156).

doomed to vicious circularity" ([Murphey&Hendrick, 1984], p.128).
Concerning the existence of objective records and sources, such as scientific journals, letters, manuscripts etc. which are independent of interpretation, Murphey and Hendrick quote Becker as follows:

> "The records are after all only paper, over the surface of which ink has been distributed in certain patterns. And even these patterns were not made by the actual occurrence, e.g., the assassination of Lincoln. The patterns are themselves only 'histories' of the event, made by someone who had in mind an image or idea of Lincoln's assassination" ([Murphey&Hendrick, 1984], p.128).

The past is indeed beyond our reach and has gone forever. In that sense the criticism is justified. The philosopher of science cannot reach out to the past, which is what Laudan's solution requires. Murphey and Hendrick oppose Laudan's proposal on a second ground. According to them, Laudan runs into the normative/descriptive paradox. HOS_1 is non-normative, and the subset of HOS_1 used to choose between competing notions of rationality also consists of non-normative descriptions. The subset can play a normative role because it consists of cases about which most scientifically educated persons have strong normative intuitions. The normative claims are traced to the decisions of a group:

> "thus the historicism of Laudan and Lakatos is in reality a marked conventionalism and, indeed, a group elitism".[15]

This also points to an element of circularity: By the "dogmatic" fixation of the PI's, all models of rationality that are not compatible with them are a priori ruled out.
Giere attacked Laudan's claim that his solution yields a valid normative theory of rationality in a similar way:

> "Does the type of metamethodology advocated by Lakatos and Laudan yield methodological principles which are genuinely normative? Not really. At its most successful, the metamethodology would tell us only that we had discovered a general *description* of situations which we intuitively regard as clear cases of rational acceptance or pursuit. We might have correctly identified the descriptive component of the methodology, without capturing its normative force.[...] In Kantian terms, Laudan's metamethodology could at most show only that we are acting in accord with his methodology, not that we are acting out of regard for that methodology. It cannot show

---

[15] [Murphy&Hendrick, 1984], p.129

that his methodology is actually embodied as a norm in our judgements" ([Giere,1985], p.338).

How serious are these criticisms? They definitely bring out weak aspects of Laudan's solution. On the other hand they are not devastating. Let us consider the first criticism concerning the distinction between HOS_1 and HOS_2. We shall consider the situation at the micro-level. When Murphey and Hendrick argue that historical facts are always value-laden and that all history of science is HOS_2 they have a point. Yet their conclusion that attempts to found notions of rationality in the historical record are doomed to vicious circularity is drawn prematurely. They would be right if it would make no sense at all to differentiate in historiography between facts and interpretation. However, it is very reasonable to suppose that the agreement that often exists among historiographers has an objective basis. Stegmüller's views are relevant in this respect. Stegmüller argued that the method of historiography sometimes *seems* circular because one cannot sharply distinguish between facts and hypothetical components in our background knowledge ([Stegmüller,1977]). In science one can usually sharply distinguish between the hypothesis H (e.g. the hypothesis that a coin is loaded) which is at issue, the facts F (the result of n tosses) and some background knowledge B (e. g. the assumption of the applicability of the binomial distribution). In philology and historiography it is often less easy to distinguish H, F and B. When we read a text we interpret the words that we read by means of our background knowledge, which may encompass the result of years of work. According to Stegmüller, in cases of difficulties with interpreting a text, this background knowledge is also used to produce hypotheses concerning the meaning of that text. In such cases one can often sharply distinguish an hypothesis H, on the one hand, from F and B together, on the other hand, but it may be difficult to reach an agreement on F, because F cannot be separated clearly from B. In the case of "Cauchy and the continuum" our background knowledge yields two different hypothesis concerning Cauchy's view of the continuum. Whenever one attempts to fix the facts so as to choose between the two hypotheses, such fixation appears to depend upon elements from our background knowledge that imply one of the two interpretations. Very often, however, no dilemmas of interpretation arise and the fact that no *sharp* borderline can be drawn should not lead us to conclude that no distinction between F and H can be made at all. From my point of view the situation at the macro-level is not much different. Historical facts concerning long term developments are theory-laden and dilemmas of confirmation may easily arise when we attempt to test theories of rationality by means of the historical material. But also here it would be very unwise to draw the conclusion that consensus among historiographers with respect to macro-level developments cannot be objective.

Also, with respect to the second criticism concerning the normative/ descriptive problem, it is important to realize that the problems arising when historical material is used as evidence for theories of rationality are analogous to the problems concerning the relationship between scientific theories and their evidence. Scientific facts are theory-laden and a fact's acceptance is based upon a "decision" of the scientific community. That is why, in order to test a scientific theory, it is important to distinguish the theory that is being tested from the theories infecting the data. In the case of the theories of rationality and their relationship with history the situation is similar. The set of historical cases against which Laudan proposes to test theories of scientific rationality is value-laden. However, as long as they are not theory-laden by the theories of rationality for which they will provide the tests, there is no circularity problem. Moreover, the normative/descriptive paradox does not occur because HOS_1 interpreted in this way is normative. We are testing a normative theory against a set of normative cases and the question is not about deriving norms from facts.[16]

IV.5 My proposal

One reason why the first circularity problem requires more attention in the case of mathematics than it does in the case of natural science is obviously the fact that it is unusual to describe the history of mathematics in terms of competition. In attempting to apply Lakatos's MSRP (or a modification of it) to mathematics, one enters a rather unexplored area in the philosophy of mathematics. If the methodology of scientific research programmes is in some form applicable to mathematics, it must be possible to identify competing programmes. Vast complexes of theories and their development have to be described by means of a relatively simple model. Lakatos's and Spalt's work concerning Cauchy show that such identification is not a (completely) trivial matter.

I will now present a modified version of Laudan's solution of the circularity problem. I will distinguish three histories of mathematics: HOM_1 is the history of mathematics as it is available in the primary sources, HOM_2 is the history of mathematics as described by historians, i.e. the history of mathematics

[16] Cf. [Maull,1976], p.125

as it is available in the secondary sources, and HOM3 is the rationally reconstructed history of mathematics. Remember that, in my sense of the phrase "rational reconstruction", the reconstruction must be based upon an explicit methodology.[17] In the relationship between HOM1 and HOM2, and also in the relationship between HOM2 and HOM3, vicious circularity can occur and should be avoided. I propose the following approach to the first and second circularity problem: in order to diminish the danger of circularity the philosopher of mathematics, when constructing HOM3, should primarily restrict himself to HOM2, i.e. the secondary sources. He must, both for the identification of competing units and for the criteria of theory acceptance and rejection, consider cases in HOM2 about which there is *considerable agreement* among historiographers, and he must discover whether or not a *factual* rational reconstruction of those cases in HOM3, on the basis of the model of rationality involved, is possible. In proposing this procedure, I assume that agreement among historiographers reflects Lakatos's "accepted basic value judgements of the scientific élite" and Laudan's "preferred pre-analytic intuitions about scientific rationality". It is reasonable to suppose that, whenever there exists consensus, historians of science or of mathematics fairly represent the pre-analytic intuitions of the scientific community as a whole in their judgement of the past. Of course, I assume that HOM2 is not accidentally infected by the methodology to be used explicitly in the reconstruction. Once a model of rationality has been corroborated by means of some rational reconstructions in the described way, it can, in principle, be used to study cases in HOM2 in which historiographers disagree. As soon as we possess different models of rationality, their different degrees of adequacy could be determined along the lines of Laudan's solution of the circularity problem: model of rationality A is superior to model of rationality B if it makes possible considerably more factual reconstructions.

My modification of Laudan's solution to the circularity problem yields a way of finding such cases, reducing the risk of vicious circularity to a minimum: *I (pragmatically) propose to handle the circularity problems by relying on consensus in HOM2*.

My proposal is to check the applicability of Lakatos's MSRP, or a modified version of it, along these lines. Unfortunately historiographers of mathematics are very much inclined to restrict themselves to micro-level developments and they hesitate to generalize. The philosopher of mathematics who wants to

[17]Cf. Chaper I, Section I.1.2

reconstruct macro-developments in terms of a rather simple model can only base himself to a limited extent upon the existing historiography. Speculations and/or, at least, many extra, particular (reasoned) judgements are probably inevitable, even if we do not require complete consensus among historiographers, and even if we interpret silence among historiographers, regarding a generalization made by one of them, as agreement. How can we provide some firm ground for such speculations? We will only see whether my solution works when we actually try to apply it. That is what we will do in the following chapters.

IV.6 Conclusion

My analysis of Lakatos's "Cauchy and the continuum" and Spalt's *Vom Mythos der mathematischen Vernunft* has revealed some meta-methodological problems that were discussed and I proposed to handle them in a pragmatic way. I am now in a position to attack the main problem of this study - the modification of Lakatos's MSRP and the corroboration of the resulting methodology by means of some rational reconstructions of developments in mathematics. In the next chapter I will consider two papers by the Italian philosopher Giorello. My criticism of these papers yields the first elements of a methodology of mathematical research traditions that I will define more precisely in Chapter VI.

Chapter V Two reconstructions of developments in mathematics by means of the Methodology of Scientific Research Programmes

V.1 Introduction

One of the few authors who has attempted to reconstruct mathematical developments in terms of the methodology of scientific research programmes is Giorello. In [Giorello,1975] he describes Archimedes's mathematical work as resulting from a research programme in the Lakatosian sense and in [Giorello,1981] he did the same with Riemann's work. Below I will first discuss some historiographical studies on the development of Archimedes's work. Then follows a discussion of Giorello's views of the work of Archimedes. My conclusion will be that it is rather artificial to describe the work of Archimedes as resulting from the execution of a research program in the Lakatosian sense. We shall see that it is more natural to consider Archimedes's work as resulting from one or more "research projects", characterized by a certain heuristic unity and executed within the much larger context of a "research tradition" that also encompasses the work of other mathematicians. In the same way I will deal with Giorello's work on Riemann: I will consider different views of the development (of parts) of Riemann's work, and then follows a discussion of Giorello's 1981 paper. Riemann's work can hardly be considered as the result of a research programme in the Lakatosian sense. Here again it is more natural to consider the work as resulting from "research projects" - characterized by a certain heuristic unity - executed within the much larger context of a "research tradition" that also encompasses the work of other mathematicians. In this way the cases of Archimedes and Riemann suggest the first elements of a methodology of mathematical research traditions. The idea is that mathematicians usually work on research projects in the context of a research tradition. A research tradition represents more or less established views on the nature of the subject and on what a good proof should consist of. The research projects primarily represent the dynamic element in mathematics. I will give a more precise definition of a methodology of mathematical research traditions in Chapter VI.

V.2 The development of Archimedes's work

V.2 1 Archimedes's method of mechanical theorems

The question that I will be primarily concerned with in this section is whether or not, in Archimedes's work or in major parts of Archimedes's work, a certain heuristic unity can be identified. In accordance with the meta-methodological considerations of Chapter IV, I will rely heavily on secondary sources.

In his *History of Greek Mathematics*, Vol. II, Heath writes on the work of Archimedes:

> "The treatises are, without exception, monuments of mathematical exposition; the gradual revelation of the plan of attack, the masterly ordering of the propositions, the stern elimination of everything not immediately relevant to the purpose, the finish of the whole, are so impressive in their perfection as to create a feeling akin to awe in the mind of the reader.[...] There is at the same time a certain mystery veiling the way in which he arrived at his steps. For it is clear that they were not *discovered* by the steps which lead up to them in the finished treatises. [...] A partial exception is now furnished by *The Method* of Ar-chimedes, so happily discovered by Heiberg. In this book Archimedes tells us how he discovered certain theorems in quadrature and cubature, namely by the use of mechanics, weighing elements of a figure against elements of another simpler figure the mensuration of which was already known [...]: 'Certain things', he says, 'first became clear to me by a mechanical method, although they had to be demonstrated by geometry afterwards because their investigation by the said method did not furnish an actual demonstration. But it is of course easier, when we have previously acquired, by the method, some knowledge of the questions, to supply the proof than it is to find it without any previous knowledge'" ([Heath,1921], Vol.II, pp. 20-21).

"The Method of Mechanical Theorems, addressed to Erathostenes" or briefly, "The Method", was discovered in 1906 by the philologist Heiberg. In Heath's view "The Method" is one of Archimedes's earlier works. Several other authors have the same opinion. The opinions of these authors concerning "The Method" differ in many details, but all agree on its importance in understanding how Archimedes's work developed. Below I will first concentrate on Dijksterhuis's views, basing myself upon [Dijksterhuis,1956].

V.2.2 The lever-principle

Dijksterhuis assumes along with Heath that the following chronology of the Archimedean corpus is a reasonable approximation of the order in which Archimedes wrote his major works[1]:

(1) On the Equilibrium of Planes or Centres of Gravity of Plane Figures, Book I
(2) The Quadrature of the Parabola
(3) On the Equilibrium of Planes or Centres of Gravity of Plane Figures, Book II
(4) The Method of Mechanical Theorems
(5) On the Sphere and Cylinder: two books
(6) On Spirals
(7) On Conoids and Spheroids
(8) On Floating Bodies: two books
(9) The Measurement of the Circle
(10) The Sand-reckoner

In the chronology, "On the Equilibrium of Planes or Centres of Gravity of Plane Figures, Book I"[2] is Archimedes's earliest treatise. It contains an axiomatic theory of equilibrium. The book also contains theorems on the position of the centre of gravity of simple geometric figures: the parallelogram, the triangle and the trapezium. Dijksterhuis argues as follows with respect to Archimedes's theory of the equilibrium. Long after Achimedes, Pappus discussed[3] the determination of centres of gravity by means of considerations of equilibrium, and argued that if a triangle is placed in horizontal position with a median on the upper edge of a vertical supporting plane, then *the figure will be in equilibrium, because the median divides the triangle in two parts of equal area.* According to Pappus, this means that the centre of gravity must be on the median. Pappus apparently did not see that parts balancing each other need not have equal weight and parts with equal weight need not balance each other. Dijksterhuis assumes that Pappus' erroneous theory dates back from before Archimedes and that Archimedes attempted to correct that theory. In other words: "On the Equilibrium of Planes, Book I" would be Archimedes's

---

[1] [Heath,1921], Vol.II, p.22 and [Dijksterhuis,1958], p.46

[2] I will indicate the book by means of the abbreviated title "On the Equilibrium of Planes, Book I".

[3] In the eighth book of Pappus's *Mathematical Collection* ([Dijksterhuis,1956], p. 299).

revision of an earlier theory to determine centres of gravity.[4] In the book, Archimedes deduces a series of propositions from seven postulates. Proposition six is the so-called lever principle for commensurable quantities:

> "Commensurable magnitudes are in equilibrium at distances reciprocally proportional to the weights".[5]

Proposition seven states the lever principle for incommensurable magnitudes. Deviating from the above given chronological order, and also from the order in which Archimedes's works appear in Heiberg's classical edition, Dijksterhuis discusses "The Method" immediately after "On the Equilibrium of Planes, book I". Dijksterhuis says that he does so because knowledge of the contents of "The Method" facilitates the understanding of the "Quadrature of the Parabola".[6] Yet when Dijksterhuis turns to "The Quadrature of the Parabola", which according to him contains the logical confirmation of an insight gained intuitively in a way described in "The Method"[7], it becomes clear that the order of treatment is for Dijksterhuis not only didactical, but that it is also the expression of a hypothesis on the development of Archimedes's work. Dijksterhuis writes:

> "The theorem on the volume of the cylinder hoof already enabled us to see how Archimedes arrived, from a surmise gained by mechanical means with the aid of the method of indivisibles, at a mathematical proof which satisfied all his requirements of exactness. An even more beautiful example of such a logical confirmation of an intuitively gained insight is furnished by the theorem on the area of any segment of an orthotome [i.e. parabola - T.K.], which formed the subject of the first proposition of the *Method*".[8]

According to Dijksterhuis, the proofs given in "The Quadrature of the Parabola" were obviously found by means of the heuristic method described in "The Method". Reading Dijksterhuis's order of treatment (first "On the Equilibrium of Planes", then "The Method" and finally "On the Quadrature of the Parabola") as corresponding to the way in which Archimedes's mathe-

---

[4]This view is not shared by everybody. Berggren wrote: "Certainly, the argument given by Dijksterhuis is plausible enough. That is to say, things could well have been as he says; however, there is no evidence that they were. in particular it is just as plausible to assume that the passage Dijksterhuis cites represents Pappos' own misunderstanding of what it means to find the centre of gravity of a plane figure" ([Berggren,1976], p.99).

[5]In Dijksterhuis's translation ([Dijksterhuis,1956], p. 289).

[6][Dijksterhuis,1956], p.313

[7][Dijksterhuis,1956], p.336

[8][Dijksterhuis,1956], p. 336

matics developed also implies the assumption that the investigations concerning "On The Equilibrium of Planes, Book I" led Archimedes to the following crucial idea[9]: in the case of homogeneous material, weights are proportional to areas or volumes, which means that considerations of equilibrium can in principle solve problems of quadrature and cubature. In "The Method" Archimedes describes how he used this idea to derive a considerable number of results. One of the first problems that Archimedes successfully attacked by means of the new idea was the quadrature of the parabola. This particular problem may have occurred to him when he tried to determine centres of gravity of curvilinear figures. It is clear from "On the Equilibrium of Planes" that Archimedes studied such matters. Because the problem of the determination of the centre of gravity is rather simple in the case of rectilinear figures and in the case of the circle, the conic sections were the natural next candidates to study.

V.2.3 "The Method"

As Dijksterhuis has pointed out, Archimedes's method consists of two methodic principles. First the *barycentric method*: geometric figures are conceived to be attached to a lever in such a way that the lever remains in equilibrium, and from that equilibrium conclusions are drawn. In the case of curvilinear figures, the barycentric method only becomes fertile in combination with the second methodic principle - the *method of indivisibles*: the area of a plane figure is looked upon as consisting of an infinite number of line segments drawn in a given direction ([Dijksterhuis, 1956], pp. 318-319).

Basing myself upon [Dijksterhuis,1956], pp. 316-318, I shall now briefly show how Archimedes applied these principles in "The Method" to the problem of the quadrature of the parabola. The application of the principles still requires considerable ingenuity. Consider a parabolic segment ABC (Figure 5.1). The line segment AC is an arbitrary cord and not necessarily perpendicular to the axis. BD is parallel to the axis of the parabola and bisects AC in D. The problem is to determine the area of the parabolic segment.

An arbitrary straight line parallel to the axis intersects AC and the parabola in respectively the points X and O. We can then consider the segment to be composed of infinitely many line segments OX parallel to BD.

---

[9]It happens frequently that the first important contributions of a mathematician play a proportionally large role in his entire work. In [Koetsier,1986] I attempted to characterize Lagrange's work, proceeding from his first two mathematical discoveries.

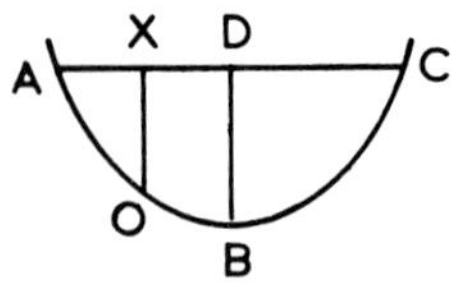

Figure 5.1

In order to "weigh" the segment using the decomposition in line segments, we need to know the length of OX, which means in the Greek context that we need to know the ratio of OX to some other segment. Archimedes now uses the following theorem that he proves elsewhere in his work (Figure 5.2).[10] Let CE (tangent to the parabola at C) intersect the line through A parallel to the axis at Z. E and M are the points of intersection of respectively XO and DB with CZ.

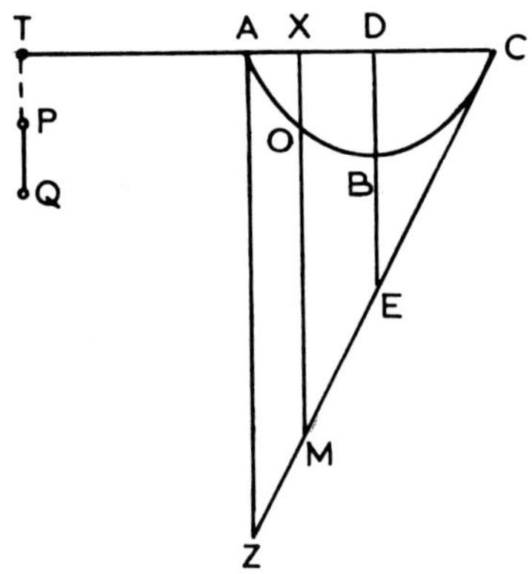

Figure 5.2

Then we know that XM : XO = AC : AX. This theorem can be applied as follows. Let us prolong CA until T such that CA = AT (Figure 5.2). We then have

[10]In section 5 of "The Quadrature of the parabola" (E.g. [Czwalina,1972], p.156).

$$XM : XO = AC : AX = AT : AX$$

which means that if one were to place at T a line segment PQ equal to XO, and if one considered CT as a lever with fulcrum A, we would have

$$XM : PQ = AT : AX$$

and PQ would balance the segment XM where it is now. Apparently the parabolic segment (considered as a sum of infinitely many line segments XO) placed with its centre of gravity in T will balance the triangle ACZ (considered as a sum of infinitely many line segments XM) where it is now. Because of the position of the centre of gravity of ACZ (on the vertical line at a distance of one third of AC from A) applying the principle of the lever we get

$$\text{Area segment} = 1/3 \text{ Area } \Delta \text{ ACZ}$$
$$= 4/3 \text{ Area } \Delta \text{ ACB}$$

Although this proof carries conviction, Archimedes did not consider it as being rigorous.

V.2.4 The mechanical proof in "The Quadrature of the Parabola"

Having discussed" The Method", Dijksterhuis turns to "The Quadrature of the Parabola" because it beautifully enables us to see "how Archimedes arrived, from a surmise gained by mechanical means with the aid of the method of indivisibles, at a mathematical proof which satisfied all his requirements of exactness" ([Dijksterhuis,1958], p.336). According to Dijk-sterhuis, it was the method of indivisibles and not the barycentric method that did not satisfy Archimedes.[11] With the method of indivisibles he touched upon the sensitive problem of atomism or continuity which had given rise to violent controversy in Greek mathematics.[12] Although he expresses the need for it, Archimedes

---

[11]According to Knorr, however, the method of indivisibles was only a simplification of the method of exhaustion (see below) and it was the barycentric method that Archimedes considered formally insecure ([Knorr,1978], pp. 247-248).

[12]The question "finds its clearest expression in the aporia that worried Democritus: if the circular sections that can be made in a cone parallel to the base are congruent, how can the

does not give the desired rigorous proof in "The Method". In "The Quadrature of the Parabola" there are, however, two different proofs of the theorem. The second proof, covering propositions 18-24, is purely geometrical. The proof given in "The Method" was presumably transformed by Archimedes into the first proof of the "Quadrature of the Parabola".[13]

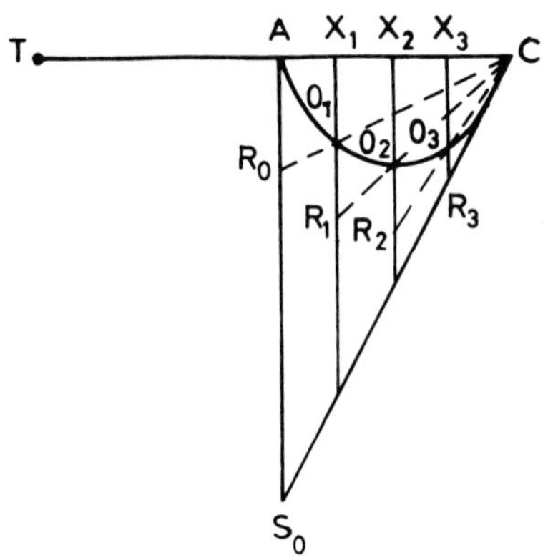

Figure 5.3

As I will show below, the first proof covering the propositions 1-17 is mechanical in the sense that the barycentric method is used (but not the method of indivisibles). I will now give a sketch of that first proof.[14] Archimedes uses the method initiated by Eudoxus, which is usually called the "exhaustion method": the magnitude to be calculated is compressed between a monotonously ascending and a monotonously descending sequence.
The proof is restricted to the case that AC and the axis of the parabola are perpendicular (Figure 5.3).The line segment AC is divided into arbitrarily many equal parts, e.g. AX_1, X_1X_2, X_2X_3 and X_3C. Through the points X_i

---

cone differ from a cylinder; and if they grow smaller towards the vertex, is not then the curved surface, which should be smooth, scalariform?" ([Dijksterhuis,1956], p.320).

13Van der Waerden expressed this most clearly: "The mechanical derivation is essentially the same as that in the "Method", but Archimedes made it into a rigorous proof by means of the exhaustion method" ([Van de Waerden,1961], p. 216).

14I base myself upon [Dijksterhuis,1953], pp.336-342.

lines parallel to the axis of the parabola are drawn intersecting the parabola in the points O_i. If we connect C with the points O_i we get a large number of trapezia: e.g. trapezium $X_1X_2O_2O_1$ (or briefly trap.O_1X_2) and trapezium $X_1X_2O_2R_1$ (or briefly trap. R_1X_2). By means of the barycentric method (using TC as a lever with fulcrum at A), Archimedes proves in the case of the division into four equal parts that if

$$S_{below} = \text{(def) trap. } O_1X_2 + \text{trap. } O_2X_3 + \Delta\, O_3X_3C$$

and

$$S_{above} = \text{(def) trap. } R_0X_1 + \text{trap. } R_1X_2 + \text{trap. } R_2X_3 + \Delta\, R_3X_3C$$

then

$$3\, S_{below} < \Delta\, ACS_0 < 3\, S_{above}$$

One gets a similar result when AC is divided into a larger number of parts. All the time we have

$$S_{below} < 1/3\, \Delta\, ACS_0 < S_{above}$$

By means of an indirect proof, Archimedes then shows that the area of the parabolic segment equals 1/3 Δ ACS_0. On the one hand, the proof is based on the fact that for all divisions the area of the parabolic segment lies between S_{below} and S_{above}. Whereas on the other hand it is based on the fact that $(S_{above} - S_{below})$ can be made arbitrarily small.

V.2.5 The geometrical proof in "The Quadrature of the Parabola"

So far, the case of the quadrature of the parabola shows us how, in Dijksterhuis's description in one particular case, Archimedes's mathematical thinking developed. Similar descriptions are given by Heath and van der Waerden. These historians assume more or less explicitly that in this case we can follow the development of Archimedes's thinking even further. They assume that Archimedes, after he had given the mechanical proof by exhaustion, devised the purely geometrical proof of propositions 18-24 of the "Quadrature of the Parabola". In my own words I will briefly describe that proof.[15]

Archimedes considers the segment ABC (Figure 5.4). D is the middle of AC, and DB is parallel to the axis of the parabola. The area Z of ΔABC can be considered as a first approximation to the area of the segment. We now draw lines parallel to the axis through the midpoints N_1 and N_2 of the cords AB and BC respectively. Those lines intersect the parabola in T_1 and T_2.

---

[15] Here again I base myself upon [Dijksterhuis,1953], pp.342-345

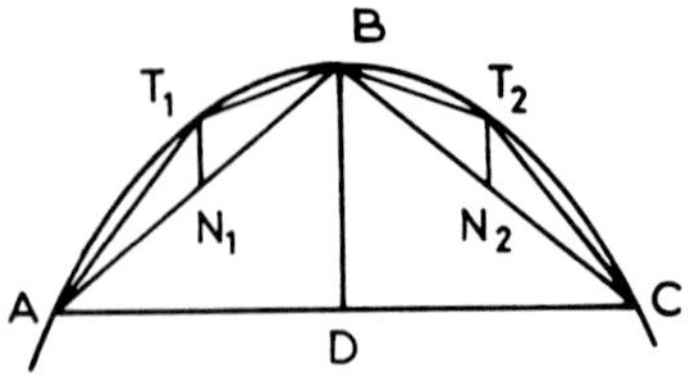

Figure 5.4

Triangle ABC extended with ΔAT_1B and ΔBT_2C yields polygon AT_1BT_2C. The area of that polygon is a second approximation to the area of the segment. Archimedes proves that area ΔAT_1B = area ΔBT_2C = 1/8 Z, which implies that area AT_1BT_2C = Z + 1/4 Z. We can get a third approximation by repeating the step from the first to the second approximation: the polygon AT_1BT_2C is extended with triangles inscribed in the remaining four segments corresponding to the AT_1, T_1B, BT_2 and T_2C. The area of the resulting polygon equals $Z + 1/4\ Z + (1/4)^2Z$. Proceeding in this way, Archimedes obtains a geometric progression with the ratio 1/4. By means of a lemma on the sum of this progression, Archimedes proves the theorem with an indirect proof.

One of the reasons why historians believe that the purely geometric proof was given after the mechanical proof is undoubtedly the fact that they occur one after another in the same treatise. But there is also the following passage from the preface of "The Quadrature of the Parabola". Archimedes writes to Dositheus as follows:

> "[...]we have taken it in hand to send to you, as we had determined to write to Conon, a certain geometric theorem which had not formerly been studied, but has now been studied by us, *one we first have found out by means of mechanics and afterwards have exhibited also by means of geometry* [...] We do not know of any earlier geometer who undertook to square the area contained by a straight line and the segment of the section of a right-angled cone, which is precisely what has now been found out by us; for it is proved that every segment bounded by a straight line and a section of a right-angled cone is four-thirds the triangle having the same base and height equal to the segment[...]. So we have written up and are sending to you the proofs of it, first as it has been studied by means of mechanics, then after

these also as it is proven by means of geometry" ([Knorr, 1978], p. 248, translation by Knorr; emphasis is mine - T.K.).

Misunderstanding seems excluded at least at first sight: Archimedes explicitly says that he first found the theorem by means of the mechanical method and only afterwards gave the geometrical proof.

V.2.6 The importance of "The Method"

"The Method" contains several other elegant results that Archimedes found by means of that method. They all concern three-dimensional situations and centres of gravity of bodies, and must have been the result of an intensive search for suitable situations, i.e. first a search for theorems on sets of indivisibles of interesting figures, followed by attempts to use those theorems in relating those indivisibles by means of a suitably chosen lever to other sets of indivisibles. The results involved encompass a considerable part of Archimedes's work.[16]

The crucial question is how important the method of mechanical theorems in the development of Archimedes's complete work actually was. How representative is the example of the parabola described above? The answer to these questions is that we do not know. Some authors place much emphasis on the role of the mechanical investigations in Archimedes. For example Berggren, after having noted that some of the works of Archimedes evidently depend on the earliest work "The Equilibrium of Planes, Book I" because the barycentric theory is used in them, writes:

"However, the real dependence of his subsequent work on P. E. I [The Equilibrium of Planes, book I -T.K.] is, I believe, even more fundamental. The crucial link between P. E. I and the rest of his work is his discovery that he could use theorems about centres of gravity to obtain results on quadrature and cubature. He described this in *The Method*, a letter to Erathostenes. Although he wrote this late in his career it is the discovery of his method, not the date of his letter, that concerns us now. The main idea must have occurred to him after he found the theorems recorded in P. E. I (for his method requires some of these theorems) and before he found the area of a segment of a parabola, his first result on quadrature. In sum, the chronology is both clear and suggestive: first came the discovery of the

[16]They concern results in at least the following four works: On the Equilibrium of Planes, Books I and II; The Quadrature of the Parabola; On the Sphere and Cylinder, Two books; On Conoids and Spheroids.

theorems in P. E. I, next the application of these to frame a method that would allow him to discover theorems on quadrature and cubature, and finally applications of this method to problems whose solutions are both surprising and subtle. We now see, in sharp relief, the role that the work P. E. I played in Archimedes's later work" ([Berggren,1976], p. 90).

The views of the historian Frajese, who distinguishes three stages in Archimedes's research, are also remarkable in this respect:[17]

"(1) pure intuitions based on the hypothesis of *simplicity*
(2) confirmation (where possible) by the mechanical method
(3) rigorous proof (mostly by means of the method of exhaustion)".

The application of the mechanical method requires a suitable choice of figures, whose area or volume are to be compared by weighing. Frajese's hypothesis is that Archimedes was in this respect guided by an intuition of simple arithmetical relations underlying the complexity of geometrical facts.[18] Frajese illustrates this idea with the following example. Let BDXV represent a cylinder with height DX and base diameter BD. Let triangle BDA and semicircle BAD represent respectively a cone and a semi-sphere with the same height and the same base as the cylinder.

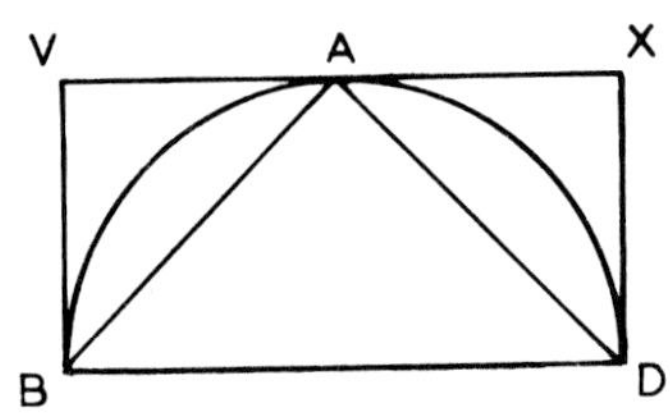

Figure 5.5

[17]"[...]lo sviluppo dei procedimenti di ricerca e di dimostrazione di Archimede può essere verosimilmente avvenuto almeno in certi casi (peraltro frequenti) attraverso tre successivi momenti: 1) pura intuizione basata sulla ipotesi della *semplicità*; 2) verifica col *metodo meccanico* (quando è possibile); 3) dimostrazione rigorosa (spesso con metodo di esaustione)" ([Frajese,1974], p.25).

[18]Frajese: "Archimede dovette pensare ad una *semplicità* delle leggi geometriche: semplicità che si traduce nella piccolezza dei numeri interi che valgono a esprimere, a coppie, i rapporti geometrici più signicativi" ([Frajese,1974], p. 23).

We know that Volume cone = 1/3 Volume cylinder. On the basis of the hypothesis of simplicity we can then guess that the volume of the semi-sphere will be precisely in between the volumes of cone and cylinder:

Volume semi-sphere = 2/3 Volume cylinder.

According to Frajese,[19] such an argument may have led Archimedes to attempt to apply the mechanical method to a cylinder with inscribed cone and inscribed sphere. In "The Method", Archimedes proves the truth of the conjecture by means of the mechanical method, and in "On the Sphere and Cylinder" he gives a geometrical proof. Frajese gives another argument to support his hypothesis of the occurrence of the three stages in the development of Archimedes's work. In the introductory letter to Dositheus in "On Spirals", Archimedes wrote that two of a list of theorems that he had sent to Konon were wrong. One of the false theorems is the following: If a sphere is cut by a plane in two unequal parts, the ratio of the volumes of the two parts equals the ratio of the squares of the areas of the two corresponding "skull-caps".[20] Frajese considers the occurrence of these two false theorems as an argument supporting his views. He supposes that Archimedes framed such conjectures on the basis of the principle of simplicity.

What is already present in the work of Dijksterhuis and others becomes even more conspicuous in the views of Berggren and Frajese: historians are inclined to attempt to define what I will call the heuristic unity in the work of a particular author. The work of a great mathematician always looks to a certain extent like a mystery or a series of mysteries. Yet apparently, historians assume that *a great mathematician repeats himself in the sense that it is possible to discern certain characteristic patterns in the way in which he attacks problems or in the way he sets new goals for himself.* Such patterns form in my definition the *heuristic unity* of the work of that mathematician. An attempt to understand the development of the work of a mathematician should encompass an attempt to characterize its heuristic unity. In the case of Archimedes, historians clearly work in this way. Unfortunately so little is known with certainty in the case of Archimedes that such attempts are bound to remain quite speculative. While Berggren and Frajese propose to understand the development of the Archimedean corpus as a whole starting from

---

[19][Frajese,1974], pp. 24-25
[20][Czwalina,1972], p.6

the contents of "The Method", Knorr attacks such attempts. Knorr rejects Heath's chronological ordering of Archimedes's works and draws conclusions with respect to the role of the mechanical writings that differ from the views of most earlier authors. According to Knorr, who extends arguments by Arendt and Kierboe,[21] "The Method" is a late work of Archimedes, and he very much restricts its heuristic role. Knorr distinguishes in Archimedes's writings an early group and a mature group. In Knorr's view, Archimedes still heavily relied on Euclidean techniques in the works belonging to the early group, while in the works of the mature group, having outgrown the technical methods of his precursors, Archimedes reaches a unique level of precision. The geometrical proof in "The Quadrature of the Parabola" belongs in Knorr's view to the early group and was given completely independently of mechanical considerations. "The Method" and the mechanical proof in "The Quadrature of the Parabola" belongs then to the mature group. Knorr's interpretation is definitely possible, but I do not feel competent to answer the question as to how probable it is. Not all of Knorr's arguments are prima facie equally convincing. For example, at the end of section V.2.5, I quoted Archimedes (in Knorr's translation) as saying about the quadrature of the parabola that he *first* found it out by means of mechanics and *afterwards* exhibited it also by means of geometry. Knorr is forced to assume that in this sentence Archimedes does not refer to the order of genesis, but to the sequential order of exposition in the treatise. I will not go further into Knorr's arguments. It is important, however, that although Knorr restricts the role of mechanical considerations in Archimedes's work, he does not doubt that the barycentric method played a considerable heuristic role when Archimedes's wrote the works from the mature group.

V.3 Giorello's views of Archimedes's work

V.3.1 Archimedes's "research programme"

Referring to the views of Frajese, Giorello has attempted to describe Archimedes's work as resulting from the execution of a Lakatosian research programme in mathematics. His interpretation shows very well the difficulties one has in applying Lakatos's MSRP to mathematics, and in using Lakatosian notions like "hard core", "protective belt" and "positive heuristic" to describe Archimedes's "research programme". For Giorello the *hard core* of the Archimedean programme consists of two principles of development:

---

[21] [Knorr,1978], p.269

A) conjectures framed by means of the hypothesis of simplicity can be confirmed or falsified by means of the mechanical method, and
B) Theorems proved by means of the mechanical method can be proved geometrically.
Those two principles each possess a *protective belt* consisting of the collections of theoretical devices that are used in the mechanical and geometrical proofs respectively. These collections of theoretical devices protect the principles of development in the sense that they made the successful application of the principles possible. As for the protective belt of the first principle, Giorello remarks that it "imparts rational substance to abandoning older working hypothesis".[22] Here he thinks of the two above mentioned false theorems that Archimedes talks of in his "On spirals". They were disproved by means of the mechanical method. As for the protective belt in connection with B) Giorello writes:

> "The idea is always present that as a matter of principle hinging upon the conception of arithmetical simplicity, every theorem able to be proved by having recourse to infinitesimal elements can also be proved without them, the surrounding protective belt being made up of geometrical devices which enable this fundamental hypothesis to be confirmed in each particular case".[23]

Giorello's Lakatosian view of Archimedes's work is not very satisfactory. It is obvious that Lakatos's MSRP is used in rather a free way. In Lakatos's views the negative heuristic of a programme, which specifies the hard core, tells us what paths of research to avoid. Moreover, in Lakatos's examples of research programmes in mathematics, the hard cores take the form of sets of commonly accepted axioms. Giorello's hard core consisting of two principles is not a set of untouchable statements. It consists of the faith that certain paths of research will yield fruit. It hardly even tells us what paths of research to avoid, the two principles telling us rather what to try next! The protective belt in Giorello's sense indeed protects his hard core: the mechanical method and the geometrical method encompass a great deal, and whenever a conjecture can be falsified or confirmed, the hard core in Giorello's sense will be strengthened. Yet Giorello's protective belt does not have much in common with a Lakatosian protective belt, i.e. a set of auxiliary hypotheses that protect a hard core in the sense that falsifications are always directed at those

---

[22][Giorello,1975], p.129
[23][Giorello,1975], p.131

distinguishes in the work of Archimedes two interacting *"lines of research"* - a geometrical line of research and a mechanical line of research. The geometrical line of research concerns Archimedes's rigorous geometrical proofs. The basic idea of the mechanical line of research is, in Giorello's view, the method of indivisibles. These lines of research extend far beyond Archimedes's work. Giorello remarks that there exists "the undeniable historical fact that the two lines are mutually opposed".[24] The geometrical line corresponds to the rejection of indivisibles and the "mechanical line" to their acceptance. Here again it becomes clear that Giorello applies Lakatos's MSRP in a peculiar way. What is a line of research? It is characteristic of Lakatos's MSRP's that research programmes do compete. The history of science is characterized by permanent competition in Lakatos's view. Giorello uses Lakatos's methodology to characterize Archimedes's work, but leaves out one of its major elements - the fact that research programmes compete with one another. On the other hand he introduces, without definition, besides research programmes other large units that do compete -"lines of research".[25]

V.3.2 The need to modify Lakatos's MSRP

Giorello's failure to apply Lakatos's terminological apparatus in a precise way suggests that the methodology of scientific research programmes must be modified in order to be applicable to mathematics. In the above I have shown that historians repeatedly gave characterizations of (elements of) the heuristic unity (of parts) of Archimedes's work. The core idea is that by means of the principle of the lever, purely geometrical problems concerning areas (and volume) can be solved. The method of indivisibilia also makes this idea a practicable method for curvilinear areas. Because Archimedes is working within a geometrical tradition in which the method of indivisibilia is seen as a dubious method of proof, after having proved a theorem by means of the mechanical method, he has to find a second proof which is in accordance with the tradition. Archimedes's solution of the problem of the quadrature of the

---

[24][Giorello,1975], p.130

[25]The two lines of research are not opposed in the sense that they yield contradictory results. They are opposed in the sense that in the course of time the use of infinitely small or infinitely large elements has frequently met with opposition. In certain periods of history this took the form of a total prohibition of the use of infinitesimals in proofs, so that they could at most play a heuristic role. Spalt's extension of Lakatos's "Cauchy and the continuum", which I discussed in the previous chapter, is in fact an attempt to describe Giorello's two lines of research as research programmes.

dubious method of proof, after having proved a theorem by means of the mechanical method, he has to find a second proof which is in accordance with the tradition. Archimedes's solution of the problem of the quadrature of the parabola, which we sketched above, illustrates the method. The development of those parts of Archimedes's work achieved in this way illustrates the following pattern: An (unexpected) combination of two rather disjunct areas (geometry and mechanics) in the form of a paradigmatic solution of a particular (geometrical) problem which functions as a guideline in a series of analogous situations, yields a considerable number of results. Given the paradigmatic example, the major heuristic rules that are actually applied are: "Find analogous situations" and "Apply analogous methods in analogous situations".

The question is whether there exists a rather simple model, comparable to Lakatos's MSRP, which is applicable to Archimedes's work. In Chapter IV, I argued that in order to identify large units like research programmes in mathematics, I would base myself primarily upon consensus among historiographers. The case of Archimedes makes clear how this idea could be used in practice. The modification of Lakatos's MSRP that we are looking for should be such that it can be considered as a *formalization of notions that are informally present in the historiography concerning Archimedes's work.* The model should allow us to identify the "mathematical process" that can partially be characterized by the heuristic pattern described above. I propose to use the term *"research project* for such processes. The notion of "research project" will be one of the elements of the modification of Lakatos's MSRP that I will propose. In the first sections of this chapter I described some attempts by historians to understand the development of Archimedes's work by means of the ideas expressed in"The Method". From the point of view of the model that I will propose, these attempts are attempts to view the development of Archimedes's work as resulting from a research project .

However, the model that we are looking for should also allow us to characterize the broader mathematical context of Archimedes's work. Archimedes based his investigations on the works of Euclid, and an understanding of the Elements of Euclid is a prerequisite to becoming acquainted with the work of a writer of the stature of an Archimedes. The following quotation is quite representative in this respect:

> "Owing to the completion of the *Elements* of Euclid, the Greek mathematicians had gained possession of a systematically arranged collection of

the fundamental mathematical propositions on which they could base their further investigations" ([Dijksterhuis, 1956], p. 49).

Historians all agree: within Greek mathematics there existed a body of mathematical knowledge, of which the fundamentals were systematically arranged by Euclid, that was extended by Archimedes (and many others). The modification of Lakatos's MSRP that we are looking for should also allow us to characterize this body of mathematical knowledge. I propose to say that Archimedes and the others worked within a *"research tradition"*, which was defined most clearly by Euclid's Elements. I will call that tradition the *Euclidean Tradition.*

I will return to the characterization of research projects and research traditions in Chapter VI when I will propose a methodology of mathematical research traditions. However, I will first discuss another paper by Giorello in which he attempted to apply Lakatos's MSRP to Riemann's work. That discussion will show that the notions of research tradition and research project can also be applied to nineteenth century analysis. Instead of describing Riemann's work as resulting from a research programme in Lakatos's sense, it is much more appropriate to say that Riemann executed research projects within the context of a research tradition in analysis.

V.4 Riemann

V.4.1 Introduction

In [Giorello,1982] Giorello has attempted to describe Riemann's work as resulting from the execution of a Lakatosian research programme in mathematics. Before I discuss that paper, I will first make some remarks on Riemann's work and consider two different attempts to understand its development. Riemann's earliest publication is his 1851 doctoral dissertation *Grundlagen für eine allgemeine Theorie der Functionen einer veränderlichen complexen Größe* ("Foundations for a general theory of functions of one complex variable quantity")[26], which is a treatise on the theory of functions of a complex variable.

A major source of Riemann's interest in the theory of functions of a complex variable was undoubtedly Cauchy's work. In 1846 Cauchy published two important papers.[27] In the first paper he proved that for a function

---

[26][Riemann,1851]

[27]My sketch of this part of Cauchy's work is based upon [Kline,1972], pp. 634-642 and in particular pp. 639-640.

$f(z)=f(x+iy)=u(x,y)+iv(x,y)$ which satisfies the equations (later called Cauchy-Riemann equations)

$$\frac{\partial u}{\partial x}=\frac{\partial v}{\partial y}, \quad \frac{\partial u}{\partial y}=-\frac{\partial v}{\partial x}$$

the integral around a curve bounding an area equals zero:

$$\int f(z)\,dz=0$$

In the second paper Cauchy shows that if the curve encloses poles, then the integral equals $2\pi i$ times the sum of the residues. He also took up the subject of multiple-valued functions. For example,what is the value of the integral $\int\omega\,dz$ along a closed path around the origin in the case of $\omega^3 = z$? Cauchy did not succeed in solving satisfactorily the problems that such multiple-valued functions offered.

In his dissertation Riemann solved these problems and at the same time generalized Cauchy's theorems much further. In the dissertation he shows special interest in the theory of algebraic functions and their integrals. This interest led to the publication in 1857 of Riemann's *Theorie der Abel'schen Functionen.*[28] Unfortunately, little is known concerning the precise development of Riemann's mathematics. There are, however, two interesting and rather speculative attempts to understand the development of his work in the theory of functions of a complex variable. The first is by Klein and the second by Brill and Noether. Klein, Brill and Noether were mathematicians with a great interest in the history of their discipline. Before I can discuss these authors I must first make some introductory remarks in the sections V.4.2 through V.4.5.

V.4.2 The problem of how to deal globally with a particular function - the input from physics

Riemann studied in Berlin from 1847 until 1849. According to Dedekind he attended a class on elliptic functions taught by Eisenstein. Later Riemann related that he had completely disagreed with Eisenstein on the way in which complex quantities should be introduced into the theory of functions:

> "Eisenstein had stuck to the formal calculation while he himself [Riemann - T. K.] had recognized in the partial differential equation the essential

[28][Riemann,1857]

definition of a function of a complex variable".[29]

Apparently Riemann at that time already was aware of the fact that analytical expressions, particularly power series, often only locally characterize a function, and he believed that the Cauchy-Riemann equations allowed for a more global approach.

If we trust Prym, a student of Riemann at Göttingen, Riemann was led to attribute decisive importance to the Cauchy-Riemann equations after he had been wondering why in so many cases one obtains correct results when operating with divergent series, as Euler repeatedly did. Prym writes:

> "He [Riemann - T.K.] then asked himself what actually made functions extend from one area into the other, and he reached the insight that it was the partial differential equation which achieves this. Dirichlet, with whom he discussed the matter, completely agreed with this opinion [...]".[30]

If Prym is right, Riemann may have reasoned as follows. For Euler a power series, convergent or not, *globally* characterizes a function in a relatively simple way. Unfortunately, this representation is not always valid. Nevertheless, the fact that Euler's risky use of the power series as a global characterization yielded so much fruit means that it is plausible to assume that there exists a reliable global characterization of a function which Riemann found in the Cauchy-Riemann equations.

The desire for a more global approach does not automatically lead to differential equations. Felix Klein has emphasized the role that physical considerations may have played in the development of Riemann's thinking. In physics the idea prevailed that in general, phenomena could be best understood by means of differential equations combined with particular boundary conditions.

Klein writes:

> "What I want to emphasize is *that the source of Riemann's pure mathematical development lies exactly here*. The banishment of action at a distance, the explanation of the phenomena by means of the internal forces of a space-filling aether, corresponds in mathematics to the understanding of

---

[29]"Eisenstein sei bei der formellen Rechnung stehen geblieben, während er selbst in der partiellen Differentialgleichung die wesentliche Definition einer Function von einer complexen Veränderlichen erkannt habe" ([Dedekind, 1876], p.512 in [Weber,1876]).

[30]"Er frug sich dann, was denn eigentlich die Functionen aus dem einen Gebiete in das andere fortsetze, und gelangte zu der Einsicht, daß dies die partielle Differentialglei-chung thue. Dirichlet, mit dem er den Gegenstand besprach, stimmte dieser Ansicht vollständig bei [...]"([Neuenschwander,1981], p.228).

functions *by means of their behaviour in the infinitely small* (in particular by means of the partial differential equations that they satisfy). And as in the area of physics the individual phenomenon depends on the general arrangement of the experimental conditions, Riemann in mathematics individualizes his functions by the *special boundary conditions* that he imposes upon them. The formula that one needs for the calculatory control of the function appears here as the final result, not as the point of departure".[31]

The point of view sketched by Klein leads rather naturally to the consideration of the Cauchy - Riemann equations, which as we have seen already played a role in Cauchy's work.

V.4.3 Conformal mappings

Gauß wrote a paper called "A general solution of the problem to map the parts of a given surface onto another given surface in such a way that the image becomes similar to the original in the smallest details".[32] Riemann studied this paper for the first time in March 1849 and re-read it later in Göttingen.[33] It is one of the two papers to which Riemann refers in his dissertation. Gauß considers an arbitrary mapping from one surface to another which maps any infinitely small triangle onto a *similar* infinitely small triangle. Nowadays such mappings are called conformal, i.e. continuous angle-preserving mappings. Gauß shows that such a mapping can be described by a function of a complex variable and, conversely, a function of a complex variable defines a conformal mapping. We must then restrict ourselves to functions f(z) with z=x+iy for which Lim $\Delta f/\Delta z$ for $|\Delta z| \rightarrow 0$ is independent of the direction of

---

31"Was ich betonen will, ist dies, *daß eben hier die Quelle von Riemanns rein mathematischen Entwicklungen liegt.* Was in der Physik die Verbannung der Fernwirkungen, die Erklärung der Erscheinungen durch die inneren Kräfte eines raumerfüllenden Äthers ist, das ist in der Mathematik das Verständnis der Funktionen *aus ihrem Verhalten im Unendlich-Kleinen*, insbesondere also *aus den Differentialgleichungen*, denen sie genügen. Und wie im übrigen die einzelne Erscheinung im Gebiete der Physik von der allgemeinen Anordnung der Versuchsbedingungen abhängt, so individualisiert Riemann seine Funktionen durch die *besonderen Grenzbedingungen*, die er ihnen auferlegt. Die Formel, deren man zur rechnerischen Beherrschung der Funktion bedarf, erscheint hier als Schlußresultat der Betrachtungen, nicht als Ausgangspunkt" ([Klein,1894], p.484, The emphasis is Klein's).

32"Allgemeine Auflösung der Aufgabe: die Theile einer gegebnen Fläche auf einer andern gegebnen Fläche so abzubilden, daß die Abbildung dem Abgebildeten in den kleinsten Theilen ähnlich wird" (Astronomische Abhandlungen (ed. Schumacher) 3, 1825, 1-40 = Gauß, Werke, Bd. 4. pp.189-216).

33[Neuenschwander,1981], p.224

$\Delta z=\Delta x+i\Delta y$. This independence of the derivative of the direction of the differentiation implies the Cauchy-Riemann equations $\{f=u(x,y)+iv(x,y)\}$:

$$\frac{\partial u}{\partial x}=\frac{\partial v}{\partial y}\ ,\ \frac{\partial u}{\partial y}=-\frac{\partial v}{\partial x}$$

Gauß' paper laid a link between two seemingly quite different notions - complex functions and conformal mappings. In Riemann's hands this link would become very fertile.

V.4.4 Potential theory

The Cauchy-Riemann equations determine a function of a complex variable uniquely only in combination with suitable boundary conditions. The formula for the total differential of v

$$dv=\frac{\partial v}{\partial x}dx+\frac{\partial v}{\partial y}dy$$

yields by means of the Cauchy-Riemann equations

$$dv=-\frac{\partial u}{\partial y}dx+\frac{\partial u}{\partial x}dy$$

which implies that

$$v_X=v_P+\int_P^X -\frac{\partial u}{\partial y}dx+\frac{\partial u}{\partial x}dy$$

Disregarding possible complications, v can be determined by integration along a suitable curve in the complex plane if u(x,y) is known and if v is known in one point P. The question then is: What kind of boundary conditions determine u?

Here a link with potential theory shows up. Differentiation of the Cauchy-Riemann equations yields (assuming that u and v are "respectable" functions):

$$\Delta u=0\ \ ,\ \ \Delta v=0.$$

Obviously u and v must be harmonic functions. This link between the theory of complex functions and potential theory became an important tool in Riemann's hands.

Potential theory was not a coherent theory in the 1840s, but existed in the form of more or less isolated papers by different authors. According to Felix

Klein, Riemann was particularly influenced with respect to potential theory by Dirichlet.[34] Dirichlet studied in three dimensions precisely what Riemann needed in two - the determination of harmonic functions by means of boundary conditions. The problems that Dirichlet dealt with are related to the following result from theoretical physics.[35] If we have a mass- (or charge-) distribution on a surface S which bounds a part Ω of $\mathbb{R}^3$, the force exerted by that distribution on a point mass (charge) at (x,y,z) is proportional to (V_x,V_y,V_z), where V(x,y,z) is called the *potential function*, which satisfies

$$\Delta V = 0 \text{ for } (x,y,z) \notin S$$

This led Dirichlet to the following mathematical problem: which mathematical boundary conditions determine V? He gave the following answer. If Ω and its boundary S are given, and a continuous function f on S is given, then in the family of all continuous functions (on $\Omega \cup S$) that are continuously differentiable in Ω and that take the values f on S, there i) exists at least one function for which the integral

$$U = \iiint_{\Omega} \{(\frac{\partial u}{\partial x})^2 + (\frac{\partial u}{\partial y})^2 + (\frac{\partial u}{\partial z})^2\} \, dt$$

attains a minimum value and ii) for this function holds

$$\Delta u = 0.$$

Dirichlet proves ii) by showing that $\Delta u = 0$ is a necessary condition for the integral U to be minimal. He finds i) evident, "because the integral is non-negative for any u belonging to the family".[36] I will call this method of proving the existence of a function satisfying the given requirements *Dirichlet's principle* in accordance with modern usage. I will call the assumption behind i) *Dirichlet's assumption.* Dirichlet's assumption was to be seriously criticized particularly in the second half of the nineteenth century.[37]

---

[34][Klein, 1979], p.262
[35][Monna, 1975], p.28
[36][Monna,1975], p.30
[37]Cf. [Monna,1975]

Riemann very successfully used the two-dimensional analogon of Dirichlet's principle.

V.4.5 Riemann-surfaces and cross-cuts

Consider the function $w = f(z) = z^2$. When this function is geometrically interpreted, it maps the ring $T = \{z \mid 1 \leq |z| \leq 2\}$ (Figure 5.6) in the complex plane U on the ring $T' = \{z \mid 1 \leq |z| \leq 4\}$in the complex plane V. The mapping is conformal. There is, however, something remarkable here: the continuous conformal mapping is everywhere locally invertible. Nevertheless it is not globally invertible. If we restrict the function $f(z) = z^2$ to $S = \{z= r\, e^{i\phi} \mid 1 \leq r \leq 2 \text{ and } 0 \leq \phi < \pi\}$, i.e. the upper half of T (Figure 5.7.a), then f(S) = T', and the restricted function is globally invertible.

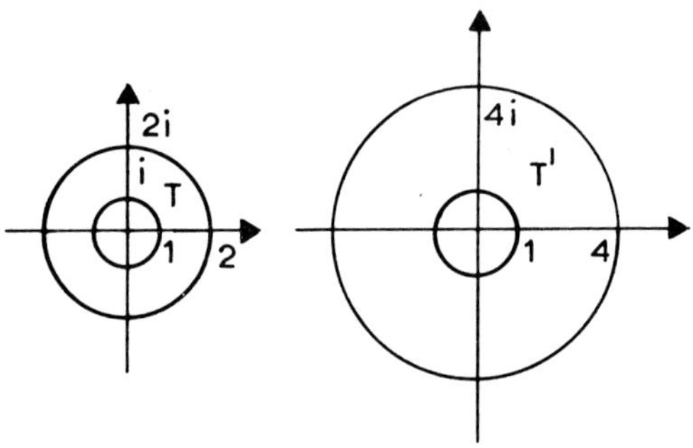

Figure 5.6

However, as soon as we extend the set S and allow ϕ to grow above π, the global invertibility is disturbed (Figure 5.7.b). This breakdown of invertibility can be avoided by letting the images f(z) for $\phi \geq \pi$ form another layer of images on top of the images for $0 \leq \phi < \pi$. For $\pi \leq \phi < 2\pi$, the images f(z) then form a ring which completely covers the ring of images for $0 \leq \phi < \pi$. If ϕ grows from 0 through π, the image f(z) then first moves anti-clockwise on the lower ring. From $\phi = \pi$ the image f(z) moves anti-clockwise on the upper ring.

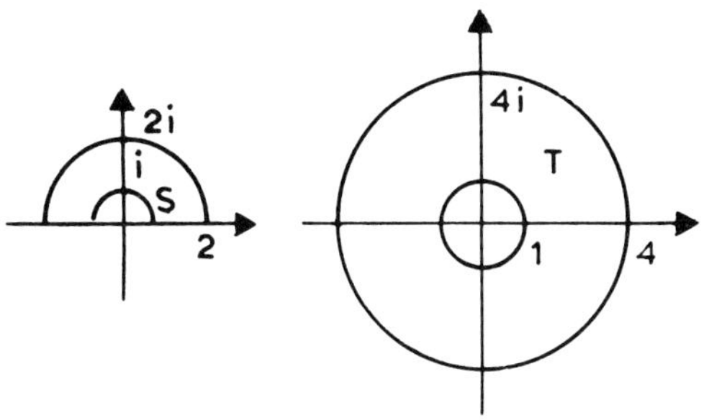

Figure 5.7a

Because for $\phi = 2\pi$ the images continuously move from the upper ring back to the lower ring, we can imagine that along the line segment connecting the points 1 and 4 in plane V the two rings are interpenetratingly connected. The two coinciding interpenetratingly connected rings form an example of what was later called "Riemann-surface" an important generalization of point sets in the plane.

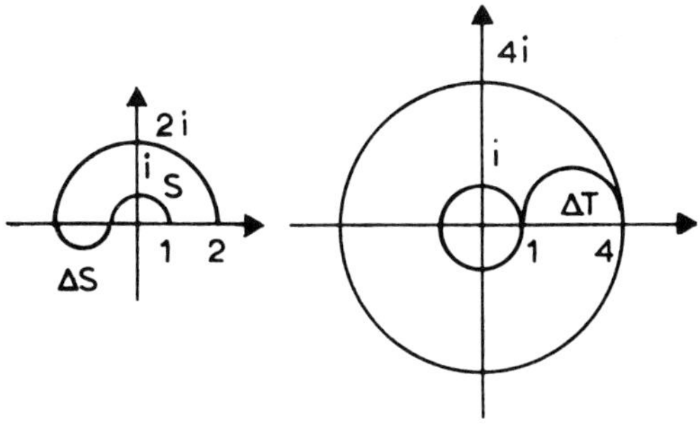

Figure 5.7b

We get another example of a Riemann-surface when we consider the function $f(z) = z^2$ on a circular disc C surrounding the origin. Then the image C' consists of two coinciding circular discs that cohere in the origin (Figure 5.7c).

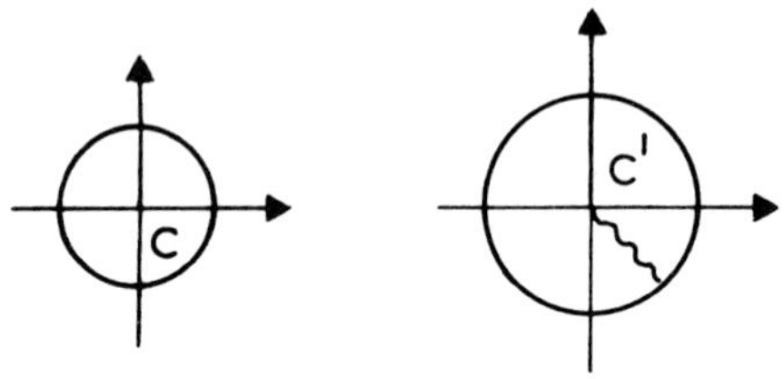

Figure 5.7c

In the case of this Riemann-surface C', the origin is a *branch point* of C'. The two circles are interpenetratingly connected along a line through the origin. The precise position of this line of intersection is irrelevant.[38]

The precise circumstances under which Riemann arrived at the idea of the Riemann surface are unknown. In his dissertation Riemann simply confronts us with the notion. Klein wrote:

> "Because the continuation of a function along different routes can yield a different u+iv for the same x+iy (and Riemann always had the conformal mapping in his mind), he hit upon the idea of a 'Riemann-surface' which covers the plane or only parts of it more than once".[39]

With the notion of Riemann - surface, the problems that other researchers had had with multiple-valued functions became in principle soluble. The possibility of generalizing Cauchy's results to complex functions mapping a Riemann-surface onto a Riemann-surface, and also the problem of the relationship between a particular function and the nature of the Riemann-surfaces that map onto one another, must have come to Riemann's mind early on.

---

[38] If Riemann had not handled the Euclidean plane in quite such a "physical" way, he would never have come up with the Riemann-surface. The rigid Euclidean plane in a platonic reality cannot, of course, be considered as consisting of several coinciding layers.

[39] "Indem die Fortsetzung einer Funktion auf verschiedenen Wegen für dasselbe x+iy verschiedenes u+iv ergeben kann und Riemann immer die konforme Abbildung vor Augen hatte, erwuchs für ihn die Idee der *'Riemannschen Fläche'*, welche die Ebene oder auch nur Teile von ihr mehrfach überdeckt"([Klein,1979], p. 256).

In his dissertation Riemann's primary goal, however, was the global characterization of complex functions by means of boundary conditions. The equations $\Delta u = \Delta v = 0$ suggested that this could be done by means of a two-dimensional potential theory. We have seen above that if u is known and if v is known in one point P, then

$$v_X = v_P + \int_P^X -\frac{\partial u}{\partial y}\,dx + \frac{\partial u}{\partial x}\,dy$$

Moreover, u could be determined by its values on the boundary of an area (a bounded Riemann-surface) on the basis of the two-dimensional analogon of the Dirichlet-principle. However, the integral is sometimes dependent on the path. This is so, for example, if we consider an area with a singularity S in it. Consider Figure 5.8. If we exclude the singularity with a very small circle and try to define a complex function u + iv uniquely on the resulting ring-shaped area by means of the values of u on the boundary and the value of v in one point, the value of v usually depends on the path. Path 1 gives a value different from that given by path 2. In order to obtain a unique function u+iv, Riemann introduces *cross-cuts*.

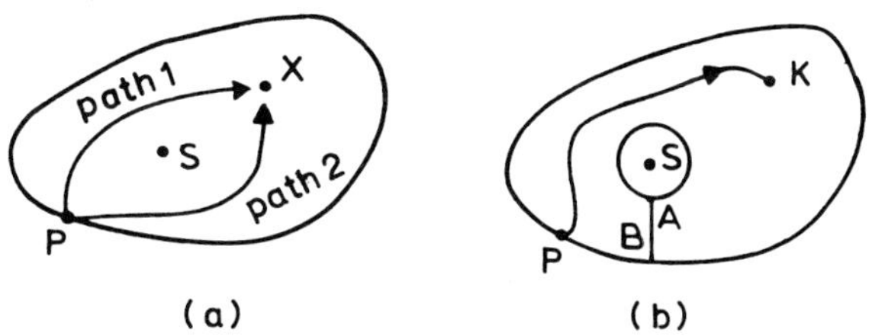

Figure 5.8

If we cut the area with a cross-cut AB from the inner circle to the outer boundary, we obtain an area for which an integral along a path PX only depends on the point P and X. For such an area the function u+iv is uniquely determined in the way sketched above. If there are more singularities more cross-cuts are needed to guarantee unicity. In his dissertation Riemann introduces the notions "cross-cut", "simply connected", "multiply-connected" and "order of connectivity" for bounded Riemann-surfaces.[40]

V.4.6 Klein's hypothesis

The two links that appeared to exist - between the theory of functions of a complex variable on the one hand and geometry and potential theory on the other - are central in the development of those parts of Riemann's work with which we are dealing. The link with potential theory enabled Riemann to follow his favoured global approach. The link with geometry yielded the Riemann-surface, which opened up impressive possibilities of generalization.

Riemann's dissertation consists of 22 parts. In Part 1 Riemann defines a function w=u+iv of a complex variable z=x+iy by the requirement that dw/dz is independent of dy/dx. He proves that this implies that the function corresponds to a conformal mapping (Part 3). In Part 4 Riemann derives the Cauchy-Riemann equations. In Part 5 he introduces the bounded Riemann-surface and it becomes clear that he intends to study functions defined on bounded Riemann-surfaces. The functions may possess discontinuities in isolated points and along isolated lines.[41] In Part 6 Riemann gives some re-

---

[40]It is interesting to compare the appearance of the notion of connectivity in Lakatos's rational reconstruction of the history of Euler's formula that we studied in Chapter I with the appearance of that notion in Riemann's work. In Lakatos's (in this respect counter-factual) reconstruction, counter-examples to a theorem bring to light hidden lemmas of simple-connectedness. It is rather obvious that the value v_x is not independent of the path if the area is such that there exist closed paths containing singularities. In principle, such areas can be considered as counter-examples to "theorems" that do not differentiate between simple- and multiple-connectedness. However, no one ever came up with such theorems. The pattern of development here is not one of bold generalizations followed by dramatic refutations. We rather have a set S (simply-connected areas, although they are not yet so called) and a particular property P (concerning unicity) that all elements of S possess. Then objects (multiply-connected areas, although they are not yet so called) appear that do not belong to S but which are related to the elements of S in the sense that they all belong to a larger set S' (the set of "connected areas"). The objects do not possess property P. The question of when P applies leads to a refinement by dissociation of the concept of connected area (Cf. my remarks in I.3.4 concerning refinement by dissociation of concepts).

[41]In a footnote Riemann remarks: "Diese Beschränkung ist zwar nicht durch den Begriff einer Function an sich geboten, aber um Infinitesimalrechnung auf sie anwenden zu können erforderlich: eine Function, die in allen Punkten einer Fläche unstetig ist [...] kann weder einer Differentiation, noch einer Integration, also (unmittelbar) der Infinitesimalrechnung überhaupt nicht unterworfen werden" ([Riemann,1851], p.9, footnote).

sults on connectivity. Then Riemann turns to Cauchy's theorems and their generalizations (Parts 7-14). In Part 15 Riemann shows that the order of connectivity is invariant under conformal mapping and he studies the degree of invariance of branch points under conformal mapping. In the Parts 16-18 Riemann concentrates on the main problem of the dissertation - finding the minimum number of conditions, including certain discontinuities (in points but also along lines), that uniquely determine a function on a Riemann-surface. While he was writing the dissertation, applications to the theory of integrals of algebraic functions[42] were undoubtedly already on his mind. This is obvious from his interest in functions that are discontinuous along lines. Integrals of algebraic functions are discontinuous along cross-cuts. This can readily be grasped from the following considerations. If we take a multiple-valued algebraic function and consider by means of the introduction of a Riemann-surface T the corresponding one - to - one mapping f: T -> S, we can integrate f on the surface T. The integral should be independent of the path. This means that we have to turn T into a simply-connected area T' by means of cross-cuts. The integral defined on T' is discontinuous along the cross-cuts. If one crosses a cross-cut, the value of the integral grows with a certain constant value depending on the cross-cut (and not on the point at which the cross-cut is crossed).

At the end of the dissertation in Parts 21 and 22 Riemann turns again to conformal mappings and proves that there always exists a conformal mapping of a simply connected surface, bounded by curves, onto the interior of a circle (the Riemann-mapping theorem).

It is clear that both links of functions of a complex variable with conformal mappings and potential theory were used by Riemann in order to effectuate a global approach to algebraic functions and their integrals. Such functions and their integrals occupy a central position in his work. Of course, one would like to know more precisely which lines of thought guided Riemann. However, not much is known about the precise genesis of Riemann's dissertation and the *Theorie der Abel'schen Functionen.*[43] Moreover, Riemann's theories were, and still are, sophisticated and difficult to understand by the non-initiated. Fortunately there exist two interesting conjectures concerning the development of Riemann's work. In an attempt to understand the dynam-

---

[42]If z and its image f(z) are connected by means of an algebraic equation, f is an algebraic function.

[43][Riemann,1857]

ics of parts of Riemann's brilliant work, Klein on the one hand and Brill and Noether on the other have come up with admittedly speculative, but possible hypotheses.

The first hypothesis is Klein's and it concerns the way in which Riemann was led to existence-theorems for functions on Riemann-surfaces. In Klein's view particular physical applications of potential theory played a major role. Klein writes:

> "First of all let a closed surface consisting of n-sheets covering the z-plane be given. The basic thought-experiment is: The Riemann-surface is imagined to be uniformly conducting electricity. That can be realized very easily if one glues tin leaf onto the surface and provides for an isolated penetration of the leaves by means of combs that grasp into each other in the branch-cuts, so that the resistance in the teeth is the same as in the homogeneous tin cover. In two points, A_1 and A_2, the poles of a galvanic battery of suitable strength are set up. This causes a current whose potential is everywhere else on the surface unique and continuous and satisfies the equation $\Delta u=0$, but becomes discontinuous in A_1 and A_2 like $\log r_1$ and $-\log r_2$, respectively. In that way we have obtained another existence theorem that can be phrased, for example, as follows: On every closed Riemann-surface there exists a continuous potential function u that becomes logarithmically infinite in a determined way that is given in advance".[44]

This real potential function can be turned into a function of a complex variable u+iv as follows:

---

[44]"Sei zunächst eine n-blättrige, geschlossene Fläche über der z-Ebene gegeben. Das grundlegende Gedankenexperiment ist: Die Riemannsche Fläche wird als gleichförmig elektrisch leitend gedacht. Das läßt sich sehr einfach realisieren, indem man die Fläche mit Stanniol beklebt und für eine isolierte Durchdringung der Blätter dadurch sorgt, daß man in den Verzweigungsschnitten Kämme ineinander greifen läßt, so daß der Leitungswiderstand in den Zinken der gleiche ist wie in der homogenen Stanniolbelegung. In zwei Punkten A_1, A_2 werden die Pole einer galvanischen Batterie von geeigneter Stärke aufgesetzt. Es entwickelt sich ein Strom, dessen Potential u auf der Fläche überall sonst eindeutig und stetig ist und die Gleichung $\Delta u = 0$ befriedigt, in A_1, A_2 aber unstetig wird wie $\log r_1$ bzw. $-\log r_2$. Damit haben wir einen weiteren Existenzsatz gewonnen, der sich etwa so formulieren ließe: Auf jeder geschlossenen Riemannschen Fläche existiert eine stetige Potentialfunktion u, die an zwei vorgegebenen Stellen in bestimmt vorgegebener Weise logarithmisch unendlich wird" ([Klein,1979], p.260).

"First of all we cut [the surface - T.K.] by means of 2p crosscuts in such a way that all further cuts cause the surface to fall apart; then we also connect A_1 and A_2 by means of a cut. In the surface thus prepared,

$$v = \int_{z_0}^{z} -\frac{\partial u}{\partial y}dx + \frac{\partial u}{\partial x}dy$$

is unique and continuous, but on the two sides of the individual cross-cuts its value differs in an additive constant and it is 2π larger on one of the sides of A_1A_2 than on the other.

In conclusion I can now phrase the theorem: If a closed Riemann-surface covering the z-plane is given and if one marks any two points A_1, A_2 of it connected by means of a curve, then there exists on the surface cut by 2p cross-cuts one and only one unique function u+iv that possesses the following properties: it is everywhere continuous, except in A_1 and A_2 where it becomes logarithmically infinite; and, whereas its real part is unique on the surface and on its boundaries, the imaginary part exhibits at corresponding points of the two sides of the connecting curve A_1A_2 the periodicity modulus 2π, and likewise at corresponding points of the two sides of each of the 2p cross-cuts periodicity moduli P_i that must be calculated. The function $\Pi A_1, A_2$ that proceeds from u+iv by arbitrary crossing of the cuts - in accordance with the terminology that was developed for elliptic integrals - is called 'an integral of the first kind'".[45]

---

[45]"Wir zerschneiden [die Fläche] zunächst durch 2p Querschnitte so, daß kein nicht zerstückender Rückkehrschnitt (=Periodenweg) mehr möglich ist; dann verbinden wir auch noch A_1, A_2 durch einen Schnitt. In der so präparierten Fläche ist v eindeutig und stetig,

$$v = \int_{z_0}^{z} -\frac{\partial u}{\partial y}dx + \frac{\partial u}{\partial x}dy$$

unterscheidet sich aber an den Ufern der einzelnen Querschnitte um je eine additive Konstante und [...]ist [...] auf dem einen Ufer von A_1A_2 um 2π größer als auf dem anderen. Ich kann jetzt zusammmenfassend den Satz aussprechen: Ist eine geschlossene Riemannsche Fläche über der z-Ebene gegeben, und markiert man auf ihr irgend zwei Punkte A_1, A_2 , die man durch eine Kurve verbindet, so existiert auf der durch 2p Querschnitte zerschnittenen Fläche ein und nur ein eindeutiger Funktionszweig u+iv, der folgende Eigenschaften hat: er ist überall sonst stetig, außer in A_1 und A_2, wo er logarithmisch unendlich wird; und, während sein reeller Teil auf der Fläche und an ihren Begrenzungen eindeutig ist, bietet der imaginäre Teil an korrespondierenden Stellen der beiden Ufer der Verbindungskurve A_1A_2 den Periodizitätsmodul 2π dar, und ebenso an korrespondierenden Stellen der beiden Ufer jedes der 2p Querschnitte Periodizitätsmoduln

Klein then goes on explaining how Riemann could likewise demonstrate the existence of other Abelian integrals and how he could prove their existence mathematically by means of the principle of Dirichlet.[46]

If Klein is right, we have the following pattern in Riemann's work. The link of the theory of complex functions with potential theory makes Riemann's desire for a global approach to the theory of functions practicable. The other link between complex functions and conformal mappings yields the notion of Riemann-surface, pointing at possibilities of generalization. But in Klein's view the "real source of Riemann's developments" ("die eigentliche Quelle von Riemann's Entwicklungen") is "the inductive-physical thinking" ("das induktiv-physikalische Denken") ([Klein,1979], p.259). In the above quotation Klein describes a paradigmatic thought experiment in mathematical physics, which shows

> "how Riemann, by the special character of his thinking, approached the above-mentioned kinds of functions from a completely different angle than had been done before. And it is precisely in this [special character] that one can view an essential element for the extraordinary achievement which he accomplished in this field".[47]

It is clear that we have here an attempt to give a characterization of the heuristic unity of (parts of) Riemann's work. Of course the characterization does not yield a complete understanding of the development of Riemann's work. There are many new elements in Riemann's work, for example the completion of each sheet of a Riemann surface with a point at infinity (thus avoiding a host of difficulties) and the important role of topological concepts, which by far surpass anything known before Riemann.

P_i, die man berechnen muß. Man nennt die Funktion Π_{A_1,A_2}, die aus dem gewonnenen u+iv durch beliebige Überschreitung der Schnitte entsteht - gemäß der Terminologie, die sich bei den elliptischen Integralen entwickelt hat - 'ein Integral dritter Gattung'" ([Klein,1979], pp. 260-261).

[46]*Abelian integrals* are integrals of rational functions on a Riemann surface. Riemann classified them as follows: those that are everywhere finite are called *integrals of the first kind*. The *integrals of the second kind* have algebraic, but no logarithmic, infinities. Integrals possessing logarithmic infinities are called *integrals of the third kind*. Abelian functions are defined by means of the inversion of algebraic integrals.

[47]"wie Riemann durch die Eigenart seines Denkens von einer ganz anderen Seite an die genannte Funktionsgattungen herantritt, als es vorher je geschehen war. Und eben hierin wird man ein wesentliches Stück für den außerordentlichen Erfolg erblicken, den er auf diesem Gebiet errang" ([Klein,1979], p.261).

V.4.7 The hypothesis of Brill and Noether

Brill and Noether objected to Klein's hypothesis.[48] They refer to Riemann's *Theorie der Abel'schen Functionen*[49] - where Riemann wrote that he was led to his results "in the autumn of 1851 and the beginning of 1852 through investigations about conformal mapping of multiply-connected surfaces"[50] - in order to support their view that conformal mappings played a much more important role than Klein suggested. According to Brill and Noether, a note out of Riemann's heritage contains a line of reasoning that must be considered as Riemann's starting point in his work on the theory of functions. I will only briefly outline their argument, since a complete treatment of the mathematics involved would lead us too far. The title of the note is "Equilibrium of electricity on cylinders with circular cross-section and parallel axes".[51] The note starts with a problem in 2-dimensional potential theory: find a function u(x,y) which satisfies $\Delta u=0$ inside an area S bounded by circles in the xy-plane. The values of u on the boundary of S are given. We can then consider u as the real part of a complex function: $w(x + iy) = u + iv$, defined on S, which assumes given real values on the boundary of S. This function corresponds to a conformal mapping w: S -> T, where T is a Riemann surface. Function w maps the boundary of S on the boundary of T. The problem now is to find w, or its inverse function w^{-1} : T -> S, which may be easier.

The problem of finding the inverse function is relatively easy if the area S consists of n disjunct circles and the original u assumes on these circles all values from $-\infty$ to $+\infty$ precisely once (walking counter clockwise on the circles). T is then a Riemann surface which covers the upper half of the complex plane n times and has 2n-2 branch points of order 1 in its interior. This follows from the fact that S and T must possess the same order of connectivity. According to Brill and Noether for n=2 the function w^{-1} could be guessed by Riemann. That solution however, in Brill and Noether's view, paved the way for the solution for higher values of n. Brill and Noether write:

> "In order to penetrate from the boundary values to those in the interior, one needs in each case again [one of] Green's auxiliary function[s] U, which possesses the value of zero at the boundary, which satisfies the equation $\Delta U=0$ in the interior, but which in an arbitrary point of the interior does

---

48[Brill&Noether,1892-93], p. 259

49[Riemann,1857]

50[Riemann, 1857], p. 95

51[Riemann,1876a]

> not this time become infinite like 1/r but like log r. However, in the case of a surface consisting of two leaves over the UV-plane this function can be guessed. It is nothing other than the imaginary part of an elliptic integral of the third kind which satisfies all the above-mentioned conditions in the twice-covered half-plane; this is because for real arguments the imaginary part of the integral is known to be zero. This paved the way towards mastering the higher cases, where the number of boundary curves is larger than two.Then the number of leaves increases and their connectivity becomes more complicated, but Green's function is always determined by an Abelian integral of the third kind".[52]

They believe that a train of thought similar to the one in the note may have led Riemann to his dissertation and his work on Abelian functions. Brill and Noether write:

> "The note not only contains in a nutshell the problems that are put in the dissertation and which are only supplemented in one direction by the theory of Abelian functions, but supplies at the same time the essential expedients for their solution".[53]

If Brill and Noether are right we have the following pattern in Riemann's work. A particular problem and its solution, in which the theory of complex functions, potential theory and conformal mappings occur interrelated, suggest further problems and the tools for their solution on the basis of analogy.

---

[52]"Um von den Randwerten zu denen im Inneren der Fläche vorzudringen, braucht man in jedem Falle wieder die Green'sche Hülfsfunction U, die [...] and der Grenze den Wert Null besitzen, im Inneren der Gleichung $\Delta^2 U=0$ genügen, aber diesmal in einem unbestimmten Punkte des Inneren nicht wie 1/r, sondern wie log r unendlich werden muss. Im Falle jener zweiblättrigen Fläche über der UV-Ebene lässt sich aber diese Function *erraten*. Sie ist nichts anderes, als der imaginäre Teil eines elliptischen Normalintegrals dritter Gattung, welches in der doppelt überdeckten Halbebene die erwähnten Bedingungen alle erfüllt; denn für reelle Argumente ist der imaginäre Bestandteil des Integrals bekanntlich gleich Null. Hiermit war der Weg gebahnt für die Bewältigung der höheren Fälle, wo die Zahl der Grenzcurven grösser als zwei ist. Es erhöht sich dann die Zahl der Blätter, ihr Zusammenhang wird complicirter, immer aber ist die Green'sche Function U durch ein Abel'sches Integral dritter Gattung bestimmt" ([Brill&Noether, 1892-93], pp. 257-258).

[53]"Ueberhaupt enthält die Note nicht nur in nuce die Aufgaben, welche die Dissertation sich stellt, und welche die Theorie der Abel'schen Functionen in einer Richtung nur vervollständigt, sondern sie stellt zugleich die wesentlichen Hülfsmittel zu ihrer Lösung zur Verfügung" ([Brill&Noether,1892-93], p. 258).

Whether it is Klein who is right or Noether and Brill, it is clear that the heart of Riemann's scientific enterprise is first of all the conviction that a very general, global approach to the theory of functions of a complex variable is possible. Secondly, the theory of complex functions was linked to geometry (conformal mappings) on the one hand and to potential theory on the other. Characteristic of the way in which the links heuristically functioned is that elements of all sorts from one of the two linked theories bring about something new with respect to the other theory. In different ways Klein, Brill and Noether point out chains of analogous situations that supposedly guided Riemann's research.

V.5 Giorello's views of Riemann

V.5.1 Three heuristic principles

On the basis of Klein's views, Giorello has attempted to describe Riemann's work on complex analytic functions as the execution of a research programme in the sense of Lakatos's MSRP. He characterized the programme by pointing out three important principles in Riemann's "programme". They are:

(a) *The Dirichlet-principle* , i.e. Dirichlet's method to prove the existence of certain harmonic functions,

(b) *The principle of the definition of mathematical entities by means of their behaviour in the infinitely small*,

(c) *A principle consisting of certain "metaphysical" theses*.[54]

The three principles have quite a different status. Principle (a) is a rather concrete mathematical tool. Principle (b) is much more general. We saw above that it was phrased by Felix Klein, who described Riemann's definition of a complex function by means of the Cauchy-Riemann equations as being an example of an application of this principle. Klein also saw the principle at work in Riemann's "About the hypotheses that underlie geometry"[55], where a quadratic form of the differentials of the coordinates is used to define a particular geometry. Principle (c) is even more general and concerns Riemann's views on the way science develops. Let us first see how Giorello characterizes this principle. Giorello starts from two quotations from Riemann's work. Principle (c) primarily concerns the conviction that the history of science shows that :

---

[54][Giorello,1981] , pp.124-125

[55] [Riemann,1854]

"The conceptual systems that nowadays underlie [the explaining natural sciences] arose from the gradual change of older conceptual systems, and the reasons that led to the new ways of explanation can always be reduced to contradictions or improbabilities exhibited by the older ways of explanation".[56]

From Riemann's point of view, progress is induced by contradictions and improbabilities in older explanations. Progress requires conceptual change and conceptual change requires new notions. Riemann also wrote in his "Ueber die Hypothesen" that the important problems in physics could only be solved

"starting from the extant empirically confirmed conception of phenomena which was originally founded by Newton and if one gradually refashions this view driven by facts that cannot be explained by it. Investigations proceeding from general notions, such as the one carried out here, can only serve the purpose of not hampering this work by conceptual limitations, and that the progress in understanding the connection of things is not curbed by traditional prejudices".[57]

In the course of such investigations, older concepts appear as special cases of more general concepts. For example, Euclidean geometry is in "Ueber die Hypothesen" one very special case of a more general notion. The new and more general point of view leads to new distinctions: for example "unbounded" has to be distinguished from "infinite". Such old and new concepts are in Giorello's terminology linked by "dictionaries" or "translation schemes".

Giorello writes:

---

[56]"Die Begriffssysteme, welche ihnen [den erklärenden Naturwissenschaften- T.K] jetzt zugrunde liegen, sind durch allmählige Umwandlung älterer Begriffssysteme entstanden, und die Gründe welche zu neuen Erklärungsweisen trieben, lassen sich stets auf Widersprüche oder Unwahrscheinlichkeiten, die sich in den älteren Erklä-rungsweisen herausstellten, zurückführen" ([Riemann,18786b], p.489).

[57]"indem man von der bisherigen durch die Erfahrung bewährten Auffassung der Erscheinungen, wozu *Newton* den Grund gelegt, ausgeht und diese durch Thatsachen, die sich aus ihr nicht erklären lassen, getrieben allmählich umarbeitet; solche Untersuchungen, welche, wie die hier geführte, von allgemeinen Begriffen ausgehen, können nur dazu dienen, dass diese Arbeit nicht durch die Beschränktheit der Begriffe gehindert und der Fortschritt im Erkennen des Zusammenhangs der Dinge nicht durch überlieferte Vorurtheile gehemmt wird" ([Riemann, 1854], p.268).

"A translation scheme then is that which shows an old and familiar theory to be a particular instance of a more abstract and more general conception".[58]

Apparently Principle (c) does not tell us how to develop general notions. Its content seems to be something like this: Progress necessarily implies the development of new and more general concepts!

V.5.2 A critique of Giorello's reconstruction

Giorello's paper is meant to show that Lakatos's MSRP is more in accordance with mathematical practice than is Popper's falsificationism[59] or Kuhn's conception of scientific growth.[60] Giorello argues that Kuhn fails to take into account that "in the growth of knowledge, *concatenations* between scientific achievements are observed which no typology of scientific change should ignore", and from Popper's perspective it is inexplicable that the awareness of certain difficulties concerning Dirichlet's assumption did not bother Riemann.[61] But in spite of the modest aims, Giorello's application of Lakatos's MSRP to Riemann's mathematics is hardly satisfactory. It appears to be difficult to characterize the heuristic unity of Riemann's programme *in a precise way* in Lakatosian terms. What precisely are the negative and positive heuristics of Riemann's programme? It seems that in Giorello's opinion, the Dirichlet-principle belongs primarily to the positive heuristic of the programme.[62] As for the principles (b) and (c) Giorello seems to assign them to the "metaphysical hard core" of the programme. He explains their function in that respect by means of the notion of translation scheme. Such translation schemes are in Giorello's view also part of the hard core of the programme.[63] Giorello opaquely writes: "principles like (b) appear to be the

---

[58][Giorello,1981], p.125

[59][Giorello,1981], p.121

[60][Giorello,1981], p.118

[61]The case study, moreover, shows that Poincaré's division of research mathematicians into two groups - the intuitive mathematicians (e.g. Riemann) and the rigorous mathematicians (e.g. Weierstraß) - is misleading. In Giorello's view the programmes of Riemann and Weierstraß both had their own level of intuition and their own level of rigor. In Giorello's view intuition "denotes that set of heuristic principles which leads to preferring one mode of 'translating' rather than another, to constructing one translation scheme rather than another" ([Giorello,1981] , p.129).

[62][Giorello,1981], pp.123-124

[63][Giorello,1981], p.126

object of the mediation of the translation scheme; assertions like (c) have the function of justifying such a mediation".[64] I understand this as follows: principle (b) concerns the way in which new notions have to be defined, and implicitly determines the way in which older notions are going to be seen from the new point of view, while (c) justifies (b). In particular the status of principle (c) bothers Giorello. On the one hand he writes that (c) concerns statements (Giorello means the statements on conceptual development by means of which (c) is expressed) that are preserved "as long as possible", and for that reason it seems to belong to the hard core. Those statements are, on the other hand, not "testable". They belong in Giorello's view to the 'rhetoric of scientific communication' and for that reason he is inclined to shift them from the 'hard core' to the farthest 'periphery' of the 'protective belt'. Yet they hardly seem to function in any protective sense.

The three principles that Giorello uses to characterize the heuristic unity of Riemann's programme define neither implicitly nor explicitly an 'irrefutable' set of axioms and they do not clearly list a chain of ever more complicated problems either. The principles characterize certain aspects of Riemann's work. Even if we forget about Lakatosian terminology, they are very unsatisfactory as a characterization of the heuristic unity of Riemann's work. Actually the very general character of Principle (c) in particular could mean that it did not play any role whatsoever in the process leading to Riemann's earliest work. Unless the opposite is proven, one cannot exclude the possibility that Riemann only became aware of the principle after his earliest publications. Then it would be a principle that was made explicit only afterwards, reflectively, and not a principle that necessarily played a role in the development of Riemann's work. One of Giorello's concluding sentences on (c) is:

> "Do we hold then that one of the most distinctive characteristics of the methodology of scientific research programmes - the distinction between the metaphysical hard core and the protective belt, in characterizing the heuristics - cannot be applied to case studies such as the one under consideration, or that it should be abandoned? As an absolute scheme, this distinction, in our opinion, leads to considerable difficulties. However, if it is assumed as a working hypothesis, in a flexible, non-schematic way, it will allow us to characterize assertions such as those of type (c), in our example

---

64 [Giorello,1981], p.130

as at once 'metaphysical' and 'auxiliary' according to our interests in the reconstruction of the research programme".[65]

Apparently Giorello was aware of the problems described above. Yet his application of Lakatos's MSRP in a "flexible, non-schematic way" remains unsatisfactory.

V.6 Concluding remarks

As far as Riemann's work is concerned we finish with a conclusion similar to our conclusion to the examination of Archimedes's work. As they have done in the case of Archimedes, historians have attempted to characterize the heuristic unity of (parts of) Riemann's work. One could use the term "positive heuristic" for such a characterization, but Giorello's attempt to view Riemann's work from a Lakatosian point of view shows that Lakatosian terminology as a whole is not very suitable to describe Riemann's work. I prefer to describe Riemann's work as resulting from one or more *research projects*. The attempts of Klein, Brill and Noether to define the heuristic unity of (parts of) Riemann's work are from my point of view attempts to distinguish a research project.

As far as the characterization of the research projects is concerned, there exists a clear analogy between the development of Riemann's work and that of Archimedes: An unexpected combination of two rather disjunct areas (theory of complex functions and potential theory in the case of Riemann, or geometry and mechanics in the case of Archimedes) in the form of a paradigmatic solution of a particular problem (in the theory of functions in the case of Riemann or in geometry in the case of Archimedes) acts as guideline in a series of analogous situations and yields a considerable number of results. Here also, as in the case of Archimedes, the major heuristic rules are then actually: "Find analogous situations" and "Apply analogous methods in analogous situations". I wrote above that Archimedes's work took place within a mathematical tradition that also encompassed the work of others. The same holds for Riemann's work. In spite of its great originality, Riemann worked within a tradition, of which Cauchy was one of the first major representatives. Riemann has been characterized more than once as "someone who followed Cauchy's way of working".[66] I will call that tradition the *Conceptual Tradi-*

65[Giorello,1981] , pp.130-131

66"jemand, der 'sich der Cauchyschen Manier' anschließt" ([Neuenschwander, 1981], p.226).

tion in analysis. In Chapter VIII, I will describe this tradition which dominated nineteenth century analysis more extensively.
In the next chapter I will give an outline of a methodology of mathematical research traditions. However, I will first compare Archimedes's and Riemann's work in still another respect. In order to do so we have to consider the fact that parts of Riemann's work were in a sense refuted by Weierstraß' criticism of the Dirichlet-principle. Klein illustrated that criticism in this way. Suppose one is to determine among all 'continuously curved' curves from point A to B via C the one whose length is minimal (Figure 5.9).

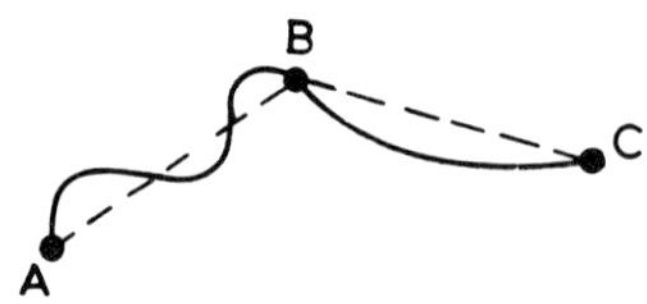

Figure 5.9

Obviously the shortest curve consists of the two straight-line segments AB and BC. This lower limit, however, does not belong to the set of curves considered, since it is not 'continuously curved' in B. Clearly the existence of a minimal 'continuously' curved curve is not at all certain. The Dirichlet-assumption consists of a similar argument. Riemann himself admitted the correctness of the criticism ([Klein,1979], p.264) and although most mathematicians after Weierstraß' criticism lost faith in Riemann's results, Weierstraß and Riemann did not ([Klein,1979], p.264). Weierstraß actually stimulated his pupil H.A.Schwarz to devise different proofs for the theorems (Ibid.). Such proofs were eventually given.
One could compare the "Dirichlet assumption" with the method of indivisibles which Archimedes used. Archimedes himself came up with the rigorous proofs of the theorems that he had found in a different way. In the case of Riemann, Weierstrassian mathematicians provided the rigorous proofs. When we look at Archimedes and Riemann within the wider context of the traditions they worked in, we can make the following observation. Both used unexpected links between mathematics and physics which led to important results in mathematics. The results were obtained in two phases: first a proof was given which did carry conviction, but which did not satisfy the criteria for a good proof that the tradition in its mature form would use. In the case of

Archimedes, who was working within the context of the already rather mature Euclidean Tradition, those criteria were known and Archimedes himself also provided the rigorous proofs. The Conceptual Tradition in analysis was still young at the time of Riemann. The work of Weierstraß and his pupils in particular would transform the Conceptual Tradition. They would also provide rigorous proofs for Riemann's theorems based upon the Dirichlet-assumption.

Chapter VI Outline of a Methodology of Mathematical Research Traditions

VI.1 Research traditions

In the preceding chapters we saw that the application of Lakatos's MSRP to mathematics requires that we modify that methodology considerably. Notions like protective belt, positive heuristic, empirical versus theoretical progress all concern natural science and they cannot be applied to mathematics without modifications. The attempts by Spalt and Giorello to describe certain mathematical developments in terms of the MSRP clearly show the difficulties involved. In this chapter I will define a *methodology of mathematical research traditions (MMRT)*. My approach owes a great deal to Laudan's *Progress and its Problems* ([Laudan,1977]), which contains a methodology of scientific research traditions.[1]

I use the following working definition of a research tradition:

> A *mathematical research tradition* is a group research activity, historically identifiable (in a certain period), characterized by common general assumptions (in the form of e.g. definitions, axioms) about the entities that are being studied in a particular *fundamental mathematical domain*, and it involves assumptions about the appropriate methods to *prove* properties of those entities.[2]
>
> A *fundamental mathematical domain* consists of the most general mathematical entities that play a role in the mathematics of the period.

A good example of a mathematical research tradition is the dominant deductive geometric tradition in Greek mathematics represented, for example,

---

[1]In the footnotes to this chapter I will state my debt to Laudan more precisely. Influential theories of scientific change have not only been advanced by Lakatos and Laudan, but also by Kuhn, Feyerabend, Toulmin, and others. With the exception of Kuhn's model (see Chapter VIII) I have not taken these models into consideration. The same holds for [Donovan e.a.,1988]. In this book a serious attempt was made to test the elements of the different theories of scientific change that I mentioned against a broad range of examples. None of the examples directly concerns the development of mathematics.

[2]Laudan gives the following preliminary definition: "A research tradition is a set of general assumptions about the entities and processes in a domain of study, and about the appropriate methods to be used for investigating the problems and constructing the theories in that domain" ([Laudan,1977], p.81). The main difference with my definition is that Laudan's definition includes not only methods to justify theories, but also heuristic methods. Although I do not deny that whole traditions can possess a certain heuristic unity, I locate the heuristics and consequently the dynamics of traditions in research projects.

by Euclid's Elements. I will call that tradition the *Euclidean Tradition.* The fundamental mathematical domain of that tradition is Euclidean space. Major entities of the fundamental domain are points, lines (straight and curved) and surfaces (plane and curved). The methodology of proof of the Euclidean Tradition is Euclid's axiomatic methodology. Another example of a research tradition is the *Structuralist Tradition* that has dominated twentieth century mathematics. The fundamental mathematical domain of the Structuralist Tradition consists of the totality of all mathematical structures, and its methodology of proof is the modern formalist axiomatic method.

In order to characterize a tradition one should ask the two classical fundamental questions in the philosophy of any science: What is the object that is being studied? What aspect of the object is being studied and what methods are being used to study that object? As regards the second question, I will interpret it as concerning the methodology of proof (this includes Lakatos's negative heuristic or hard core). In order to identify research traditions, in mathematics I will give priority to the ontological question: "What is the object that is being studied in the tradition?".[3] The views on the object under investigation that prevail within a tradition imply, of course, certain border conditions as for the methodology of proof. But within the border conditions created by a particular view of the object, a considerable variation in methodology is still possible. If such differences are considerable, I will distinguish more than one tradition.

In a tradition conjectures are framed with respect to the entities involved and attempts are made to prove those conjectures.

> A *conjecture* expresses a particular presumed state of affairs concerning the objects that are being studied in the tradition.
>
> A *proof* of a conjecture is an embedding of the conjecture in the already established knowledge belonging to the tradition by means of a successful application of methods belonging to the tradition.

A proof is always a proof relative to a tradition. Usually traditions encompass several theories that concern different *subdomains* of the fundamental domain. For example, Euclid's theory of the circle (Books III and IV of Euclid's Elements) concerns the subdomain consisting of all circles in the Euclidean plane. The last book of the Elements contains another example - it

[3] I suspect that essentially different views of the object necessarily imply important methodological differences, while one and the same object may be approached by means of essentially different methodologies.

is dedicated to the properties of the regular solids. It is important to emphasize that mathematical research traditions are different from mathematical theories. I use the following definition of a theory:

> A *theory* belonging to a tradition is a set of connected mathematical statements that are considered to be true of (a subdomain of) the fundamental domain of that tradition encompassing at least one theorem, i.e. a proven conjecture.

Theories exemplify and illustrate traditions, and a tradition in general encompasses more than one theory. Yet my definition does not exclude the possibility that a young tradition encompasses only one theory.

VI.2 Examples of traditions

Let us tentatively try to list some research traditions in mathematics. Above I mentioned the Euclidean Tradition in Greek mathematics. The great Greek mathematicians, as for instance Apollonius and Archimedes, all worked within this tradition. In Chapter VII, I will discuss the emergence of this tradition and show that there are good reasons to suppose that it was preceded by another Greek mathematical tradition: the *Demonstrative Tradition.*

One may also say that the 17th century is dominated by a tradition whose object is Euclidean space, but whose methodology of proof is different from that of the *Euclidean Tradition* in the sense that it encompasses more. The methodology is less rigid making the borderline between an acceptable and an unacceptable proof less clear. The introduction of powerful algebraic methods by Fermat and Descartes was of crucial importance. Although the objects that are being studied in this tradition are in Euclidean space and the results of the Euclidean Tradition were not rejected by this tradition, one must distinguish this tradition from the Euclidean Tradition. I will call it the *17th Century Geometric Tradition.*

Within the 17th Century Geometric Tradition, Leibniz and Newton developed the theories of differential calculus and the calculus of fluxions. The tremendous success of Leibniz's calculus in particular led to what I will call the *Formalist Tradition* in analysis, of which Euler and Lagrange are major contributors. This tradition was replaced in the nineteenth century by the *Conceptual Tradition* in analysis. Cauchy and Weierstraß were important representatives of this tradition, but Riemann also worked within the Conceptual Tradition. I will study these two traditions below in Chapters VIII and IX.

New traditions are usually born within other traditions. The 19th century shows important developments in this respect. Cantor's set theory grew out of the 19th century tradition in analysis, but soon began to deviate considerably from this tradition in its object and method. The appearance of non-Euclidean geometry within the Euclidean geometric tradition eventually contributed to a new view of the object of geometry.

History also shows traditions being united in new traditions. For example in the first half of the twentieth century, both the Conceptual Tradition in analysis and the Euclidean Tradition in geometry were absorbed by the *Structuralist Tradition* in mathematics. The Bourbaki - enterprise was a major and quite successful attempt to found that tradition.

VI.3 Research projects

Progress in a tradition consists in the production of conjectures and theorems. In order to frame conjectures and to find proofs, different heuristic rules are used. There are more or less tradition-independent rules like Polya's rules and the rules belonging to Lakatos's MP&R. There are, however, also tradition-dependent rules defined by the general assumptions concerning the objects that are being studied and by the methodology of proof of a tradition. For example, in the Conceptual Tradition in analysis we have the rule that the infinitely small and the infinitely large ought to be dealt with in terms of limits. The application of such tradition-independent and tradition-dependent rules leads to conjectures and proofs. Yet the examples of Archimedes and Riemann discussed in the previous chapter suggest that progress often takes place as follows: a new paradigmatic solution of a problem leads to a series of conjectures and theorems on the basis of analogy. I will use the word research project for such concatenations of conjectures and theorems.

> A *research project* consists of a number of research goals together with a set of hints as to how one can reach the goals. The project includes a paradigmatic solution of a problem that shows the kind of goals and the effectiveness of the hints with respect to the goals. Large projects may very well encompass subprojects.[4]

The research projects belonging to a tradition represent the dynamic element of that tradition. An important research project is always the result of a 'stroke of genius' and represents something that is completely new.

---

[4]The notion "research project" does not occur in Laudan's methodology of research traditions.

It will happen often that in a small, young tradition one particular project dominates. If in such a case there is also a stable hard core consisting of a set of assumptions that are not subjected to criticism, the tradition can be considered as a research programme in Lakatos's sense. Obviously the notion "mathematical research tradition" is more general than the notion "mathematical research programme".

There are at least two different kinds of research projects that have to be distinguished. First, there are projects that fit wholly within a particular research tradition. Secondly, there are projects that, although they spring from a particular tradition, nevertheless deviate from that tradition. Within a tradition there are always attempts to give the tradition as rigorous a foundation as possible. Two nice examples of *foundational project* are Lagrange's attempt to give a formalist foundation to eighteenth century analysis without the use of infinitesimals, and Weierstraß' consistent application of ε - δ methods in nineteenth century analysis. A successful foundational project may lead to modifications of the tradition to which it belongs, and also, if it deviates a great deal, to a new rival of the tradition from which it proceeded. There are also projects that, although their goal was not to modify the tradition they spring from, lead to the introduction of new notions that influence the ontology of that tradition. Such projects may also lead to rival traditions. One notices that a mathematical research tradition is not rigid, that it can be modified in the course of time.[5] Also in this respect the notion of research tradition is more general than the notion of research programme. As I said above, two traditions have to be distinguished if their domain and/or their methods differ drastically.

VI.4 Progress and rationality in mathematics

When one looks at the list of traditions given in VI.2 it is clear that in the course of time older traditions are replaced by newer traditions. Some traditions, however, possess a very long life. Although the Euclidean Tradition did not receive much attention for centuries because of a lack of successful research projects, it continued to exist as a tradition in which mathematicians sometimes worked. Only in the course of the 19th century did the Euclidean Tradition become subject to revision. And even nowadays the Euclidean tradition has not totally disappeared. In a paedagogical context, for instance,

---

[5]Also in Laudan's methodology of scientific research traditions, traditions are not rigid but can be modified in the course of time.

examples from that tradition could still play a role. Something similar holds for the Conceptual Tradition in analysis. The Conceptual Tradition was not discarded in the twentieth century Structuralist Tradition but subsumed under it in a modified form.

Although there is competition in mathematics it is different from the competition found in natural science. A major reason is undoubtedly the following. Unlike the theories of natural science, whose justification is primarily restricted to the empirical data and whose starting point may be completely hypothetical, in mathematics the starting point of a theory concerning a specific domain will always be based upon statements considered to be true of the domain.[6] In the case of the Euclidean Tradition, this starting point was made quite explicit in the form of definitions, postulates and axioms. Because of the fact that mathematical theories, right from the start, are based upon "direct evidence" concerning their domain, older mathematical theories are hardly ever completely rejected. Usually they are absorbed by the newer theories in a modified form.

I will now concentrate on the problems of progress and rationality in the development of mathematics. Mathematicians generally agree that over the centuries mathematics progresses. It is less easy to say in general terms (without giving examples) what this progress consists in. I claim that if one wants to study the rationality of its development, mathematics is better subdivided and structured in terms of research traditions and research projects than in terms of research programmes. The next question then is: What constitutes progress in mathematics? And a related question is the following: What choices should a rational mathematical community make with respect to competing research traditions and competing research projects?

In Lakatos's MSRP a research programme is theoretically progressive if it predicts new facts. It is empirically progressive if these facts are also confirmed by experience. According to the MSRP, rational scientists prefer progressive programmes to degenerating programmes. At first sight this notion of progress seems inapplicable to mathematics, because in mathematics there is no clear analogon of empirical confirmation. An idea by Marchi ([Marchi,1976]), however, suggests a modification of the MSRP in this respect such that it becomes applicable to mathematics. Marchi argues that mathematical theorems should be regarded analogously to facts in science. Let

[6]In principle a mathematical theory could be developed from a completely arbitrary (consistent) set of axioms without an intended model. However, this hardly ever happens.

us first consider Marchi's argument. In what sense is a mathematical theorem a fact? In three senses:

> "(1) a theorem is something universally recognized and accepted and is in this sense a fact; (2) a theorem is a task and is in this sense a fact; (3) a theorem's acceptance rests on a provisional compromise between dogmatism, infinite regress and psychologism and is in this sense a fact" ([Marchi,1976], p.391).

The first sense does not need much explanation. Marchi compares facts such as the fact that the atomic weight of water is 18 with, for example, the fundamental theorem of algebra. Continued failure to prove the theorem was not taken as evidence that the theorem was false. Instead, the theorem was so widely recognized and accepted that mathematicians continually sought new proofs of it.[7]

As for the second sense, Marchi compares proofs of theorems to the explanations of facts. Referring to Agassi, Marchi argues that a fact in science is a task because it requires an explanation, or a proof that in the experimental procedure a mistake has been made. In the same way a theorem in mathematics is a task, because it requires a proof or a refutation.[8]

As for the third sense Marchi refers to Popper. According to Popper a scientific fact is based on our experience. We do not, however, dogmatically accept our experience. If we have reasons to doubt our experience we start testing. We can go on testing, but this would lead to an infinite regress if we insist on testing the test to test the test etc. At a certain moment we stop because we are satisfied. In the words of Marchi:

> "Mathematical theorems are expressions of our experience, our mental experience in this case rather than physical experience. We do not accept our mental experiences dogmatically. We question them when we have causes to do so. But we don't feel compelled to question our experiences until we reach a final justification. That is, in accepting mathematical theorems, we rely neither on psychologism, nor on the justification of reasoned

---

[7]The intermediate value theorem for continuous functions is also a good example. Spalt nicely shows how the intermediate value phenomenon was first simply taken for granted, then phrased and proved and repeatedly rephrased and reproved more precisely in the form of a theorem ([Spalt,1988]).

[8]From my point of view, Marchi here insufficiently distinguishes conjectures from theorems. A conjecture can clearly be considered as a task. Within a particular tradition an established theorem loses much of its task character. Yet even such theorems can often be considered as tasks in the sense that improvement of the existing proofs is never excluded.

arguments nor on dogmatism. Thus we accept theorems as facts in mathematics on the same basis as we do facts in science"([Marchi,1976], 393)

The point is that the acceptance of a theorem in mathematics, as the acceptance of a scientific fact, rests on a provisional compromise between dogmatism, infinite regress and psychologism: mental experiences are repeatedly tested until the mathematical community "dogmatically" finds that it has been enough.

One can further clarify the proposal to conceive of mathematical theorems as analogous to facts and of proofs as analogous to explanations, as far as the second sense of fact is concerned, by means of a model of explanation given by Van Fraassen ([Van Fraassen,1980], pp. 134-153). Van Fraassen defends a pragmatic view of explanation: all kinds of things, not just facts, can be the object of explanation. According to Van Fraassen an explanation is an answer to what he calls a why-question. A why-question in a given context is determined by three factors: (1) the topic of the question, (2) the contrast-class of the question and (3) a relation of explanatory relevance. In Van Fraassen's model, a why-question can only be understood in a particular context. The topic of the question is one particular proposition with respect to the context. The contrast class consists of the proposition supposed to be true plus a set of alternatives that are false. Finally the relevance relation determines what will count as a possible explanation. Let us consider an example given by Van Fraassen. "Why is this man called Hillary?" The topic of the question is the proposition "This man is called Hillary". The contrast class could consist of all sentences of the form "This man is called <arbitrary name>" and the relevance relation could allow as relevant different events leading up to the usage of the name Hillary. A possibly valid answer to the why-question would then be: Already as a little girl his mother dreamt of having a son called Hillary and since Hillary was her first son, he is called Hillary.

We can apply Van Fraassen's analysis as follows in order to show analogy between facts and theorems. The why-question concerning any theorem T is "Why is statement T true?" The contrast class consists of the two accompaniedd "Statement T is true" and "Statement T is false". The relation of explanatory relevance is determined by the methodology of proof of the mathematical research tradition involved. A proof of statement T then counts as an explanation.[9]

[9]Cf. [Resnik&Kushner,1987], p.153

The fact that proofs of theorems can be viewed as explanations of the truth of mathematical facts suggests the following tentative (partial) notion of mathematical rationality.

> A mathematical research project or research tradition *progresses heuristically*[10] if it produces conjectures (theorem candidates) of weight.[11] Apart from heuristic progress there is *absolute progress*. If the project or the tradition succeeds in proving or refuting the conjectures resulting from its heuristic progress, it progresses absolutely. But we have also absolute progress, if the project or tradition succeeds in proving (or refuting) conjectures or theorems produced by a competing project or tradition. (Because a project producing interesting conjectures very often also produces proofs of special cases or analogous cases, heuristic progress will usually be accompanied by absolute progress.)
>
> The preference of a rational mathematical community for a research project or a research tradition is proportional to its expected progress.[12]

Often when traditions or projects are competing, one of them clearly progresses heuristically and absolutely much more than the other(s). It is easy to find examples of competition where the behaviour of mathematicians was in accordance with the model of rationality which I developed above. The 17th Century Geometric Tradition was considerably less rigid regarding methodology than was the Euclidean Tradition, but it succeeded in proving many new theorems. Undoubtedly that is why the mathematicians involved turned to the more liberal methodology. As regards competition between projects within the 17th Century Geometric Tradition, take the competition between Leibniz's differential calculus and the preceding methods represented, for example, by Huygens. Very soon the high degree of progress of the differential calculus became clear. Consider, for example, the problem of

---

[10]As we have seen in II.3 Lakatos speaks of heuristic progress when the modifications of a protective belt are in accordance with the positive heuristic of a programme. I obviously use the term heuristic progress in a different way.

[11]A theorem is by definition a proven conjecture. Proofs are, however, tradition - dependent. A theorem belonging to tradition A may become a conjecture in tradition B because the proof of the theorem is unacceptable in tradition B. Conjecture and theorems possess weight. Below, in section VI.5, I will consider the problem of the weighing of conjectures and theorems.

[12]Compare my definition with Laudan's "rationality consists in making the most progressive theory choices" ([Laudan,1977], p. 6).

the catenary, which was easily solved by means of the differential calculus and only with a lot of effort by means of the older geometric method. Bos writes:

> "In fact, the catenary was a test-case between the old and the new style in the study of curves, and only because the champion of the old style was a giant like Huygens, the test-case can formally be considered as ending in a draw" ([Bos,1980], p.80).

Eighteenth century mathematicians turned to the differential calculus because it enabled them to work in highly progressive projects. The "new style in the study of curves" would soon lead to the Formalist Tradition.

Very much in the style of Lakatos's "Proofs and Refutations", Marchi does not seem to distinguish between a conjecture and a theorem. Within a research tradition one does, however, entertain such a distinction, a theorem being a conjecture proven within that tradition. Obviously, whenever we study a particular tradition, we will have to distinguish between the two. Moreoever, in spite of the analogy between scientific facts and mathematical theorems, I will mostly avoid the use of the notion "mathematical fact" below and only speak of conjectures and theorems. The reason is the following. From the perspective created by Marchi's remarks, a research tradition in mathematics is a way of viewing mathematical facts and a way of explaining them by relating them to each other. This particular point of view together with the way of relating the facts to each other are characteristic of the research tradition. In the outline of a methodology of mathematical research traditions that I gave above competition between traditions or projects that should be regarded as dealing with the same or closely related facts is of particular importance. Speaking of different research traditions dealing with the same facts, however, has a Platonist ring. It suggests that there exist mathematical facts independent of all traditions. And this suggestion deserves to be taken seriously. For example, speaking of the same mathematical fact viewed by different traditions is in accordance with mathematical practice. That is how mathematicians behave. This is undoubtedly related to the fact that the competition between mathematical research traditions is less fierce than between scientific traditions. Usually the basic ideas of a new tradition grow out of an older tradition and do not contradict all the core ideas of and the "facts" studied by the old tradition. Often the basic ideas of a new tradition concern something like a shift of emphasis which leads to a rather different way of viewing the facts. And although a fact is viewed differently, this shift is hardly ever seen as replacement of the old fact by a completely new fact. Moreover, the frequent application of mathematical facts in science or real life is to a large extent independent of the mathematical tradition which explains these facts. Mathematical facts indeed appear to be tradition-independent to a certain extent. Yet it is not my intention now to choose for or against Platonism, and I will leave the difficult questions concerning the object of mathematics aside. In Chapter X, I will return to these matters, but for the time being I will avoid the notion of mathematical fact. As long as we are willing to accept the fact that mathematicians working in different traditions believe that they are studying the same mathematical facts and do not question that belief, an evaluation of the methodology of mathematical research traditions without use of the notion of mathematical fact is quite possible.

VI.5 Some problems

The weighing of conjectures and theorems[13]

Not all theorems are equally important or interesting. Moreover, judgements about degrees importance are never precise. This means that the progress of a project or a tradition cannot be determined exactly. Mathematical rationality is not a computable rationality, but is more like the rationality of the court room where pros and cons are weighed. It is not only the number of theorems which counts, but also their quality. From now on I will use "weight" as a general term to denote "quality", "importance", etc. of conjectures and theorems. In determining the weight of the theorems that a project or tradition produces, or is expected to produce, one can do no more than give a list of factors that influence that weight. In order to gain some insight into the way mathematicians weigh conjectures and theorems, I will consider some examples.

Example 1

In a paper dating from 1979 Hallett tackles the problem of specifying criteria for progress of one version of a mathematical theory with respect to its predecessor, and in this paper he proposes the following criterion for progress in mathematics:

> "the setting up of a new mathematical theory T_{m+1} constitutes progress with respect to its predecessor T_m if T_{m+1} is used in the solution of at least one problem P which T_m did not solve, provided that P is not of T_{m+1} 's

---

[13]This section also shows that I am considerably influenced by Laudan's methodology of research traditions (Cf. [Laudan,1977], Chapters 1 and 2).

> own making and that the statement solving P was not used in the construction of T_{m+1}" ([Hallett,1979], p.10).

Hallett calls this *Hilbert's criterion* (with reference to a similar idea uttered by Hilbert). The condition "provided that P is not of T_{m+1} 's own making" should guarantee that the new theory's contribution is significant. The condition "that the statement solving P was not used in the construction of T_{m+1}" should exclude ad hoc solutions of problems. Hallett illustrates his proposal by means of a partial reconstruction of the development of Cantor's set theory. He considers in particular the transition from Cantor's early theory of linear point sets (die *Punktmannigfaltigkeitslehre*) to the set theory of *Grundlagen einer allgemeinen Mannigfaltigkeitslehre.* The major achievement of the *Grundlagen* was the presentation of the transfinite ordinal numbers as an autonomous and systematic extension of the natural numbers. Hallett mentions the fact that in 1884 Mittag-Lefler made substantial use of Cantor's new results in extending certain fundamental theorems concerning the representation of analytic functions. Hallett also shows that Borel, in his 1895 proof of the Heine-Borel theorem, used Cantor's transfinite ordinals. Set theory in the form that it has in the *Grundlagen* was not designed for these applications. Hallett argues on the basis of several quotations that the applications influenced mathematicians in their appraisal of Cantor's work, and he concludes that they do tacitly apply something similar to Hilbert's criterion.

Let us consider Hilbert's criterion. From Hallett's point of view Hilbert's criterion definitely constitutes a *sufficient* criterion for progress. I agree. Yet there are some problems. In order to distinguish degrees of progress, which is inevitable, one needs some criterion to decide about the *importance* of problems.

Hallett writes:

> "in historical investigations of theories it may be possible to discover *historical* evidence for the importance of a problem, particularly if mathematical theories are indeed arranged as programmes with long term research aims" ([Hallett,1979], p.9).

Hallett creates the impression that for him a problem is important if prominent mathematicians consider it to be important (Cf. [Hallett,1979], p. 20; remark on Weierstrass). One crucial question then remains unanswered: How do prominent mathematicians recognize an important problem? By means of Hilbert's criterion, Hallett really only shifts the problem of what constitutes progress to the problem of the importance of problems.

Hallett also presents Hilbert's criterion as a *necessary* condition for progress. This, however, is not at all obvious. If a theory solves an important problem which so far went unsolved, then it can be considered as *proven* that the theory constitutes progress. But this means that the theory already constituted progress at the moment of its appearance, i.e. already *before* its application to the problem so far unsolved. Hilbert's criterion will sometimes be one way of recognizing progress in retrospect. I argue that a theory may generate and at the same time solve such "interesting" problems that even without definite proof of its value through Hilbert's criterion, the theory should be considered valuable.

From Hallett's research we can draw the conclusion that *external successes* are a weighty form of absolute progress.[14] By definition an *external success of a project or tradition* is the application of a theorem belonging to the project or tradition for the solution of a problem outside of the project or tradition.

---

[14]In the Notices of the American Mathematical Society **28**, 1981, p. 507 Oscar Zariski was quoted as follows: "In 1950 I gave a lecture at the International Congress of Mathematicians at Harvard; the title of that lecture was "The fundamental ideas of abstract algebraic geometry". This lecture is a good illustration of how relative in nature is what we call "abstract" at a given time. [....] Because it dealt only with projective varieties, that lecture, viewed at the present time, after the great generalization of the subject due to Grothendieck, appears to be a very, very concrete brand of mathematics. There is no doubt that the concept of "schemes" due to Grothendieck was a sound and inevitable generalization of the older concept of "variety" and that this generalization introduced a new dimension into the conceptual context of algebraic geometry. What is more important is that *this generalization has met what seems to me to be the true test of any generalization, that is, its effectiveness in solving, or throwing new light on old problems* by generalizing the terms of the problem (for example: the Riemann-Roch theorem for varieties of any dimension; the problem of the completeness of the characteristic linear series of a complete algebraic system of curves on a surface, both in characteristic zero and especially in characteristic $p \neq 0$; the computation of the fundamental group of an algebraic curve in characteristic $p \neq 0$)" (Emphasis is mine - T.K.). This quotation illustrates the point made by Hallett. It is also interesting to consider the next sentences in Zariski's text: "But a mathematical theory cannot thrive indefinitely on greater and greater generality. A proper balance must ultimately be maintained between the generality and the concreteness of the structure studied, and usually this balance is restored after a period in which it was temporarily (and understandably) lost. There are signs at the present moment of the pendulum swinging back from "schemes", "motives", and so on, toward concrete but difficult unsolved questions concerning the old pedestrian concept of a projective variety (and even of algebraic surfaces). There is no lack of such problems."

Example 2

A famous problem in 20th century group theory is the problem of the classification of all finite simple groups. A finite simple group is a finite group which has no normal subgroups except the trivial ones (the whole group or the identity element). The theory of finite groups is to a certain extent analogous to the theory of prime numbers. An arbitrary natural number has a unique factorization into a product of primes. Similarly an arbitrary finite group can be built up out of finite simple groups. A crucial theorem in this area says: every finite group can be represented as a product of a normal subgroup and a quotient group.

Between 1963 and 1981 a considerable number of mathematicians have been working on a large research project to solve the problem. It was discovered that most finite simple groups fall into a number of infinite families. In the mid-1950s these families were quite well described. Five special finite simple groups, the Mathieu groups, were not covered by that classification. Such groups came to be called "sporadic" groups. In 1963 Feit and Thompson proved the theorem: "A finite simple group is either cyclic or its order is even".[15] Their work initiated the project. In the course of the period under survey, 21 new sporadic groups were discovered and it was finally proved in 1981 that there are no more sporadic groups.

Regarding the heuristic unity of the project[16] I will restrict myself to one quotation. Daniel Gorenstein, one of the leading experts involved in the project, writes:

> "In a number of cases (primarily but not exclusively those in which computer calculations were ultimately required) "discovery" did not include the actual construction of a group - all that was established was strong evidence for the existence of a simple group G satisfying some specified set of conditions X. The operative metamathematical principle is this: if the investigation of an arbitrary group G having property X does not lead to a

---

[15]They actually proved "All finite groups of odd order are solvable" ([Feit&Thompson, 1963]).

[16]I want to emphasize that the solution of the problem of the classification of all finite simple groups was a very impressive achievement. Compared to what actually took place, the following characterization of part of the heuristic unity of the research project involved is extremely superficial and inevitably leaves out very much of what is essential. It would be completely wrong to suppose that a characterization of the heuristic unity of a project is more than the phrasing of some characteristic principles that guided the research in question. It can never characterize the quality of the total mathematical effort involved.

> contradiction but rather to a "compatible" internal subgroup structure, then there exists an actual group with property X. In all cases the principle has been vindicated; however, the interval between discovery and construction has varied from a few months to several years" (Quoted in [Franklin,1987], pp.10-11).

One sees how the project produced conjectures that were then turned into theorems. In other words, heuristic progress was followed by absolute progress. In 1972 Gorenstein set forth a sixteen-point programme for completing the classification. At that time there was scepsis but by 1976 this had disappeared. Gorenstein writes about that year:

> "The turning point undoubtedly occurred at the 1976 summer conference in Duluth, Minnesota. The theorems presented there were so strong that the audience was unable to avoid the conclusion that the full classification could not be far off. From that point on, the practising finite group theorists became increasingly convinced that the "end was near" - at first within five years, then within two years, and finally momentarily. Residual skepticism was confined largely to the general mathematical community, which quite reasonably would not accept at face value the assertion that the classification theorem was almost proved" (Quoted in [Franklin, 1987], p.11).

As regards the weight of the mathematical results involved, *the desire for a complete classification is a factor* from my point of view, adding extra weight to the results that bring such a classification nearer. *Thus independently of external successes, systematicity factors also determine the weight of mathematical conjectures and theorems.*

Example 3

A paper by Hopf, "Differenzialgeometrie und topologische Gestalt" ([Hopf,1932]), shows how factors concerning harmony contribute to the weight of problems. In this expository paper Hopf presents a number of solved and unsolved problems which are all related to one general problem. This problem concerns the relation of the topological properties of a surface on the one hand, and the differential-geometrical properties of this surface once it has been turned by means of a differential-geometrical metric into a differential-geometrical surface, on the other. Hopf writes:

"Which connections and bonds exist between the topological properties on the one hand, and the differential-geometrical properties on the other?" [17]

The general problem consists of two sub-problems;

i)"Given the topological surface F: to introduce onto it a differential geometry free from singularities. What possibilities exist for such a metric?".[18]

ii)."I am given a small piece from a surface that I can investigate with all desired precision [...]. What kind of conclusions can I then draw, from my knowledge of the small piece of the surface, concerning the global properties of the whole surface, particularly concerning its topological form?".[19]

Hopf expresses in his paper the desirability of a theory that connects existing global and local theories. Global and local theories should be in "harmony".

Example 4

In an interesting paper Bos has investigated the rise and decline of a mathematical theory - the "construction of equations" ([Bos, 1984]. The paper concerns a theory within the 17th century Euclidean tradition in our terminology. For our purposes Bos' spaper is particularly interesting because he explicitly states that in his opinion the Lakatosian approach by means of the methodology of scientific research programmes is not applicable in that case ([Bos,1984], p.375). According to Bos, the Lakatosian model does not take into account factors on the level of *meaning and motivation* (ibid.), which in his opinion played a crucial role in the case under consideration.

I shall first briefly describe the events. Within the Greek geometrical tradition it had been held that constructions should preferably be executed by ruler and compass. Yet in cases where such a construction could not be found, the Greeks had also considered other methods of construction. However, they had not succeeded in reaching agreement on the means were and which means were not allowed in such cases. Descartes in his Géométrie embarked upon a

---

[17]"Welche Zusammenhänge und Bindungen bestehen zwischen den topologischen Eigenschaften einerseits und den differentialgeometrischen Eigenschaften andererseits?" ([Hopf,1932], p.210).

[18]"Gegeben ist die topologische Fläche F; man soll sie differentialgeometrisch metrisieren, d.h. auf ihr eine singularitätenfreie Differentialgeometrie einführen. Welche Möglichkeiten bestehen für eine solche Metrik?"([Hopf,1932], p.210).

[19]"Aus einer Fläche wird mir ein kleines Stück vorgelegt, das ich mit jeder beliebigen Genauigkeit untersuchen kann[...] Was für Schlüsse kann ich dann aus meiner Kenntnis des kleinen Flächenstückes auf die Eigenschaften im Großen der ganzen Fläche, insbesondere auf ihre topologische Gestalt ziehen?" ([Hopf,1932], p.210).

project to solve that methodological problem: he laid down which constructions were acceptable and which were not, and he gave an order of preference among the available means of construction. *This order of preference in due course turned out to constitute a major problem.* Descartes's central ideas are the following:

i) A construction problem can be reduced by means of analytic geometry to solving an irreducible polynomial equation H(x)=0 in one unknown, the coefficients of H(x)=0 being given or constructible by ruler and compass.

ii) Construction problems can be classified according to the degree of the corresponding irreducible equations and constructions can be found for the roots of the irreducible equations through the intersection of well chosen constructing curves. It is inevitable that one should accept constructing curves other than straight lines and circles. However, one should only use the simplest possible curves, and "simple" meant for Descartes "algebraically simple".

Descartes showed that for degrees 1 or 2, a construction of the roots of the equation is possible by means of ruler and compass. For degrees 3 or 4 he showed that the roots could be found by intersecting a circle and a parabola. For degrees 5 or 6 Descartes had to introduce a new kind of curve - the (subsequently styled) "Cartesian parabola" - which is a curve of degree 3. He showed that for degrees 5 or 6 the roots could be constructed by intersecting a circle and a suitable Cartesian parabola.

Although Descartes thought that he had solved the problem in principle, his legacy generated several questions which later writers attempted to solve. For example, techniques had to be found to pinpoint the constructing curves. The choice of constructing curves other than the parabolas used by Descartes was taken into consideration, and subsequently also the construction of equations of higher degree than 6 came under discussion. Van Schooten,Wallis, Sluse, Fermat, Huygens, De la Hire, l'Hôpital and Newton, amongst others, worked on the subject.

In the 18th century the interest in the subject vanished. Euler in his *Introductio ad analysin infinitorum* (1748), and Cramer in his *Introduction à l'analyse des lignes courbes algébriques* (1750), still devoted substantial sections to the theory. But after 1750 the subject fell into oblivion. According to Bos one may speak here of the death of a mathematical theory.

By 1670 Newton had come to the conclusion that Descartes' approach had run into problems. He started working on an alternative project. Descartes had used algebraic criteria of simplicity (in particular the degree of curves) in order to determine the order of preference: a construction should preferably be executed by means of straight lines and circles, and only when this was inevitable a parabola or, in even more difficult cases, a suitable Cartesian parabola was allowed. Newton attempted to use geometric criteria of simplicity to determine such an order of preference. But Newton did not succeed in producing a really workable alternative ([Bos,1984], p.366).

The project started as a project within geometry with a geometrical motivation. Because of its algebraic nature, however, it generated its own algebraic problems. Bos' paper clearly shows the opposition that developed between the algebraic theory and its original geometrical background. A mathematician like De la Hire either forgot or did not understand Descartes' original motivation, and Van Schooten and Huygens discovered that a geometrical classification of problems as to their constructibility does not coincide with Descartes' classification of the corresponding equations as to their degree ([Bos,1984], p.357). Jakob Bernoulli also discovered that the simplest possible constructing curves in the algebraic sense are not always the simplest possible constructing curves in a geometrical sense ([Bos,1984], pp. 358-359). The opposition between algebra and geometry is the clue to the disappearance of the theory. Bos sums up his conclusion as follows:

> "This is the basic contradiction, and in it we can sum up the cause for the death of the construction of equations: the subject had a natural tendency to become algebraic, but it could not pass on its original geometrical criteria to algebra, and so it lost its motivation and it died." ([Bos,1984], p. 373)

The methodology of mathematical research traditions that I developed above is quite adequate to describe the developments studied by Bos. We are dealing with theories that belong to the 17th Century Geometric Tradition. Why were the theories born and why did they die? Descartes's approach corresponds to a research project. The resulting theory appeared in Descartes's *Géométrie* and undoubtedly owed part of its initial weight to the weight of the other parts of that book. Descartes' analytical methods successfully solved geometrical problems that were considered important. Descartes' project on the construction of equations also originally defined and solved relatively interesting problems. Yet ultimately the theory died, although Newton defined a different, competing research project. It is very interesting that it clearly did not die owing to a lack of solvable problems ([Bos,1984], p.372). It is easy to define problems concerning the construction of equations that 18th century mathematicians could have studied. Yet they didn't. Why? Bos says:

> "The causes that made the construction of equations disappear lay in the sphere of motivation and method, rather than in the sphere of the mathematical problems and techniques. The causes were connected with the reasons why mathematicians considered certain problems as meaningful, and with the criteria of adequacy which mathematicians set for the solutions of these problems. Such reasons and criteria are very essential in the development of mathematics" ([Bos,1984], p.373).

I argue that the theories involved lost weight, and they lost weight because all of the different possible directions of past and future research appeared to be equally *arbitrary* than the others. In particular the determination of an order of preference for the choice of constructing curves in order to construct the root of a particular equation appeared to be a very serious problem: it could not be solved without an *arbitrary* decision. Without the connection with geometry the whole problem lost its sense. In connection with geometry, algebraic criteria of simplicity appeared to clash with geometric criteria. Newton's attempt to use geometrical criteria also failed. Bos writes:

> "Newton could not pursue that line for higher-order equations in a natural and convincingly geometrical way" ([Bos,1984], p.362).

I would add to this that the theories that were developed did not possess any external relevance. External successes would have added weight to the theorems produced.

The case studied by Bos shows that the conjectures and theorems that a research project produces, or is expected to produce, are weighed by mathematicians. Arbitrariness in the foundations of a theory is an example of a meta-problem that mars that theory. Such a meta-problem influences the weight of conjectures and theorems negatively.

As for the weight-reducing influence of meta-problems, it is interesting to quote Scharlau, who writes that in the beginning of the 19th century it was not so much a desire for rigor that made mathematicians turn to the new analysis, but rather the interesting new theories that could be developed. He adds:

> "[...] in the entire philosophical and historical discussion of foundations of this century the same above mentioned mistake has been repeated. The great uprising of mathematical logic and proof theory, as well as related areas (set theory, model theory), had very little to do with an intensified need of mathematicians for an "exact foundation" of their subject. These areas were undertaken because they presented many difficult and interrelated unsolved problems" (W. Scharlau in [Jahnke & Otte, 1981], p.338).

Apparently the influence of meta-problems should not be overestimated: in the beginning of the 19th century the new analysis did not owe its progress to the fact that 18th century theorems lost weight, but to the weight of its own results.[20] I will return to the revolution of rigor at the beginning of the 19th century in Chapter VIII. Scharlau's remark suggests that in the case of the constructibility of equations, the arbitrariness in the foundations of the theory will only have led to the disappearance of the theory because of the absence of other factors adding weight to the results obtained - for example external relevance, or factors concerning simplicity and elegance.

Factors influencing the weight of conjectures and theorems.

I will now give a tentative list of factors that influence the weight of a theorem in mathematics. The list is undoubtedly not incomplete.

1. If a theorem allows for the solution of a problem outside the domain of the project, this contributes considerably to the weight of the theorem. I will call such theorems *externally relevant* (Cf. example 1 above). Unexpected external relevance of a theorem particularly adds weight .

2. If the theorem only solves a problem within the project, connections with other problems, logical or by analogy, bear weight. For example, the theorem may lead to the proof of another theorem which was in need of proof. But connections of analogy are also of importance. A theorem may bring forth new conjectures and thus add to the heuristic progress of the project.

3. If several theorems of a theory gain in weight, then other theorems belonging to the theory also gain in weight.[21]

4. In situations of competition, theorems that have been proved in the context of one project (or tradition), and not by a competing project or tradition, gain in weight.

5. If on the basis of analogy a conjecture is expected to be easily provable, but in fact turns out to be very difficult to prove, it gains in weight.

6. If others in the past tried in vain to prove a conjecture, that conjecture gains in weight.

7. Some "meta-problems" vexing a tradition or a project have a negative influence on the weight of theorems with respect to such a project or tradition.

---

[20]Mehrtens expressed a similar view: "In the case of the value of rigor it bears the more weight the easier it is to exert rigor in mathematical work" ([Mehrtens,1976], p.309).

[21]According to Bernstein the interest in graph theory since the 1950s is due to its many applications ([Bernstein, 1979], p.249).

i) A theory may not be rigorous, or else not rigorous enough. It may even be inconsistent.
ii) A theory may be based upon one or more dubious axioms.
iii) Arbitrariness in the foundations of a theory can be a serious meta-problem and affects the weight of the results that the theory produces (Cf. example 4).
8. Factors concerning economy, simplicity, elegance etc. influence the weight of conjectures and theorems. If a project succeeds in producing much shorter proofs of theorems than another project, this will contribute to the weight of the theory that it produces.[22]
9. Factors concerning harmony, systematicity and completeness of classification influence weight (Cf. examples 2 and 3).

Incommensurability

In an interesting paper Glas dwells at length on the different views of mathematics entertained by Lagrange and Laplace on the one hand and Monge on the other hand ([Glas,1986]). Lagrange and Laplace represented the predominant late 18th century Formalist Tradition in analysis. They saw a geometrical proof as a mere illustration of an analytical argument. For Monge, however, geometry represented what mathematics was all about. Comparing the diverging underlying overall views on the nature of mathematics Glas argues that we have here *a case of incommensurable approaches where there is no common ground for judging the value of the results obtained.* Glas states:

> "Because they perceived mathematics so differently, they entertained incompatible views on what were the right questions to ask, the right paths towards the answers, and the true standards against which mathematical achievement had to be measured. We clearly have here a case of incommensurable mathematical programmes [...]" ([Glas,1986], p.267).[23]

In terms of the MMRT the quotations seem to imply that a determination of the weight of mathematical results could be essentially *tradition-dependent*. If that were so, it would be impossible to compare traditions in a tradition-independent way with respect to their progress. Let us have a look at Glas's

---

[22]Cf. Van Bendegem, who argues convincingly that if a proof has a simple proof-outline, then the quality of that proof is considered to be high by mathematicians ([Van Bendegem,1988], p.252).

[23]In a 1989 paper Glas repeats the same message: "Our analysis so far has shown that we do have here a case of incommensurable approaches lacking a common basis for judging the relative merits of the alternatives" ([Glas, 1989^{I}], p. 128)..

argument. Glas shows extensively that Lagrange and Monge expressed themselves quite differently with respect to the nature of mathematics. However, Glas then draws the improbable conclusion that Lagrange and Monge possessed *no common ground whatsoever* for judging the value of mathematical results. To a certain extent Glas is undoubtedly right. Whenever two mathematicians judge a mathematical result, they will interpret it from their own point of view and they will weigh it differently. Monge was, in particular, inclined to interpret results in geometrical terms. Lagrange was inclined to interpret the same mathematical results analytically. However, to what extent did these different interpretations imply a radically different weighing of mathematical results? In Taton's excellent comprehensive account of the work of Monge, the name Lagrange occurs repeatedly. We learn that in 1778 Lagrange used methods very close to those of Monge's descriptive geometry ([Taton,1951], p. 70) and, referring to a situation where Monge created new methods in analytical geometry without recourse to any figures, Taton even wonders whether Monge may in that particular case have identified with the so totally different personality of Lagrange (Op. cit., p. 363). Monge also frequently dealt with problems that Lagrange had studied earlier. Obviously Monge and Lagrange understood each others contributions to mathematics perfectly well and must have repeatedly appreciated each others results. Lagrange and Monge were on friendly terms with each other whether doing mathematics or not (Op. cit., p.32). According to Boyer, Lagrange is even said to have exclaimed about Monge: "With his application of analysis to geometry this devil of a man will make himself immortal".[24]

If Glas is right, the absence of open disagreement between Lagrange and Monge is a highly peculiar circumstance. Glas is aware of this problem and he relates the absence of controversies to the fact that research was "centered around prestigious individuals who could not easily be attacked" ([Glas,1986], p.249). However, this is not a satisfactory explanation. Prestige and an established position of course discourage criticism by others but they also make it easier for those who have an established position to criticise others. If Glas is right it remains highly remarkable that Lagrange and Monge did not disagree more openly on the value of mathematical theories.

In my view Glas underestimates the considerable common ground for judging results that existed at the time among French mathematicians and he overestimates the influence on the weighing of results of the ideological opposition that he perceives between Monge and Lagrange.

---

[24] [Boyer,1968], p. 521

It is also good to consider in this context the work of Monge's pupil and protégé Lacroix. Lacroix's famous treatise *Traité du calcul différentiel et du calcul intégral* was not primarily inspired by Monge, as Glas has it ([Glas, 1986], p. 264), but a very successful synthesis of the works of Euler, Lagrange, Laplace, Monge, Legendre, Poisson, Gauss, and others. Lacroix's book and its favourable reception illustrate that in the mathematical community very often very soon consensus is reached about the value of results.

VI.6 About the next two chapters

In section IV.5 I pointed out that in this study my goal is rather modest . It is not primarily a universal theory of mathematical rationality that I am after, but a motivated modification of Lakatos's MSRP together with some rational reconstructions of developments in mathematics which are in accordance with that modified MSRP. The methodology defined in this chapter can be considered as a modification of Lakatos's MSRP under influence of Laudan's 1977 model. The next thing to do now is to give some factual reconstructions of developments in mathematics on the basis of the MMRT. It is very important that the reconstructions are factual because the reconstructions should *directly support the methodology*. Only if we were to possess several clear examples of competing rival research traditions in mathematics could we apply to mathematics the ideas set forth in Lakatos's "History of science and its rational reconstructions", and counter-factual reconstructions might then become interesting.[25] However, we do not possess those clear examples. In the next two chapters I will give two rational reconstructions on the basis of my MMRT. In accordance with the meta-methodological remarks in Chapter IV, I will reconstruct the history involved attempting to base myself on considerable agreement among historiographers. I will thereby inevitable rely heavily on secondary sources. In both reconstructions the emphasis will be on the descriptive part of the MMRT. The identification of the traditions involved will receive particularly attention. As regards the normative part of the MMRT, by means of a limited number of examples I will show on the basis of the existing historiography that in both reconstructions the transition to the succesful tradition was accompanied by progressive projects (in the sense of the MMRT) representing that tradition. In both cases the entire mathematical community, over a period of time, went over to the succesful tradition. This was a rational move from the point of view of the MMRT when judged after the event.

---

[25] Cf. Chapter IV.

Chapter VII Two research traditions in Greek mathematics

VII.1 A thesis of Lakatos

In this chapter I shall present my first rational reconstruction of a development in mathematics on the basis of the MMRT. Historiographers of mathematics do not usually use explicit global models when they describe the development of Greek mathematics. I will show, however, that some of the implicit notions and criteria that they use when they study the development of pre-Euclidean Greek mathematics can be made explicit by means of my methodology of mathematical research traditions. The MMRT enables us to distinguish in this way two rival mathematical research traditions in pre-Euclidean Greek mathematics.

My point of departure is Lakatos's historiographical thesis concerning the origin of the *Elements*. The thesis consists of three parts:

(i)"The Greeks inherited from the Babylonians and Egyptians a body of naive *conjectures* with relatively high truth content (and they invented others)",

(ii)"Then came test - and proof-thought-experiments, mainly analyses[1], without any *known* lemmas, without any safe axiomatic systems",

---

[1]In order to give an idea of the notions of analysis and synthesis prevalent at the time, we can rely on Pappus' definitions. About 300 A.D. Pappus defined analysis in the 7th book of his main work, the *Synagoge* (= collection), as follows: "*Analysis*, then, takes that which is sought as if it were admitted and passes from it through its successive consequences to something which is admitted as the result of synthesis: for in analysis we assume that which is sought as if it were already done [...], and we inquire what it is from which this results, and again what is the antecedent cause of the latter, and so on, until by so retracing our steps we come upon something already known or belonging to the class of first principles, and such a method we call analysis as being solution backward [...]. But in *synthesis*, reversing the process, we take as already done that which was last arrived at in the analysis and, by arranging in their natural order as consequences what before were antecedents, and successively connecting them one with another, we arrive finally at the construction of what was sought; and this we call synthesis" (Quoted and translated by Heath in [Heath,1921],Vol.II, pp. 400-401). Pappus distinguishes between two kinds of analysis (corresponding respectively, by the way, to the propositions and the problems in Euclid's *Elements*): theoretical analysis and problematical analysis. The problematical analysis begins with the assumption of the existence of what is required to be constructed and works backward from there. The theoretical analysis begins with the statement which is supposed to be true and then works backward (ibidem).

(iii)"It was only after hundreds of successful analyses and syntheses, after hundreds of 'proof-procedures' (in the sense of my *Proofs and Refutations*) that certain lemmas kept cropping up, became 'corroborated' (their alternatives remaining sterile) and finally were turned into the hard core of a research programme (an 'axiomatic system') by Euclid" ([Lakatos,1978b], pp.99-100).

I will first critically examine this Lakatosian thesis on the basis of the extant historiography. Although Lakatos's thesis is to a large extent correct, I will show that Euclid's *Elements* rather represent a research tradition (in my sense) than a research programme. The extant historiography also suggests that early Greek mathematics was dominated by another research tradition, which I will call the *Demonstrative Tradition*. This older tradition was replaced in the course of time by the *Euclidean Tradition* in a way that is in accordance with the methodology of mathematical research traditions. In order to support the reconstruction by way of a critical argument, I shall also make use of Van Hiele's theory of developmental levels.

VII.2 Lakatos's thesis is to a large extent correct

It is generally assumed that the Greeks inherited from the Babylonians and Egyptians a "body of naive and less naive conjectures". (They presumably became only conjectures, in the sense of statements in need of proof, from the critical Greek point of view). As for what happened to this heritage in pre-Euclidean times very little is known with certainty. Most mathematical works from that period are lost. We possess only two pre-Euclidean mathematical works intact, both by Autolycus (who flourished in the last decennia of the 4th century[2], i.e. immediately before Euclid), and as for the other works, when we know of their existence at all, we possess only fragments or references to them by later authors[3]. Inevitably the reconstruction of the development of pre-Euclidean mathematics is a rather speculative affair.

Lakatos's thesis is certainly correct in a global sense. Book I of Euclid's *Elements* begins with three kinds of unproven statements which form (in Lakatosian terminology) part of the hard core of the Euclidean research programme - definitions, postulates and axioms. Greek mathematics did not develop deductively from these unproven statements. Instead it began with

---

[2]Cf. e.g. [Van der Waerden,1961], pp.193-195.

[3]One reason, presumably, is that Euclid's *Elements* made many of the other works superfluous.

conjectures, followed by a period of proof-experiments, new conjectures, proof-analysis, which only in the course of time led to the axiomatic foundations given in Euclid's *Elements* A brief look at the extant historiography is enough to recognize this. Let us have a look at the final result - the definitions, the postulates and the axioms - and at the way in which they could have been arrived at.

The definitions

Not all definitions are used in the *Elements*. Some of them seem to be only an attempt to characterize the subject matter of geometry: e.g. Elements I, Def.1:

> "A *point* is that which has no part"([Heath,1926], Vol.I, p.153)

and *Elements I*, Def. 2:

> "A *line* is breadthless length" ([Heath,1926], Vol.I, p.153).

Some of them were perhaps inserted only out of respect for tradition. For example, the definition of rhombus (*Elements I*, Def.22) plays no role whatsoever in what follows. On the other hand the definition of a circle (*Elements I*, Def.15),

> "A *circle* is a plane figure contained by one line such that all the straight lines falling upon it from one point among those lying within the figure are equal to one another" ([Heath,1926],Vol.I,p.153),

is used (e.g. in the proof of *Elements I*,1)[4], which implies that it could at least be considered as proof-confirmed if not proof-generated.

The postulates (aitemata)[5]

Let us take, for example, the famous postulate 5 of book I of the *Elements*. It says:

> "That, if a straight line falling on two straight lines makes the interior angles on the same side less than two right angles, the two straight lines, if produced indefinitely, meet on that side on which are the angles less than the two right angles" ([Heath,1926],Vol.I,p.155).

Euclid uses postulate 5 for the first time in order to prove *Elements* I.29. Proposition I.29.1[6] reads:

> "A straight line falling on parallel straight lines makes the alternate angles

---

[4]Elements I,1 denotes proposition 1 of Book I of the *Elements*.

[5]According to Szabó the name "aitemata", i.e. demands, refers to the fact that the postulates are assumptions which were not accepted without reservation ([Szabó,1978], p.279).

[6]Elements I.29.1 denotes the first part of proposition 29 of Book I of the *Elements*.

equal to one another" ([Heath,1926],Vol.I,p.311).

According to Tóth ([Tóth,1965]), proposition Elements I.29.1 was proved by the Greek geometers of the first half of the 4th century B.C., i.e before Euclid, *without* using the 5th postulate or an equivalent statement. In Tóth's reconstruction, Elements I.29.1, which was used to prove the theorem that the three interior angles of a triangle are together equal to two right angles, followed immediately from a proposition that must have run roughly as follows:

"*The* line through a given point parallel to a given line can be drawn by drawing a line from the point to an arbitrary point of the line and constructing two equal alternate angles".[7]

This reasoning only looked acceptable because *implicitly* the unicity of a line through a point parallel to a given line was assumed ([Tóth,1965], pp. 276 - 277).

In Tóth's reconstruction, somewhere in the middle of the 4th century critical proof-analysis by mathematicians under the influence of Eudoxus revealed that the assumption of unicity was based on observation. This led to attempts to prove the assumption. A possible attempt, which may be of an earlier date, consists of deriving the unicity from the transivity-theorem Elements I.30, which may have looked rather evident. That theorem says:

"Straight lines parallel to the same straight line are also parallel to one another" ([Heath,1926],Vol.I,p.314).

Of course the problem then arises how to prove Elements I.30. During such attempts, postulate 5 must have showed up as one of a set of equivalent assumptions of which at least one could not be missed in order to develop the theory of parallels.

As for the postulates 1 through 3 the situation must be similar. Postulate 1 states:

[It is possible] "To draw a straight line from any point to any point" ([Heath,1926], Vol.I, p.154).

Postulate 2 states:

[It is possible] "To produce a finite straight line continuously in a straight line" ([Heath,1926], Vol.I, p.154).

---

[7]The construction returned later in Elements I.31 because it is needed to prove I.32 concerning the sum of the three angles of a triangle. A proof of the proposition showed up in Elements I.27 as part of Euclid's theory of parallels.

Postulate 3 states:

> [It is possible] "To describe a circle with any centre and distance" ([Heath,1926], Vol.I, p.154).

Most authors regard the postulates as being designed to guarantee the existence of certain basic geometrical objects from which the other objects could be constructed, in other words, by means of which the existence of other figures could be guaranteed. Presumably *Oenopides* in the 5th century made conscious use of the first three postulates (Cf. [Szabó,1978], p.275). In this case, it must have been repeated problematical analysis which led to lemmas that eventually became the postulates 1, 2 and 3.[8] Some form of ruler and compass as simple instruments with which to draw decent geometrical figures must have appeared very early.[9] It is probable that the choice for these particular three postulates was also influenced by the fact that adding other assumptions would have been arbitrary.[10]
The conclusion is that regarding the postulates Lakatos's thesis is correct.

The axioms

Euclid uses the term *koinai ennoiai* (i.e. common notions).[11] The axioms are used in Euclid's proofs. E.g. Euclid's proposition Elements I.4. says:

> "If two triangles have the two sides equal to two sides respectively, and have the angles contained by the equal straight lines equal, they will also have the base equal to the base, the triangle will be equal to the triangle, and the remaining angles will be equal to the remaining angles respectively, namely those which the equal sides subtend"([Heath,1926], Vol.I, p.247).

---

[8]Cf. footnote 1. Cf. also [Hankel,1965], p.141 and further.

[9]Perhaps peg and cord were used. If Seidenberg is right Greek geometry has a ritual origin and in the ritual context the constructions were carried out by peg and cord ([Seidenberg,1959], [Seidenberg,1972/73]). Approvingly Seidenberg quoted Heath, who wrote:" The *centre* is keéntron, the *circumference* périfereia, the former word having doubtless been suggested by something, e.g. a cord stretched tight, *carried around* the center as fixed point and describing a circle with its other extremity" ([Seidenberg,1959], p.117).

[10]Cf. [Steele, 1936], pp.177-181. It is also clear from the existence of proposition I.2 for example, that Greek mathematicians tried to use a minimum number of assumptions (Cf. e.g. [Dijksterhuis,1929], p.142).

[11]According to Szabó the pre-Euclidean term for such a common notion was *axioma,* meaning exactly the same as aitema, i.e. demand ([Szabó,1978], pp.282-287).

It is proved by means of the method of superposition, i.e. by putting the two triangles whose equality (congruency) should be proved on top of each other. The general assumption on which the method is based is Elements I, Axiom 7:

> "Things which coincide with one another are equal to one another" ([Heath,1926],Vol.I,p.155).

It is indeed very reasonable to suppose that this axiom also appeared as a lemma in the course of proof-experiments. The same holds for the other axioms.

VII.3 Two phases in Greek pre-Euclidean mathematics

Roughly speaking the Lakatosian thesis seems correct. There is, however, one complication. In the reconstructions of mathematical developments before Euclid that have been given by several authors, at least two different kinds of proof occur. The *epagogic* method[12], consisting of visually showing the truth of a theorem, was in particular developed in Ionia. The *apagogic* method[13], consisting of deductive argumentation, was in particular developed in Southern Italy ([Von Fritz,1971],pp.211). The epagogic method was used first and came subsequently to be replaced by the apagogic method after a period of transition during which the two methods were both used (ibid.). The view that such a transition took place in the course of pre-Euclidean Greek mathematics can be defended by means of several, more or less speculative arguments.

Von Fritz argued as follows. Elements I, Def. 17 says:

> "A diameter of the circle is any straight line drawn through the center and terminated in both directions by the circumference of the circle, and such a straight line also bisects the circle" ([Heath,1926],Vol.I,p.154).

Von Fritz remarked that the second part of the definition, which says that the circle is divided into two equal parts by its diameter, was called a *theorem* by Eudemus (second half of the 4th century B.C.).[14] Moreover, Eudemus mentions Thales as the first to prove this particular theorem. Von Fritz assumes that Thales' proof must have been unacceptable to Euclid and that Euclid as a result "hid" in a definition what actually had been a theorem. Thales' proof would have been based upon the epagogic method of

---

[12] epagoge (Greek) = leading nearer (literally).

[13] apagoge (Greek) = leading away (literally)..

[14] The source is Proclus (4th century A.D.), who is considered as reliable. See Proclus's commentary to Elements I, Def.17 (E.g.[VerEecke,1948], p.139).

superposition. Euclid's very limited use of that method (e.g. to prove Elements I.4) would be the last, irreplaceable, rest of the epagogic method that Thales would have used without inhibitions ([Von Fritz,1952],pp. 202-207).

More or less the same distinction is made by Szabó who explicitly distinguishes two phases in the history of pre-Euclidean Greek mathematics corresponding to these two different methods ([Szabó,1978]).[15] During the first phase in Greek mathematics a proof consisted of *showing or making visible* the truth of a statement. This first phase was followed by an apagogic or deductive phase. During this phase visual evidence was rejected and Greek mathematics became *a deductive system.* Szabó gives three arguments to support his assumption of the existence of two phases in the development of pre-Euclidean Greek mathematics.

a) The first argument concerns the meaning of the Greek word for proof: "deiknumi". The original meaning is probably "to show", "to point out" in the literal sense of the word ([Szabó,1978], p.188).

b) The question is then whether the word could already have lost this original meaning in early Greek mathematics. After all in the Odyssey the word sometimes means "to make known by words" ([Szabó,1978], p.188). However, in Plato's Meno we find the well-known passage in which Sokrates asks a slave how to double a square. The passage offers us a beautiful example of a pre-Euclidean proof by literally making the truth visible. The example shows that the first argument holds water, while at the same time the mere fact of the occurrence of the example also points at the existence of a non-deductive phase in pre-Euclidean mathematics.

c) The etymology of the Greek word for proof, together with one example, is, of course, not enough to conclude that initially in Greek mathematics proving consisted of literally making visible, while after some time the "demonstrative" proofs were replaced by deductive proofs. However, it seems that Euclid in the *Elements* consciously avoided proofs that were not deductive.[16]One of the indications is that in a considerable number of cases it is possible to reconstruct from Euclid's own words predecessors of the proofs in the *Elements* that are based upon literally showing the truth of the statements involved and in that sense similar to the proof in the Meno ([Szabó,1978],pp. 189-191). Further on I will give some examples.

---

[15]Cf. also [Mittelstraß,1965].

[16]Cf. also [Dijksterhuis,1929],pp.147-151).

VII.4 Van Hiele's levels of thinking

The thesis that in pre-Euclidean mathematics two phases must be distinguished remains rather speculative. Szabo's arguments are not fully convincing. It is after all imaginable that in Greek mathematics both deductive and demonstrative arguments were used together right from the start. Yet the thesis can also be supported by means of an argument that I have not come across in the literature. The argument is based upon Van Hiele's paedagogical theory of levels. Van Hiele developed his theory of levels of thinking, through which any mathematics student inevitably has to pass, in the 1950s. Van Hiele himself, and a considerable number of other authors, repeatedly have written on the theory. Various versions and interpretations of it are now extant. I shall refer to [Van Hiele, 1973] (p. 91 and further) and [Van Dormolen,1977]. Van Hiele in his book distinguishes between three levels of argumentation: *the ground level*, *the first level* and *the second level*. At the *ground level* objects have a name. The learner can recognize a trapezium, a square, an angle, a line. At this level a square for example will not yet be seen as a particular trapezium. At the ground level the subject judges on the basis of observation - this is a square because one can see that it is a square. At the *first level* objects become concepts, which are bundles of properties. Van Hiele says that the learning subject possesses a net of relations. In such a net of relations words like 'parallelogram', 'square', 'angle', etc. are vertices, connected by different relations. At this level the subject judges on the basis of his net of relations - this is a square because it is a rectangle with equal sides. At this level the learner knows for example a number of true statements with respect to a parallelogram, but he or she does not yet connect these statements deductively. At the *second level* the subject reasons deductively. He or she infers properties from other properties. There is a shift of emphasis: the primary question is no longer whether or not a particular statement is true, but whether it follows logically from other statements. This level is characterized by a net of relations in which the vertices are statements and the connections are logical.[17]

---

[17]Van Hiele pays particular attention to the question as to whether there are precisely three levels or more. He argues that there are precisely three different levels corresponding to three different forms of judgement. The impression that there are more than three levels can have different causes. Van Hiele mentions three of them. (1) For example the generalization of the theory of quadratic equations to equations of higher degree is not a shift of level: the same net of relations is being used. (2) When we formalize a deductive theory, giving formal derivations in the formal system is a first level activity. Van Hiele calls this a level

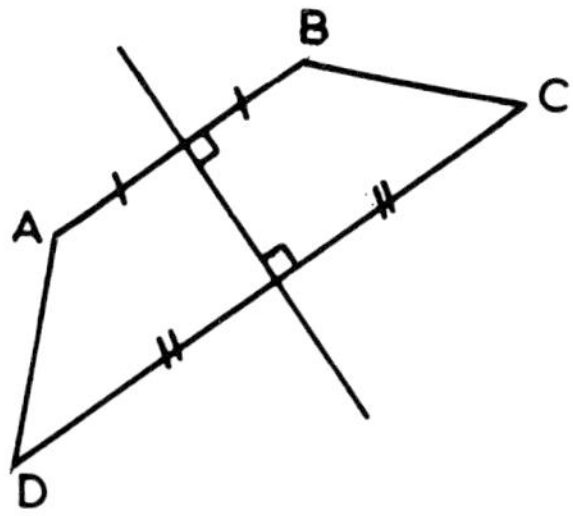

Figure 7.1

In 1977 Van Dormolen illustrated Van Hiele's theory by means of some examples. I shall reproduce one of the examples. Van Dormolen shows how a student could solve the problem - " Prove that the diagonals in an isosceles trapezium are equal" - at the three different levels.

At the *ground level* the student could measure the diagonals with a ruler, find the same results and draw the conclusion of the equality of the diagonals.

At the *first level* the student could mentally cut out the trapezium, turn it about and put it back in the hole. Because this can be done (given the definition of the trapezium), the diagonals are equal as one can easily see.

At the *second level* a student might reason as follows. The isosceles trapezium ABCD is by definition a quadrangle with an axis of symmetry that does not pass through a vertex. Consequently there is a reflection S such that S(ABCD)=BADC. We have S(AC)=BD. Because under reflection a segment and its image are congruent, the diagonals have the same length.

Of course one might wonder what a paedagogical theory has got to do with the development of Greek mathematics. The point is, however, that Van Hiele's theory is of a very fundamental kind: according to the theory an individual student of mathematics must *inevitably* pass through the lower levels before a higher level can be reached. The theory is, moreover, based upon thorough

---

reduction: by means of the formalization a second level activity is reduced to a first level activity. The proving of metamathematical theorems is then again a second level activity. (3) The transition of one level to the next level is not something that takes place all at once. In the process of transition different phases can be distinguished. This can create the impression that there are more than three levels.

research and deserves to be taken seriously.[18] Deductive thinking in mathematics is a Greek discovery. It follows from Van Hiele's theory that the first Greek mathematicians who discovered deductive thinking in mathematics, i.e. the first mathematicians who reached Van Hiele's second level of reasoning, could only make this discovery after having reached considerable mathematical knowledge at the first level and after considerable activity at this level. This considerable "first level knowledge" must have been the result of a collective enterprise. In other words, these mathematicians could only reach the second level because they came at the end of a phase in Greek mathematics characterized by "first level mathematics", and they initiated the deductive phase, represented for example by Euclid's *Elements*.

VII.5 Two traditions in Greek mathematics

Van Hiele's theory supports the view of Szabò and of others that in pre-Euclidean Greek mathematics two phases should be distinguished. The two phases then correspond to the first and second level of thinking in Van Hiele's theory respectively. It is tempting, moreover, to identify the phase of naive conjectures - which in Lakatos's view the Greeks at least partially inherited from Babylonians and Egyptians - with Van Hiele's ground level. The process of proof-experiments, proof-analysis etc. described by Lakatos, which finally yielded the "hard-core of the Euclidean research programme", corresponds to the (beginning of the) deductive phase in Greek mathematics and not to the demonstrative phase.

I propose to describe the developments as follows. The demonstrative phase corresponds to a non- deductive mathematical research tradition - the *Demonstrative Tradition.* The second, deductive phase corresponds to the formation of the deductive *Euclidean Tradition.* I will now give three examples that illustrate and support this proposal. At the same time the three examples show that not only Von Fritz and Szabo, but several other historiographers of Greek mathematics, assume that in Greek mathematics proofs were originally demonstrative and only afterwards became deductive. The first example illustrates this in the case of the historical studies of VerEecke, the second example illustrates it in the case of those of

[18]Cf. [Freudenthal,1973], pp. 116 - 117. The theory also drew international attention. Cf. [Fuys,1985] and in particular [Bell e.a.,1983], Ch. 8. Other research evidence concerning stages in the understanding of proof confirms that the ability to reason deductively is always preceded by a phase of substantial concrete reasoning ([Bell e.a.,1983], Ch.8).

Neuenschwander and finally, the third example shows that Becker also in fact assumed the existence of a period in pre-Euclidean Greek mathematics in which proofs were demonstrative.

Example 1

Tradition attributes several theorems to Thales. In his commentary on Elements. I, Def.17 Proclus writes:

> "It is related that Thales was the first to prove that the circle is cut into two equal parts by the diameter".[19]

Proclus then proceeds to prove this property of the circle. He folds the circle along a diameter and states that the two halves cover each other completely. Proclus writes:

> "Because, if one part was unequal to the other it would fall inside or outside the other and, whatever the case might be, it would follow that a smaller straight line would be equal to a greater straight line, as all those that leave from the centre to the circumference are equal; so the one that goes to the exterior circumference would be equal to the one that goes to the interior circumference; which is impossible".[20]

Proclus' proof is deductive, based upon a reductio ad absurdum. VerEecke remarks that this proof cannot have been Thales' proof because it is apagogic.[21] A demonstrative proof by Thales or some other early Greek mathematician might have run essentially as follows (Figure 7.2). After folding the circle along a diameter TS the half TPS will coincide exactly with the other half TP'S. This will be so because the definition of the circle implies that an arbitrary point of the circumference P coincide with some other point P' of the circumference will after the folding. This is visually evident. One can literally see it. (Of course one assumes implicitly that during the folding the distance of P to the center M of the circle does not change).

---

[19]"On rapporte que Thalès a été le premier à démontrer que le cercle est coupé en deux parties égales par le diamètre" ([VerEecke,1948],p.139).

[20]In VerEecke's French translation the text runs as follows. "Car, si une partie n'était pas égale à l'autre, elle tomberait à l'intérieur ou à l'extérieur de l'autre, et, quel que fût le cas, il s'ensuivrait qu'une plus petite ligne droite serait égale à une plus grande, puisque toutes celles qui partent du centre à la circonférence sont égales; donc, celle qui va à la circonférence extérieure serait égale à celle qui va à la circonférence intérieure; ce qui est impossible" ([VerEecke,1948], p.139).

[21][VerEecke,1948], p.139, footnote.

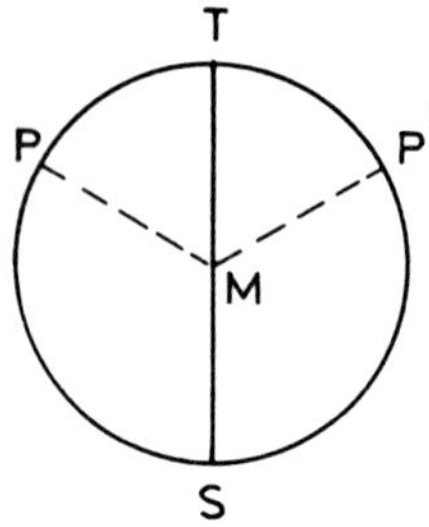

Figure 7.2

Proclus' proof is deductive. It is a second level proof in the sense of Van Hiele. To see this, one should consider the fact that the reductio ad absurdum begins with the remark: suppose that a point P after folding lies inside or outside the circumference. On the first level this remark would be met with the comment: that is unimaginable because of the definition of the circle. On the second level the unimaginable is expressed in language and a contradiction can be deduced from it. On this level it is no longer what one sees that convinces, but logic.

Example 2

Proclus writes the following in his commentary on Euclid's proposition Elements I.5, which says that the base angles in an isosceles triangle are equal:

> "it is said that he (i.e. Thales - T.K.) was the first to think and to state that the angles situated at the base of an isosceles triangle are equal".[22]

How could Thales have proved this theorem? In Anal. Prior. I 24 41b 13-22 Aristotle writes that in order that a syllogism may be valid *one* of the premisses must be affirmative and *one* must be universal. Aristotle writes:

> "This is more manifest in geometrical propositions, e.g. the proposition that the base angles of an isosceles triangle are equal. Suppose the straight lines A, B have been drawn to the center. Then, if one should assume that the angle AC is equal to the angle BD without having claimed generally that

[22]In VerEecke's French translation: "on dit qu'il fut le premier à penser et à affirmer que les angles situés à la base de tout triangle isocèle sont égaux" ([VerEecke,1948],p.217).

the angles of semicircles are equal, and again, if one should assume that the angle C is equal to the angle D without making the additional assumption that every angle of a segment is equal to the other angle of that segment, and if one should then, lastly, assume that, the whole angles being equal and the angles substracted being equal, the remaining angles E, F are equal, he will beg the question unless he assumes (generally) that 'if equals be substracted from equals the remainders are equal'" ([Heath,1949], pp.23-24).

Heath reconstructed the figure belonging to this passage in this way (Figure 7.3). Three geometrical theorems are involved in the example:

(1) The angles of semicircles are equal.

(2) The two angles of a circle segment are equal.

(3) The base angles in an isosceles triangle are equal.

Neuenschwander, who discussed this passage, assumes that the theorems (1) and (2) are old and that they were originally proved by superposition ([Neuenschwander, 1972], p.361). These original demonstrative proofs could have run as follows (Figure 7.4).

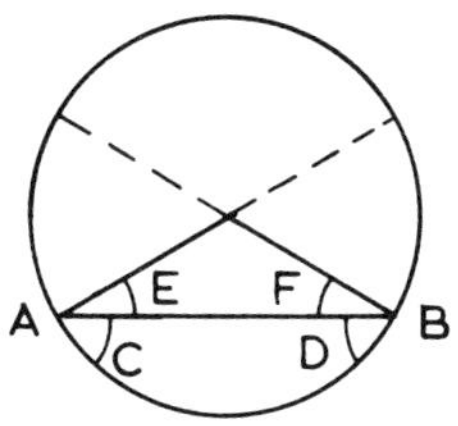

Figure 7.3

One folds a circle along the diameter TS. Let the point P after folding coincide with P'. The point of intersection Q of PP' and TS does not change position during the folding, so PQ after folding coincides with P'Q. From this follows directly

$\angle TPM = \angle TP'M$ theorem (1)

$\angle TPQ = \angle TP'Q$ theorem (2)

$\angle QPM = \angle QP'M$ theorem (3).

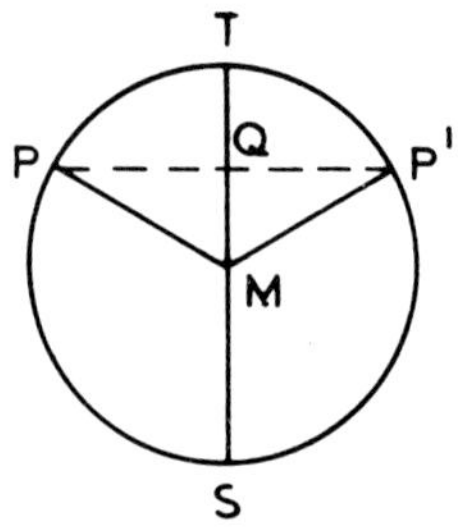

Figure 7.4

It is remarkable that in Aristotle's proof, (3) is derived from (1) and (2) and statement (4) :"If equals are substracted from equals, the remainders are equal" (In Euclid's *Elements* this statement returns as Axiom 3 of Book I!).
I suppose that in the Demonstrative Tradition in Greek mathematics the theorems (1) through (3) were proved by superposition.The proofs were level 1 proofs in the sense of Van Hiele. The insight that (3) can be derived from (1), (2) and the general statement (4) concerning equality then belongs to the (early) deductive phase. The result is a level 2 proof in the sense of Van Hiele. We also see how Axiom 3 may have been proof-generated.

Example 3
Becker's 1934 reconstruction of the demonstrative Pythagorean theory of odd and even ([Becker,1965],pp.125-145) can also show us something about the transition from the Demonstrative Tradition to the Euclidean Tradition. The reconstruction is based upon the propositions IX,21 through IX,34 together with IX,36 in Euclid's *Elements*. In Euclid's *Elements* this theory has a deductive character. On Becker's view numbers in this theory originally were represented by pebbles. Central notions were "even number" and "odd number".
Elements VII, Def.6 says:

> "An *even number* is that which is divisible into two equal parts" ([Heath, 1926], Vol.2, p.277).

E.g.

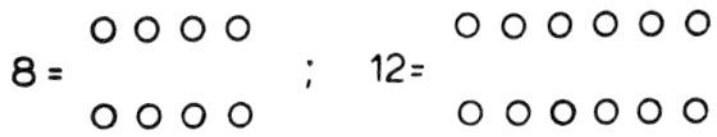

Odd numbers differ one pebble from an even number. Elements VII, Def.7 says:

> "An *odd number* is that which is not divisible into two equal parts, or that which differs by an unit from an even number" ([Heath, 1926], Vol.2, p.277).

E.g.

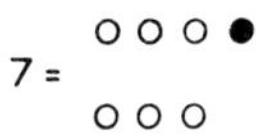

A simple theorem from the theory is Elements IX, 22:

> "If as many odd numbers as we please be added together, and their multitude be even, the whole will be even" ([Heath,1926],Vol.2,p.413).

The truth of the statement can be shown (and presumably was shown) by means of a representative example.: e.g. $7 + 9 + 13 + 5$.

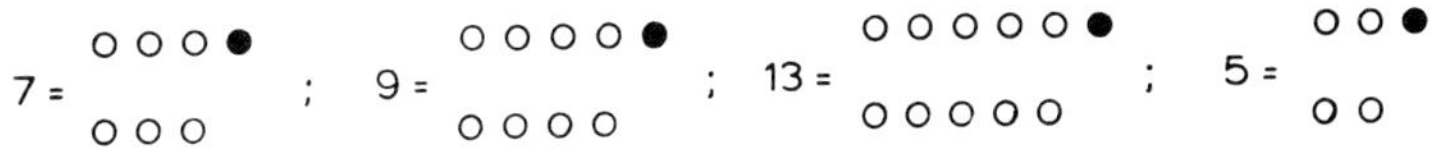

The "even parts" of the odd numbers are added first and the even number of remaining units is added afterwards. Clearly the sum consists of two equal parts and is even as a result of this.

The climax of the theory of odd and even is the theorem which asserts that numbers of the form

$$2^n(1+2+2^2+\ldots\ldots\ldots\ldots 2^n)$$

are perfect[23], provided

$$p=1+2+2^2+............2^n$$

is a prime number.
In the *Elements* this is proposition IX,36:

> "If as many numbers as we please beginning from an unit be set out continuously in double proportion, until the sum of all becomes prime, and if the sum be multiplied into the last make some number, the product will be perfect" ([Heath,1926], Vol.2, p.421).

Becker has shown how Elements IX,21 through IX,34 can all be proved demonstratively, and also how Elements IX,36 can be demonstratively derived from them.[24]
In order to see what this theory looked like in Euclid's *Elements*, let us have a look at Euclid's proof of the theorem that says that the sum of an even number of odd numbers is even. In Heath's translation this proof runs as follows:

> "For let as many odd numbers as we please, AB, BC, CD, DE, even in multitude, be added together; I say that the whole AE is even. For, since each of the numbers AB, BC, CD, DE is odd, if an unit be substracted from each, each of the remainders will be even (Book VII, Def.7); so that the sum of them will be even (Book IX, 21) [This proposition says: "If as many even numbers as we please be added together, the whole is even"- T.K.]. But the multitude of the units is also even. Therefore the whole AE is also even (Book IX, 21). Q.E.D." ([Heath,1926], Vol.2, pp.413-414).

The numbers AB, BC, etc. are represented by segments on a line as follows.

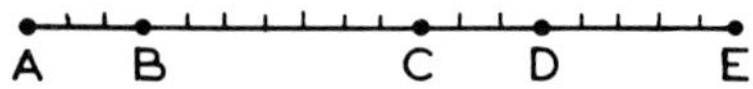

Figure 7.5

The reasoning nevertheless clearly is meant to be abstract and is an attempt to derive the theorem deductively within a larger deductive theory.

---

[23]A perfect number is a number that equals the sum of its divisors (the number itself should not be considered as divisor).

[24]The Pythagorean theory of odd and even is actually quite a sophisticated piece of constructivist mathematics. It shows that level 1 reasoning in the sense of Van Hiele may become sophisticated in different ways. Only one of those ways is the transition to deductive reasoning.

VII.6 The transition in terms of the methodology of mathematical research traditions

The three examples show what kind of transition took place. They do not show *why* it took place. If the methodology of mathematical research traditions is applicable to the transition, it is to be expected that one or more progressive research projects involving deductive elements must have brought about the transition.

It is Szabó's opinion that (i) Without any influence from outside, Greek mathematics could never have become deductive, and (ii) The transition to the deductive phase took place in the 5th century B.C. when mathematics became anti-empirical and abstract under the influence of Eleatic philosophy. According to Szabó the method of indirect proof first occurred in Eleatic philosophy and found its way from there into mathematics. It is not my intention here to discuss extensively Szabó's views on this particular point. They are very controversial among historians. Some authors agree with Szabó. From the fact that at the beginning of the 5th century B.C. in Southern Italy abstract philosophical reasoning appeared in Epicharm's comical sketches, Von Fritz has drawn the conclusion that large sections of the population must have been acquainted with this reasoning ([Von Fritz,1971], pp.210-211). He assumes that the mathematics of that time was also influenced by it. Von Fritz writes:

> "Szabó himself derives the abstract deductive method of proof in mathematics from the Eleatic philosophy, which is presumably to a large extent correct".[25]

On the other hand, Szabo's thesis has been criticized vehemently by Knorr. Without denying some influence of philosophy on mathematics Knorr argues, that

> "the origin and elaboration of formal methods in mathematics, both in the pre-Euclidean period and afterward, are best understood as a development internal to the mathematical discipline"([Knorr,1981], p.179).

Moreover, Knorr assumes that the crucial changes took place in the 4th century, instead of in the 5th century ([Knorr,1981a], p.195). In view of the speculative nature of all considerations concerning pre-Euclidean Greek mathematics, it is risky to entertain too strong convictions. In spite of

---

[25]"Szabó selbst leitet die abstrakt deduktive Beweismethode in der Mathematik aus der eleatischen Philosophie ab, was wohl weitgehend richtig ist" ([Von Fritz,1971], p. 210, Footnote 438).

differences of opinion about when and how it occurred, there seems to be consensus concerning the fact that the discovery of incommensurable magnitudes played an important role. Knorr:

> "the notion of incommensurable and the confirmation that given constructed lines are incommensurable are *intrinsically* theoretical and demand deductive arguments of the indirect kind; no concrete or practical procedure can do more than suggest that such lines evade description in terms of ratios of whole numbers" ([Knorr,1981], p.156; the emphasis is Knorr's).

An indirect proof requires that a situation which cannot be visualized - because it cannot exist - is nevertheless visualized "vaguely" or only phrased, i.e. expressed in language. A shift to some form of deduction is then necessary.[26]

The question whether Knorr is right is of minor importance for my purposes. If the transition from the Demonstrative Tradition to the Euclidean Tradition took place in accordance with the Methodology of Mathematical Research Traditions, it must be possible to identify one or more progressive research projects within the Demonstrative Tradition, whose progress depended more and more on the use of deductive level 2 reasoning in the sense of van Hiele. In this respect projects executed by the Pythagoreans could very well have played a role. In this connection it is interesting that, although most of Becker's reconstruction of the Pythagorean theory of odd and even consists in

---

[26] It is remarkable that, although students generally seem to experience a great deal of difficulty with indirect proofs (e.g.[Fremont,1969],p.342), in paedagogical literature not much attention is paid to the problem. Fremont advises teachers to show their pupils that daily life is full of indirect proofs. For example a mechanic, who can think of only three possible causes for a malfunction of a car, and after having excluded two of them concludes that it must be the third cause, in Fremont's opinion uses an indirect proof. Formally this is correct. Yet the example makes it even less clear why students have difficulty with indirect proofs. I can think of two reasons: a) a teacher wants to prove indirectly a statement which is obviously true to his pupils. This can occur when the pupil reasons on level 1 in the sense of Van Hiele and the teacher on level 2. In this case the pupil is asked to imagine what is for him at that moment unimaginable right at the start of the proof. b) It is also possible that the pupil, although the negation of the statement to be proved is not obviously absurd, simply has difficulties with deductive reasoning. For example, because he does not realize that deductive reasoning may be valid while premisses and/or conclusion are untrue or even absurd. If this is the cause, the example of the mechanic is not a good example, because eliminating two of the three possibilities need not involve any sophisticated (non-everyday) logical reasoning.

literally showing the truth of the propositions by means of representative examples, at one point Becker's reconstruction necessarily requires an indirect proof. If one assumes that $2^n p$ has no other divisors than 2, 2^2,..., 2^n and 2p, 2^2p,....2^{n-1}p, then the truth of the above mentioned theorem concerning perfect numbers can be shown literally by laying down pebbles in an exemplary case. However, there is no way in which one can *directly* show that there are *no* other divisors. Becker's proof of that part of the theorem is based upon the theorem Elements IX,30:

> "If an odd number measures an even number, it will also measure the half of it" ([Heath,1926],Vol.2,p.417).

In the reconstruction this theorem is used to prove by "finite descent" that $2^n p$ has no other odd divisors than p. Suppose the odd number in IX.30 is a and the even number is b then the proof of IX, 30 requires that one first proves that a.c = b with even c. This can only be done indirectly: if c would be odd, then b would also be odd (this is based upon Elements IX, 29: "If an odd number by multiplying an odd number makes some number, the product will be odd" [Heath,1926],Vol.2, p.416), which is not the case, so c must be even. Once one knows that c is even, the truth of IX, 30 can be shown by means of pebbles. Let a=3, c=8 and b=24. Then the matrix below shows the truth of the theorem: c being even implies that the matrix can be halved, which means that a is a divisor of the half of b.

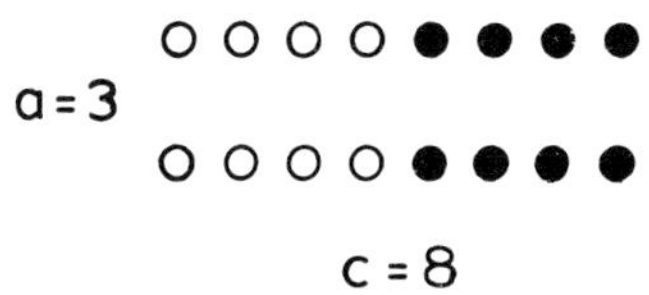

In two ways the theory of odd and even reconstructed by Becker goes beyond reasoning on level 1 in the sense of van Hiele. Although the truth of statements is shown literally and there is no use (or an extremely limited use) of logic, the theory is an ordered whole of statements. In this respect there is definitely a shift to the level of language. On the other hand the indirect proof, needed to prove that certain divisors *do not exist*, also goes beyond level 1. Although the mere existence of the theory of odd and even did not bring about the transition to the deductive Euclidean tradition, this shift to the level of language and the inconspicuous occurrence of indirect proofs indicate that the demonstrative tradition was subject to change.

Worth mentioning in this context is also Hippocrates (ca. 420 B.C.). The work on the quadrature of lunes attributed to him can be considered as a research project belonging to the early Euclidean Tradition and contributing to the degree of progress of that tradition. That work shows proofs of theorems consisting of chains of geometrical statements. Hippocrates is also reported to have written an *Elements of Geometry*. Knorr's views on the roots of Euclid's deductive mathematics also support the hypothesis that the transition from the Demonstrative Tradition to the Euclidean Tradition took place in accordance with the Methodology of Mathematical Research Traditions. Knorr finds the roots in Theodorus' work on incommensurables in the second half of the 5th century B.C. ([Knorr,1981], p.154). He has given an interesting reconstruction of some of Theodorus' work. In the terminology of the methodology of mathematical research traditions, Knorr reconstructed a research project that was born within the demonstrative tradition, but which contained important deductive elements.

Knorr assumes that Theodorus knew the following theorem on Pythagorean triples:

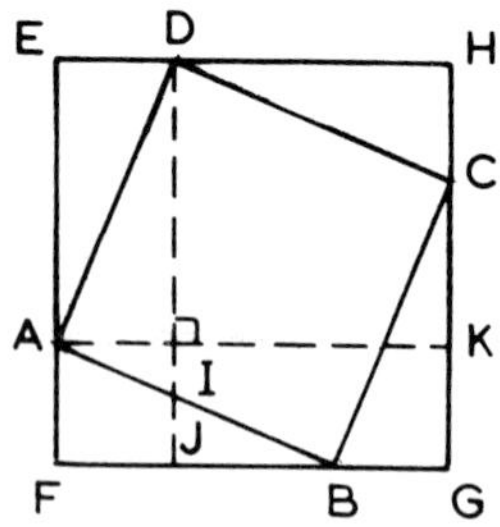

Figure 7.6

Theorem: If in a given Pythagorean triple the largest number C is even, then all three terms are even ([Knorr, 1975], p.158). This theorem can be shown demonstratively.[27] Knorr supposes also that a variant of the following demonstrative dissection proof of Pythagoras' theorem was part of the Demonstrative Tradition ([Knorr,1975], pp.176-177). Consider Figure 7.6. A, B, C and D respectively divide EF, FG, GH and HE according to the same ratio.

Obviously

square ABCD = square EFGH - 4 times triangle AFB

while also

square AFJI +square DIKH = square EFGH - 4 times triangle AFB.

Conclusion: square AFJI +square DIKH = square ABCD.

The discovery of the incommensurability of the hypothenuse and the sides in a right isosceles triangle (in fact the discovery of the irrationality of $\sqrt{2}$) is reconstructed by Knorr as follows. Attempts to find Pythagorean triples corresponding to the right isosceles triangle would have been the point of departure. Let the numbers A and B correspond respectively to the hypothenuse and the sides. Then $B^2+B^2=A^2$. Obviously, if A is even, B is also even. Then we can take twice as large a common measure and we end up with the numbers A_1 and B_1 corresponding to hypotenuse and the sides. If A_1 is even we take again a common measure twice as large. This process must end. We may then assume A_n to be odd. It follows that if $A_n^2$is odd, then the square of one of the two legs must be odd and the square of the other leg even.

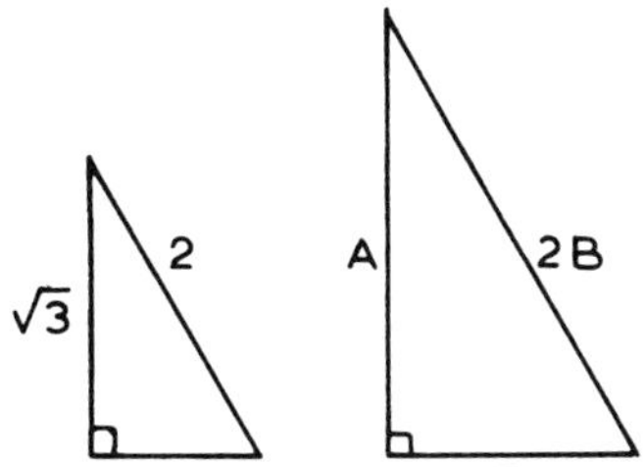

Figure 7.7

[27]One needs to show that squares of odd numbers are a multiple of four plus unity and that squares of even numbers are multiples of four.

Since the two legs are equal we have a contradiction ([Knorr, 1975], pp. 179-180). The conclusion must be that hypotenuse and legs are incommensurable. According to Plato, Theodorus did prove the incommensurability of a series of segments of length $\sqrt{n}$ and the unit of length. Plato mentions n=3, 5 and 17 and says that Theodorus encountered difficulties in the case of n=17.

In Knorr's reconstruction, Theodorus constructed rectangular triangles with one leg equal to $\sqrt{n}$ for n=3, 5, etc., followed by an examination of the triangles involved with respect to the oddness and evenness of the terms of the Pythagorean triples. As for the case of $\sqrt{3}$ the reasoning is the following. Assume commensurability. Then the hypotenuse is even, so A and B are both even. We can double the common measure repeatedly until one of the legs corresponds to an odd number. Because the hypotenuse remains even one of the two legs then corresponds at the same time to an odd and to even number. In a similar way Theodorus would have dealt with the cases n = 5, etc. For the complete reconstruction, I refer to Knorr's book.[28] I will not go into Knorr's reconstruction of Theaetetus' and Eudoxus' work. The core idea is that the work on incommensurables by these mathematicians essentially could not be based upon practical value or "graphic" evidence, but had only one criterion of validity: logical consistency. Knorr writes:

> "Hence, if within such a theory one has to call for geometric constructions, for theorems on proportionality or for other mathematical materials, these must be brought into conformity with the logical standards of the whole theory before their introduction is possible. For the technical invalidity of any step must invalidate the whole argument in which it occurs" ([Knorr,1975], p.4).

From the point of view of the methodology of mathematical research traditions Knorr's reconstruction of the development of Theodorus' work is a nice example of a simple research project leading to new mathematical theorems that were considered to be weighty. The heuristic unity can be characterized easily: one particular method to prove incommensurability is applied in a number of analogous situations. The project started in the Demonstrative Tradition, but clearly contains important deductive elements.

---

[28]Knorr shows that in the case where n = 17 the method breaks down, because, if we consider $\sqrt{17}$ as the leg of a right triangle whose other leg is 8 units long and whose hypothenuse is 9 units, commensurability is seen to be compatible with the conditions imposed on Pythagorean triples that led to a contradiction in the other cases ([Knorr,1975], p.191).

Greek culture is characterized by a quest for justification. The grounds of knowledge were a central issue: what is true? How does one distinguish what is true from what is false? As soon as it became clear that certain theorems could be proved on the one hand by demonstration and on the other hand by deduction from other theorems, this attitude must inevitably have led to the question as to which of the two methods was the more reliable. It is easy to imagine how the successes of the deductive method in the theory of irrationals combined with possible doubts with respect to visual evidence (Cf. my remarks above on the possible origin of Postulate 5) led to an axiomatic ideal of grounding mathematical knowledge only upon first principles or deductions from them.

VII.7 Conclusion

The above-given reconstruction shows that the way in which historiographers view the history of Greek mathematics is compatible with the assumption that in pre-Euclidean mathematics a transition took place from a Demonstrative Tradition to an Euclidean Tradition in accordance with the MMRT. The two traditions can be clearly distinguished, particularly in terms of their completely different methodologies. The appearance of the new tradition coincides, moreover, with the occurrence of progressive research projects in the sense of the MMRT.

Because of a lack of primary sources, the existing historiography - and as a result my reconstruction - is rather speculative. That is one of the reasons why in the next chapter I will present a reconstruction on the basis of the MMRT of another, more recent development in mathematics.

Chapter VIII Two research traditions in analysis

VIII.1 Introduction

I will now give a second rational reconstruction involving a transition from one research tradition to another in accordance with the MMRT. The transition studied in the previous chapter concerned all of Greek mathematics and the entire mathematical community of that time. The transition studied in this chapter concerns the development of analysis in the eighteenth and nineteenth centuries and it does not embrace mathematics as a whole. Yet, because of the dominant position of analysis in the mathematics of those two centuries, we are dealing with a transition of major importance involving the whole mathematical community of that time.

In this chapter the emphasis will be on the descriptive part of the MMRT. As regards the normative part of the MMRT I will restrict myself to some remarks on Cauchy's foundational project, which show that prominent mathematicians and historiographers have judged the transition from eighteenth century Formalist Tradition to the nineteenth century Conceptual Tradition in accordance with my MMRT.

VIII.1.1 Eighteenth century analysis versus nineteenth century analysis

In his book on the conceptual development of the calculus, Boyer has characterized the eighteenth century as a "period of indecision" which was followed in the nineteenth century by "rigorous formulation" ([Boyer,1959], Chapters VI and VII). This rigorous formulation, particularly by *Cauchy* and after him by *Weierstraß*, meant a complete revision of the eighteenth century calculus. Other historians of mathematics besides Boyer have contrasted the nineteenth century with the eighteenth century and have emphasized the great differences between the analysis of the two periods. Historians, however, usually restrict themselves to pointing out particular aspects of the transition. Attempts to describe the transition in terms of a general model or scheme are rare. Historians of mathematics apparently hesitate to describe such comprehensive developments as the history of analysis over a period of more than one century in terms of a rather simple model. An exception is Grabiner's suggestion that the complete revision of analysis brought about by Cauchy's work in particular was a *revolution in Kuhn's sense* ([Grabiner,1981], p.186). In this chapter, I take Grabiner's suggestion seriously and consider the following questions: How can we characterize eighteenth century analysis? How

can we characterize nineteenth century analysis? What was the nature of the transition? I will first briefly consider the idea that Cauchy's work inaugurated a revolution in Kuhn's sense and I will finally argue that a description of the transition in terms of the MMRT is more adequate.
Eighteenth century analysis encompasses a great deal and mathematical production in the nineteenth century was enormous. Throughout this chapter, I inevitably have to focus primarily on secondary sources. My references to primary sources mainly concern illustrations of general statements. The description of the transition from eighteenth to nineteenth century analysis that I develop, in terms of the methodology of mathematical research traditions, must be considered as a sort of extrapolation of views that are frequently expressed in the existing historiography of mathematics. I do not exclude the possibility of different extrapolations, but I do claim that mine encompasses major aspects of the developments as historians of mathematics have seen them. The implicit criteria that they use are compatible with the methodology of mathematical research traditions.

VIII.1.2 A Kuhnian proposal and its dismissal

Mathematics can be compared to a game whose goal is the invention and solution of mathematical puzzles, and in any particular period of history mathematicians will usually agree as to whether or not a proposed puzzle makes sense or whether a proposed solution of a puzzle is correct or not. If for a moment we do consider mathematics as a game, then there should be a set of rules according to which the game should be played in the period involved. The history of mathematics shows that over time the rules of the game of mathematics change. Kuhn has used the game metaphor to describe the development of natural science (Cf. [Kuhn,1970]). In Kuhn's terminology the rules of the game are determined by a *disciplinary matrix* (The word "matrix" is not used in a mathematical sense. See below). Such a disciplinary matrix characterizes a period of *normal science*. Sometimes during a period of normal science, a pronounced failure in the normal problem-solving activity occurs: some puzzles resist any attempt at solving them. Such puzzles are called *anomalies*. A *revolution in Kuhn's sense* occurs when one or more anomalies evoke a *crisis*, which is terminated by a switch to a new disciplinary matrix that will dominate during a new period of normal science. Kuhn very much emphasizes the fact that *succeeding disciplinary matrices are fundamentally incompatible* and he stresses the *irrational* character of the transition. In Kuhn's view the transition requires a *gestalt-switch* on the level of the individual. He also uses the term *conversion* to stress its irrational character.

On the one hand it seems reasonable to describe the eighteenth century development of analysis as a period of normal science vexed by foundational anomalies. These foundational problems slowly gained weight because of several factors. One of those factors, for example, was the fact that the growing teaching duties of mathematicians forced them to face again and again the problem of the foundations ([Grabiner, 1974]). On the other hand, it also seems reasonable at first sight to consider Cauchy's textbooks as inaugurating a new period of normal science dominated by an entirely different disciplinary matrix. That at the individual level something like a conversion was needed to switch from the old eighteenth century ways to the new approach is suggested by Abel's statements to Holmboe:

> "Divergent series are in general something quite fatal, and any demonstration based upon them is a shame. By using them one can prove anything one wants [...]. *At last my eyes have been opened in a striking manner*, because with the exception of the simplest cases - for example the geometric series - there are almost no series in mathematics whose sum has been determined in a rigorous manner; in other words the most essential part of mathematics is without foundation" (Emphasis is mine - T.K.).[1]

However, on further examination a characterization of the transition as a revolution in the sense of Kuhn is unsatisfactory. I will give three arguments to support this.

1) Kuhn has listed four different sorts of components of a disciplinary matrix: (i) Symbolic generalizations: i.e. fundamental laws of nature or definitions; (ii) Beliefs in particular models: these models supply a group of scientists with preferred or permissible analogies; they also help to determine what will be accepted as a puzzle-solution; (iii) Values; (iv) Exemplars: i.e. shared examples which show how the business of science is to be carried out ([Kuhn,1970], pp.182-187). Kuhn had disciplinary matrices in mind that pertain to natural science (e.g. Ptolemaic versus Copernican astronomy, or

---

[1]"Les séries divergentes sont en général quelque chose de bien fatal, et c'est une honte qu'on ose y fonder aucune démonstration. On peut démontrer tout ce qu'on veut en les employant [...]. Enfin mes yeux se sont dessillés d'une manière frappante, car à l'exception des cas les plus simples, par example les séries géométriques, il ne se trouve dans les mathématiques presque aucune série infini dont la somme soit déterminée d'une manière rigoreuse, c'est-à-dire que la partie la plus essentielle des mathématiques est sans fondement" ([Abel,1881], Vol.2, pp.257-258).

Newtonian versus Einsteinian dynamics), and it is not obvious what the components of a disciplinary matrix in mathematics are. Grabiner seems to apply Kuhn's terminology rather loosely. She does not attempt to define precisely the components of the disciplinary matrices corresponding to eighteenth and nineteenth centuries.
2) The fact that the problem of the foundation of the differential and integral calculus could not be solved satisfactorily during the course of the eighteenth century undoubtedly played a role in the shift from eighteenth century analysis to nineteenth century analysis. Strictly speaking, however, the problem cannot be considered as an anomaly in Kuhn's sense. If one wants an example of an anomaly in Kuhn's sense in eighteenth century mathematics, one should consider for example the problem of finding an algorithm to solve the general equation of the fifth degree. That is a clear example of a puzzle that stubbornly resisted all attempts at solving it. *In the case of the problem of the foundations of the differential calculus, no one had any idea as to how the puzzle should be solved.* Most mathematicians seem to have been reasonably satisfied with, for example, Euler's remarks on the problem. Only during the course of the century did the problem become a challenge to the mathematical world. Of course, if one wishes one can apply Kuhnian terminology here. This would require, however, that one stretches the notion of "anomaly".
3) I will show below that, although 18th century and 19th century analysis are to a certain extent incompatible, it is inappropriate to characterize the transition by means of notions like gestalt switch or conversion. The mathematical community of Europe as a whole behaved quite rationally when it dropped the 18th century ways of doing mathematics.

One could attempt to modify Kuhn's model in such a way that it becomes applicable to mathematics. This has been done by Mehrtens, who argues that a modified notion of disciplinary matrix could be used in mathematics, but that the concepts of revolution and crisis must be rejected ([Mehrtens,1976]). Yet Mehrtens, whose aims are primarily historiographical, does not go further than a few suggestions concerning the way in which the various elements of the disciplinary matrix could be applied to mathematics. In this chapter, I will argue that it is much more appropriate to describe the transition from 18th to 19th century calculus in terms of the methodology of mathematical research traditions than in terms of the Kuhnian model. I will characterize 18th and 19th century analysis as being dominated by different research traditions - the *Formalist Tradition* and the *Conceptual Tradition* - and I will show that the transition was rational in the sense of the methodology of mathematical research traditions.

VIII.2 The Formalist Tradition

VIII.2.1 The object of eighteenth century analysis

In the thirties of the eighteenth century, in Euler's work, analysis came to be considered as a discipline in its own right, independent of geometry and mechanics. In Euler's view the object of investigation in analysis consisted of *analytical expressions*. The most important analytical expressions were functions. Thus the word "function", used by Leibniz to denote variable geometrical quantities, acquired a different meaning. In the words of Euler in one of his influential textbooks, the *Introductio in analysin infinitorum*:

> "*A function of a variable quantity is an analytical expression composed in whatever way of that variable and of numbers and constant quantities.* So each analytical expression that contains, besides the variable quantity z, only constant quantities is a function of this z.
> E.g. a+3z; az-4zz; az+b√(aa-zz); c^z etc. are functions of z"
> ([Euler,1748], Par.4; The emphasis is mine - T. K.).

In Euler's work this notion of function dominates.[2] Half a century later, also in a textbook, Lagrange gave a similar definition:

> "One calls a *function* of one or of several quantities, any analytical expression in which these quantities occur in any way whatsoever, mixed or not with other quantities that one considers as having set and invariable values, while the quantities of the function can receive all possible values".[3]

In this way Euler and Lagrange defined the object of analysis, and although these definitions as such play hardly any role in Euler's and Lagrange's mathematics, they do express a major aspect of 18th century analysis. Historians have repeatedly emphasized the formalist character of 18th century analysis. Kline writes:

> "Though a controversy about the notion of a function did arise in connection with the vibrating-string problem and caused Euler to generalize his own notion of what a function was, the concept that dominated the eigh-

---

[2]It is only in connection with physical problems that sometimes a different concept of function appears in Euler's work (Cf. [Youschkevitch,1976/77], pp. 69-70).

[3]"On appelle *fonction* d'une ou de plusieurs quantités, toute expression de calcul dans laquelle ces quantités entrent de manière quelconque, mêlées ou non avec d'autres quantités qu'on regarde comme ayant des valeurs données et invariables, tandis que les quantités de la fonction peuvent recevoir toutes les valeurs possibles" ([Lagrange,1797], p.1).

teenth century was still the notion of a function given by a single analytic expression, finite or infinite" ([Kline,1972], p.406).

Boyer writes:

> "Functionality was a matter of formal representation, rather than conceptual recognition of a relationship. The almost automatic development of the calculus during the eighteenth century was largely the result of this formalistic view, to which the notation of Leibniz was so remarkably well adapted" ([Boyer,1959], p. 243).

And Grabiner writes about the eighteenth century:

> "[...] mathematicians placed great reliance on the power of symbols. Sometimes it seems to have been assumed that if one could just write down something which was symbolically coherent, the truth of the statement was guaranteed"([Grabiner,1974], p. 356).

VIII.2.2 The inductive character of eighteenth century analysis.

I now turn to the methods by means of which functions were studied. It is clear that Euler's and Lagrange's definitions of a function are rather *imprecise*. Algebraic expressions in general - and different kinds of infinite series, for example power series, and also expressions containing operations like differentiation and integration - were all considered to be functions. A precise borderline between a proper function and a meaningless composition of variables and constant quantities was not given. It is therefore understandable that in practice the meaning of the analytical expressions played an important role. For example when Euler, in the *Introductio in analysin infinitorum* ([Euler,1748]), defines the exponential function a^z for $a>1$, he stipulates that although a^z has multiple values for non-integer values of z, the function should be considered as one-valued; that is to say for any z only the main (i.e. real and positive) value should be considered.[4]

The absence of a precise definition had its consequences for the methods by means of which functions were studied in the eighteenth century. From a modern point of view the proofs of almost all theorems which, from the eighteenth century point of view, allegedly hold for all functions, are not rigorous: there is always (in the background) an unproved generalization because the boundary of the content of the notion 'function' was not precisely defined. This is one reason why eighteenth century analysis, in spite of the many "deductive" formal derivations, has essentially an inductive character. I

---

[4] See also section VIII.2.3 Example 4.

shall illustrate this by means of an example concerning the total differential from Euler's *Institutiones calculi differentialis* ([Euler,1755]).

Euler shows that the (total) differential of an arbitrary function of three variable quantities x,y and z has the form pdx+qdy+rdz. How does he do this?

After having proved the well-known fundamental rules of the ordinary differential calculus like

$$d(pq) = pdq + qdp; \; d\frac{p}{q} = \frac{qdp - pdq}{q^2}$$

([Euler,1755], Pars. 163-164) etc.[5], and after having applied these rules in several examples, Euler remarks ([Euler,1755], Par.170) that a comparison of these results shows that the differential of an *algebraic function* (an expression involving the operations +, -, x and :) always equals the sum of the differentials that one obtains if one subsequently considers each of the quantities involved as being variable and all the others as being constant. Somewhat further on ([Euler,1755], Pars. 208-216) Euler attempts to prove the rule for *all* functions. He defines the differential dF of a function F(x,y,z,...) of two or more independently variable quantities x, y, z, etc. as follows:

$$dF = F(x+dx,y+dy,z+dz,...) - F(x,y,z,...)$$

First he proves the rule for the case that

$$F(x,y,z)=X(x)+Y(y)+Z(z)$$

([Euler,1755], Par. 211) and he then considers the case that

$$F(x,y,z)=X(x).Y(y).Z(z).$$

In this case we have, say, dX=Pdx, dY=Qdy and dZ=Rdz.

So

$$\begin{aligned} F(x+dx,y+dy,z+dz)&=(X+Pdx).(Y+Qdy).(Z+Rdz)= \\ &=XYZ +PYZdx+XQZdy+XYRdz+ZPQdxdy+ \\ &\quad +YPRdxdz+XQRdydz+PQRdxdydz. \end{aligned}$$

And after dropping higher order terms one gets

$$dF=YPZdx+XQZdy+XYRdz$$

Then Euler remarks:

---

[5]From Euler's point of view differentials are special differences, which means that one can apply the normal rules of algebra to them. We then have: d(pq) = (p+dp)(q+dq)-pq = pdq +qdp+dpdq. Moreover, differentials are infinitely small, so that higher order terms, like dpdq in this case can be neglected. Ergo, d(pq) = pdq+qdp.

"These examples of functions of three variable quantities x,y and z, whose number can be arbitrarily multiplied, sufficiently show that the differential of an arbitrary function of three variable quantities has the form pdx+qdy+rdz" ([Euler,1755], Par. 212).

After having thus established the form of the total differential, Euler finally shows that the coefficients p, q and r are partial differential coefficients. For example dV=pdx+qdy+rdz implies for constant y and z that dV=pdx, so that p is a partial differential quotient ([Euler,1755], Pars. 213-214).

It is clear that Euler's proof concerning the form of the total differential of a function of several variables is a *proof by induction*, i.e. a proof on the basis of a (large) collection of examples.

VIII.2.3 The principle of the generality of algebra

In order to prove nowadays the equality of two functions f and g of one variable, one must prove that $\forall(x)\ f(x) = g(x)$. In the 18th century the equality of the two functions f and g was proved differently: one of the two functions had to be turned into the other function by means of a series of "analytical transformations". The collection of transformations allowed was not defined precisely. Cauchy later criticized the eighteenth century mathematicians for the use of arguments drawn from the *"generality of algebra"*. This principle of the generality of algebra cannot be precisely defined. I will roughly define the principle as follows: it consists in the assumption that algebraic rules which hold obviously for a certain class of objects, also hold generally, i.e. also for a larger class of objects where their validity is not so obvious. The principle of the generality of algebra implies, for example, that one can calculate with infinite series in the same way as one calculates with finite expressions. *The applications of the principle show clearly the inductive character of eighteenth century mathematics.* I shall give several examples of its application.

Example 1.

At times the (finite) quotient of two algebraic expressions can be determined by means of elementary division. Sometimes, however, the division does not end. In such cases eighteenth century mathematicians often assumed that the quotient of the two algebraic expressions equals the infinite series that the division of the two expressions yields.

The following quotation from Lacroix is quite illustrative of the way in which eighteenth century mathematicians reasoned. Lacroix remarked that, *independent of convergence or divergence of the series,* "the expression $1+(x/a)+(x/a)^2+(x/a)^3$+etc. [...] is in such a way connected with the fraction a/(a-x), that if a question led us to [that expression] we would have the right to draw from that fact the conclusion that the wanted function is no other than a/(a-x)".[6]

Example 2.

In a letter to Goldbach, dated 7th of August 1745, Euler remarked that for

$$y = \frac{1}{1 - \ln x}$$

we have

$$(1) \quad \int y\, dx = xy - 1.xy^2 + 2.xy^3 - 6.xy^4 + \dots$$

([Juškevič&Winter,1965], pp.218-219). Euler simply states (1), but I assume that he thought of a proof by means of repeated partial integration as follows. We have

$$\int y\, dx = xy - \int x\, dy$$

The function y, however, is such that

$$\frac{dy}{dx} = \frac{y^2}{x}$$

So we have

$$\int y\, dx = xy - \int y^2 dx = xy - xy^2 + \int x\, dy^2$$

and so on.

Example 3.

The series obtained in this way were freely subjected not only to algebraic manipulations but also to differentiation and integration. The differentiation and integration could also take place with respect to a parameter. For example the following expressions, considered as functions of r, are equal in the eighteenth century sense:

---

[6]"l'expression $1+(x/a)+(x/a)^2+(x/a)^3$+etc. [...] est tellement liée avec la fraction a/(a-x) que si une question nous conduisait à la série $1+(x/a)+(x/a)^2+(x/a)^3$+etc. nous serions en droit d'en conclure que la fonction cherchée n'est autre que a/(a-x)" ([Lacroix,1810], pp.4-5).

$$(2) \qquad \frac{1 - r\cos x}{1 - 2r\cos x + r^2} = 1 + r\cos x + r^2\cos 2x + r^3\cos 3x + ...$$

Euler substituted r = 1 and obtained

$$(3) \qquad 0 = \frac{1}{2} + \cos x + \cos 2x + \cos 3x + ...$$

This series diverges for all real values of x. Nevertheless Euler integrated (3) and got

$$(4) \qquad \frac{\pi - x}{2} = \sin x + \frac{1}{2}\sin 2x + \frac{1}{3}\sin 3x + ...$$

([Burkhardt,1911], pp.171-173). (N.B. (4) is correct from a modern point of view for $0<x<\pi$.).
Faith in these results was not unfounded. Euler had a habit of checking his results in different manners. For example, the series resulting from an infinite division appeared to be completely in accordance with what one obtained by means of the method of the undetermined coefficients. In [Euler,1755] (Pars. 86, 87 and 92), the equalities (3) and (4) are derived in a different way. From (1) Euler drew the conclusion that

$$(5) \qquad \int_0^1 \frac{1}{1 - \ln x}\,dx = 1 - 1 + 2 - 6 + 24 - 120 + 720 + ...$$

He remarked that this implies that the value of the integral, and at the same time the sum s of

$$(6) \qquad 0! - 1! + 2! - 3! + 4! - 5! + 6! + ...$$

equals approximately 0.5963475922. He added that he had obtained that value by transforming the series *in different ways* into convergent series ([Juškevič & Winter, 1965], pp.218-219).
I argue that *checking results in different ways is an essential element in eighteenth century methodology*, and that it is related to the inductive character of eighteenth century analysis. It served not only to check calculations, but it could also reveal risky applications of the principle of the generality of algebra .

Example 4.

The principle of the generality of algebra is intimately related to a principle that was expressed by Leibniz in 1702 as follows: "the rules of the finite also succeed in the infinite" ([Leibniz,1971], p.93)). In Euler's work, and also in the work of other eighteenth century mathematicians, we find that this principle is applied in the sense that *one calculates quite freely with infinitely small and infinitely large quantities.*

Consider for example Euler's derivation of the series expansion of a^z in the *Introductio in analysin infinitorum* ([Euler,1748]). Euler first introduces the exponential function $y=a^z$ for a>1 in a way that reminds the modern reader of a secondary school book: a definition is given for rational z, while a similar behaviour of the function for irrational z is assumed (*induction*!); continuity of the function - in the modern sense of the word continuity - is implicitly assumed. Then Euler derives the series expansion for a^z as follows ([Euler,1748], Pars 114-125). Let a>1 and let ω be infinitely small. Then we have for all i

$$a^{i\omega}=(1+k\omega)^i ,$$

or in other words

$$a^{i\omega} = 1 + \frac{i}{1}k\omega + \frac{i(i-1)}{1.2}k^2\omega^2 + \frac{i(i-1)(i-2)}{1.2.3}k^3\omega^3 + \ldots$$

For positive integer values of i this is perfectly legitimate and the binomial expansion is finite. Euler, however, immediately applies the equation for infinite i. If z is a finite positive number then i=z/ω is infinitely large. Because ω=z/i we have

$$a^z = 1 + \frac{kz}{1} + \frac{1(i-1)}{1.2i}k^2z^2 + \frac{1(i-1)(i-2)}{1.2i.3i}k^3z^3 + \ldots$$

However, because i is infinitely large

$$\frac{i-1}{i} = \frac{i-2}{i} = \frac{i-3}{i} = \ldots = 1$$

and we have

$$a^z = 1 + \frac{kz}{1} + \frac{k^2z^2}{1.2} + \frac{k^3z^3}{1.2.3} + \frac{k^4z^4}{1.2.3.4} + \ldots$$

This implies that the relation between a and k is now known, because putting z=1 yields

$$a = 1 + \frac{k}{1} + \frac{k^2}{2!} + \frac{k^3}{3!} + \ldots$$

If a is chosen such that k=1 we obtain the natural logarithm. Euler introduces the notation e for that number and finds

$$e^z = 1 + \frac{z}{1} + \frac{z^2}{1.2} + \frac{z^3}{1.2.3} + \ldots$$

Examples 1 through 4 illustrate the principle of the generality of algebra. The principle was a major methodological principle in eighteenth century analysis. Many historians of mathematics have more or less clearly expressed it. I will restrict myself to one quotation. Kline writes:

> "In view of the almost complete absence of any foundations, how could the mathematicians proceed with the manipulation of the variety of functions? In addition to their great reliance upon physical and intuitive meanings, they did have a model in mind - the simpler algebraic functions, such as polynomials and rational functions. They carried over to all functions the properties they found in these explicit, concrete functions [...]" ([Kline, 1972], p. 434).

I disagree with the first sentence. Eighteenth century analysis did possess a foundation: it was based upon algebra by means of the principle of the generality of algebra. The second part of the quotation, however, expresses precisely that principle.

VIII.3 Problems that vexed the eighteenth century tradition

Boyer rightly characterizes the eighteenth century as a period of uncertainty. And although most mathematicians involved do not seem to have worried about it, in retrospect this is a correct characterization. Eighteenth century analysis, as regards object and method, was as a whole vexed by uncertainty. This uncertainty is mainly related to the principle of the generality of algebra. Questions concerning the status of infinitely small elements in particular arose in connection with the differential calculus. Although the validity of the bulk of the methods of the differential and integral calculus was never seriously doubted, their foundation by means of infinitely small elements or by means of a vague limit concept was a cause of concern. Yet only at the end of the eighteenth century did the problem receive considerable attention. Lagrange especially, whose *Théorie des fonctions analytiques* ([Lagrange, 1797]) is an attempt to give a foundation of the calculus that avoids the use of infinitesimal elements, played an important role in this respect (Cf. [Grabiner,1981a]). As we have seen, checking results in different ways in order to make sure that they are correct is an important element in the methodology of eighteenth century analysis. This is not surprising since it was not clear as to precisely what the principle of the generality of algebra did and what it did not allow. Sometimes, however, a check failed and contradictions appeared. Such contradictions were dealt with in an ad hoc way. The following example illustrates Euler's way of handling contradictions. It also illustrates other problems that vexed eighteenth century analysis: problems concerning the use of divergent series. Euler considers in his *Institutiones Calculi Differentialis* the following equality between analytical expressions ([Euler,1755], Par. 102).

$$\frac{1}{1-x} = 1 + x + x^2 + x^3 + \dots$$

Euler now substitutes for x the numbers 1, 2 and 3 ([Euler,1755], Par. 103). This yields:

(A) $1 + 1 + 1 + 1 + \dots = \frac{1}{1-1} = \infty$

(B) $1 + 2 + 4 + 8 + \dots = \frac{1}{1-2} = -1$

(C) $1 + 3 + 9 + 27 + \dots = \frac{1}{1-3} = -\frac{1}{2}$

Because the terms of the series A are all smaller than the corresponding terms of the series B, one could argue that the sum of series A is smaller than the sum of series B. This would, however, mean that the infinite is smaller than minus 1, i.e. smaller than something which is smaller than nothing. This Euler finds unimaginable. The fact that the sums of B and C are negative while all the terms of the series are positive Euler also finds hard to reconcile with mathematical common sense . The problem that Euler faces is to draw ad hoc a borderline between what use of (divergent) series is allowed and what not. Euler proposes the following solution: it is necessary to state that in case of divergence the sum given by the expression 1/(1-x) is no *true sum.* If one states that the true sum $1+x+x^2+x^3+.....+x^\infty$ equals 1/(1-x) then from Euler's point of view one makes a mistake equal to $(x^\infty+1)/(1-x)$, which is for x>1 a very large mistake ([Euler,1755], Par. 105). Euler does not reject the use of divergent sums, because

"these sums with all their deviation from truth never lead to a mistake; they actually put us in a position of being able to find many of the most important theorems, theorems which one would not be able to find at all without those sums. Now if these sums were wrong, then they would not lead us to true theorems all the time, but would instead cause very big mistakes, because they do not differ insignificantly but infinitely from the truth" ([Euler,1755], Par. 109).

Euler's solution of the problem is simple. He stretches the meaning of the word "sum":

"So we want to call the expression whose expansion yields an infinite series, the sum of this series. In this sense 1/(1-x) is indeed the sum of the series $1+x+x^2+x^3+$etc., because the expansion of this ratio yields the series, no matter which value one substitutes for x. So if the series converges, the new notion is in agreement with the usual one; and as for the divergent series, if one assumes that explanation, one avoids with them all difficulties. By means of that explanation one is at last in a position to state the utility of the divergent series and to defend them against all objections" ([Euler, 1755], Par. 111).

Not all eighteenth century mathematicians appreciated Euler's solution. A few objected to the use of divergent series, for example Nikolaus I. Bernoulli and d'Alembert. Yet they seem to have been unable to support their criticism by means of arguments convincing others. As we have seen above (in examples 3 and 4), Euler substitutes without hesitation numerical values for the variables in equations containing (divergent) infinite expressions. In the 1890s, however, Callet derived a contradiction ([Bossut&Lagrange, 1799]). Daniel Bernoulli had argued, following Leibniz and Grandi, that the substitution of x=1 in

$$1 - x + x^2 - x^3 + x^4 - ... = \frac{1}{1+x}$$

makes sense, because the result, i.e.

$$1 - 1 + 1 - 1 + 1 - 1 + ... = \frac{1}{2}$$

can be understood as follows: the value 1/2 is precisely the average of the two different values, i.e. 0 and 1, of the partial sums of the series.

Callet wrote that this point of view, and obviously also Euler's, leads to contradictions since we can get quite a different result from

$$1 - x^2 + x^3 - x^5 + x^6 - \dots = \frac{1+x}{1+x+x^2}$$

by substitution of x=1:

$$1 + 0 - 1 + 1 + 0 - 1 + 1 + 0 + \dots = \frac{2}{3}$$

The way Bossut and Lagrange dealt with the objection is worth describing. It is a good example of a clever ad hoc solution. Bossut and Lagrange argued that Callet's objection does not hold water because one should also consider the zero-coefficients in the expansion of $(1+x)/(1+x+x^2)$ If one does so, one gets

$$1 + 0.x - x^2 + x^3 + 0.x^4 - x^5 + x^6 + \dots = \frac{1+x}{1+x+x^2},$$

which yields after substitution of x=1

$$1 - 1 + 1 - 1 + 1 - 1 + \dots = \frac{2}{3}$$

and the average of the partial sums on the left hand side is indeed $\frac{2}{3}$.
Yet Bossut and Lagrange were not fully satisfied with this ad hoc solution. This is apparent from the fact that they went beyond it and proved the following theorem. The theorem is one of the earliest results from what is nowadays called the theory of summable series. As multiplication of numerator and denominator by (1-x) shows, expressions like 1/(1+x) and $(1+x)/(1+x+x^2)$ are special cases of the more general expression $(a+bx+cx^2+\dots ux^{n-1})/(1-x^n)$. It is also easy to see that division yields

$$\frac{a+bx+cx^2+\dots+ux^{n-1}}{1-x^n} = a+bx+cx^2+\dots+ux^{n-1}+ax^n+bx^{n+1}+\dots\text{etc.}$$

(Nota bene: The right hand side is a geometrical series with first term $a+bx+cx^2+\dots+ux^{n-1}$ and ratio x^n). The *theorem* that Bossut and Lagrange proved says:
If $a + b + c +\dots+ u = 0$ then

$$\lim_{x \to 1} \frac{a+bx+cx^2+\dots ux^{n-1}}{1-x^n} = \frac{na+(n-1)b+(n-2)c+\dots+u}{n}$$

or, in other words: the value for x=1 of (the continuous extension of) $(a+bx+cx^2+\dots ux^{n-1})/(1-x^n)$ equals the value after substitution of x=1 of the

series $a+bx+cx^2+...+ux^{n-1}+ax^n+bx^{n+1}+...$etc. if the infinite sum is considered to be equal to the average of the periodic partial sums.
Proof: Bossut and Lagrange determined the value of the limit by means of substitution of $x=1/z$ followed by an application of De l'Hôpital's criterion.[7] Moreover, since $a+b+c+...+u=0$, the partial sums of the series

$$a+b+c+...+u+a+b+c+...+u+a+b+c... \text{ etc.}$$

can only assume the following n values: a, a+b, a+b+c, ..., a+b+c+...+u whose average is precisely $(na+(n-1)b+(n-2)c+...+1.u)/n$.
As for the substitution of particular values for x in a divergent expansion of a particular function, Bossut and Lagrange were more reserved than Euler. They wrote:

"[the theorem] can only be considered as a purely analytical and curious truth"

and

"because periodical series are not at all convergent the notion of sum or limit does not suit them".[8]

Yet for Bossut and Lagrange too, the series into which a function can be developed is *representative* of that function, whether it converges or not.
The example of divergent series nicely illustrates how eighteenth century mathematicians attempted to save their results in an ad hoc manner. Below I will now first turn to the characterization of the dominant research tradition in nineteenth century analysis and then I will consider the transition from the eighteenth to the nineteenth century tradition.

VIII.4 The 19th century Conceptual Tradition in analysis

VIII.4.1 The object of nineteenth century analysis

Nineteenth century analysts also studied *functions*, but for them the notion "function" usually had a different meaning than for eighteenth century mathematicians. For example, in the Introduction of his *Cours d'analyse*, Cauchy condemned the tendency to attribute

"to the algebraic formulae an undetermined extension, while, in reality, the

---

[7]This substitution is unnecessary. Immediate application of de L'Hôpital's criterion to the quotient in x also yields the desired result.

[8]"Comme les séries périodiques ne sont point convergentes, l'idée de somme, ni même de limite, ne peut leur convenir, et le théorème de Bernoulli ne peut être regardé que comme une vérité purement analytique et curieuse" ([Bossut&Lagrange,1799], p.10).

majority of these formulae are only valid under certain conditions, and for certain values of the quantities that they include".[9]

He added that as a result of his work it is clear that

"the different formulae are nothing more than relations between real quantities, relations that one can always easily verify".[10]

Cauchy defined a function as follows:

"If variable quantities are connected with each other in such a way that, when the value of one of them is given, the values of all the others can be determined, then one usually considers these different quantities as being expressed in terms of one of them which takes the name of *independent variable*, and the other quantities that are expressed in terms of the independent variable are the so-called *functions* of this variable".[11]

For Cauchy a function of one variable is a formula expressing the value of a dependent variable in terms of an independent variable. In the course of the nineteenth century the analytical expression of the function by means of a formula would come to be seen as secondary, and a function would become an *relation* between variables. Historians of mathematics have repeatedly emphasized this. I will illustrate it with one quotation. Boyer writes about the nineteenth century:

"The formalistic view that a function was a simple analytic expression gave way to the understanding that it was any relationship between variables" ([Boyer, 1959], p.277).

Fourier's *Théorie analytique de la chaleur* from 1822 shows the transition in a different way:

"In general the function f(x) represents a sequence of values or ordinates, each of which is arbitrary".[12]

---

[9]"aux formules algébriques une étendue indéfinie, tandis que, dans la réalité, la plupart de ces formules subsistent uniquement sous certaines conditions, et pour certaines valeurs des quantités qu'elles renferment" ([Cauchy,1821], p. iii).

[10]"les différentes formules ne présentent plus que des relations entre les quantités réelles, relations qu'il est toujours facile de vérifier" ([Cauchy,1821],p. iv).

[11]"Lorsque des quantités variables sont tellement lieés entre elles que, la valeur de l'une d'elles étant donnée, on puisse en conclure les valeurs de toutes les autres, on conçoit d'ordinaire ces diverses quantités exprimées au moyen de l'une d'entre elles, qui prend alors le nom de *variable indépendente*; et les autres quantités exprimées au moyen de la variable indépendante sont ce qu'on appelle des *fonctions* de cette variable" ([Cauchy,1821], p.31).

[12]"En général, la fonction f(x) représente une suite de valeurs ou ordonnées, dont chacune est arbitraire" ([Fourier,1822, p.500).

VIII.4.2 The methods of nineteenth century analysis

As far as nineteenth century methodology is concerned I will be brief. *Compared to eighteenth century analysis, nineteenth century analysis is primarily deductive.* Grabiner wrote:

> "Nineteenth century analysts, beginning with Cauchy and Bolzano, gave rigorous, inequality-based treatments of limit, convergence and continuity, and demanded rigorous proofs of the theorems about these concepts. We know what these proofs were like; we still use them" ([Grabiner,1974] p.358).

I will give some illustrations from Cauchy's work. In the introduction to the *Cours d'analyse* Cauchy rejects eighteenth century methodology. He wrote:

> "As for the methods, I have tried to give them all the rigor that one demands in geometry, in such a way that one never has to fall back upon reasons drawn from the generality of algebra. Such reasons, although generally admitted, particularly in moving from convergent series to divergent series and from real quantities to imaginary expressions, can only be considered, it seems to me, as inductions that are suitable for giving now and then a presentiment of the truth, but have little in common with the exactness so much praised in the mathematical sciences".[13]

The central concept in Cauchy's calculus is the limit concept. Cauchy defined the limit concept as follows:

> "When the successively attributed values of one variable approach indefinitely a fixed value by as little as desired, that fixed value is called the *limit* of all the others".[14]

Grabiner has rightly pointed out that Cauchy's successful rigorization of the calculus was due to his realizing two facts. Although the definitions employed by Cauchy also occur in the earlier literature, he first of all realized that fundamental notions, such as the convergence of a series, continuity, the

---

[13]"Quant aux méthodes, j'ai cherché à leur donner toute la rigueur qu'on exige en géométrie, de manière à ne jamais recourir aux raisons tirées de la généralité de l'algèbre. Les raisons de cette espèce, quoique assez communément admises, sur-tout dans le passage de séries convergentes aux séries divergentes, et des quantités réelles aux expressions imaginaires, ne peuvent être considérées, ce me semble, que comme des inductions propres à faire pressentir la vérité, mais qui s'accordent peu avec l'exactitude si vantée des sciences mathématiques" ([Cauchy,1821], pp. ii-iii).

[14]"Lorsque les valeurs successivement attribuées à une même variable s'approchent indéfiniment d'une valeur fixe, de manière à finir par en différer aussi peu que l'on voudra, cette dernière est appelée la *limite* de toutes les autres" ([Cauchy,1821], p.19).

differential quotient and the integral, could all be defined by means of the limit concept. Secondly, Cauchy realized that the limit concept could be handled by means of inequalities.[15] I shall illustrate this by means of a simple example from the sixth chapter of the *Cours d'analyse*. There Cauchy derives several criteria for the convergence of series. For example, Cauchy proves the well known theorem that a series $u_0+u_1+u_2+\ldots$ $(u_i>0)$ is convergent if

$$\lim \sqrt[n]{u_n} = k < 1$$

His proof runs as follows. There exist U and N such that for n >N:

$$\sqrt[n]{u_n} < U < 1$$

So that for all $n > N$: $u_n < U^n$ with $U < 1$. (Here Cauchy interprets the limit definition in terms of inequalities). Then

$$u_n+u_{n+1}+u_{n+2}+\ldots\ldots u_{n+p} < U^n+U^{n+1}+U^{n+2}+\ldots\ldots U^{n+p}$$

The geometric series converges, and this means that the first expression can be made arbitrarily small for all p by taking n large enough. (Here Cauchy handles convergence of series in terms of inequalities.) In fact Cauchy applies the so-called *Cauchy-criterion* for the convergence of a series $u_1+u_2+u_3+u_4+\ldots$: The condition - that for increasing values of n all sums

$$|u_n+u_{n+1}|,\ |u_n+u_{n+1}+u_{n+2}|,\ |u_n+u_{n+1}+u_{n+2}+u_{n+3}|, \text{ etc.}$$

ultimately assume values less than any assignable limit[16] - is not only necessary for convergence but is also a *sufficient condition* for convergence ([Cauchy,1821], pp. 115-116). Cauchy does not prove this. He must have found it obvious: if the condition holds, one can calculate the "sum" as accurately as one wishes. In the background is the deep assumption of the completeness of the real numbers.

I wrote above that the nineteenth century tradition in analysis was primarily

---

[15]As we have seen in Chapter III, Cauchy attempted to avoid the extensive use of inequalities by means of the use of infinitesimals defined in terms of variables converging to zero.

[16]Cauchy basically says: For all $\varepsilon > 0$ there exists an N such that for $n > N$ we have for all p: $|u_n +\ldots u_{n+p}| < \varepsilon$.

deductive. This does not, however, mean that the building of analysis was erected on the basis of axioms and definitions that were all clear from the start. In the course of time critical analysis of Cauchy's proofs revealed hidden assumptions. An example is the assumption of the completeness of the real numbers in the case of the Cauchy criterion for the convergence of a series. There is in this respect considerable similarity between the Greek deductive tradition in geometry and the nineteenth century tradition in analysis[17]: they are examples of traditions that nicely show how definitions and axioms are sometimes proof-generated in the sense of Lakatos. I will illustrate this as regards the nineteenth century tradition in analysis by means of Cauchy's definition of continuity.

VIII.4.3 Cauchy and continuity

As we have seen in Chapter III, Cauchy defined the notion of continuity in the *Cours d'analyse* as follows. A function f(x) is continuous on an interval if for all x the numerical value of the difference $f(x+\alpha)-f(x)$ decreases indefinitely with the value of α. Where does this definition of continuity come from? My answer is that it is a proof-generated notion in the sense of Lakatos.

In an appendix to the *Cours d'Analyse*, Cauchy deals with methods to approximate roots of equations ([Cauchy,1821], pp.378-380). Cauchy takes a real continuous function f(x) on an interval $x_0 < x < X$, which is such that $f(x_0)$ and f(X) have opposite signs.[18] Cauchy proves the existence of a root as follows. He divides the interval in equal parts and, going from left to right, the first sub-interval $x_1 < x < X^1$ is determined (in thought) for which $f(x_1)$ and $f(X^1)$ again have opposite sign. With this sub-interval the process is repeated. It is divided in equal parts etc. All $f(x_i)$ have the same sign as do all $f(X^i)$. In this way one obtains sequences x_i (2) and X^i (3) with

$$x_0 < x_1 < x_2 < \ldots\ldots\ldots < x_n < X^n < X^{n-1} < \ldots\ldots\ldots < X^3 < X^2 < X^1 < X$$

while $X^i-x_i<(1/m)^i(X-x_0)$. Going from left to right guarantees that all $f(X^i)$ have the same sign as do all $f(x_i)$, but that the two signs are opposite.[19] Because the terms of the decreasing sequence (3) "will finish by differing as little

[17]These two traditions have been compared frequently. Struik: "What Eudoxus had done in the period after the fall of Athenian democracy, Cauchy and his meticulous contemporaries began to accomplish in the period of expanding industrialism"([Struik, 1967], p.149).

[18]I use Cauchy's notation.

[19]Cauchy does not mention the trivial case that for some i: $f(x_i) = 0$ or $f(X^i) = 0$.

as one would want" from the terms of the increasing sequence, one can conclude that "the general terms of the series (2) and (3) will converge to a common limit". Cauchy calls this limit a. Cauchy's definition of continuity implies that $f(a+(x_i-a))-f(a)$ converges to zero because (x_i-a) converges to zero and also that $f(a+(X^i-a))-f(a)$ converges to zero because (X^i-a) converges to zero. Therefore

$$\lim_{x_i \to a} f(x_i) = f(a) \quad \text{and} \quad \lim_{X^i \to a} f(X^i) = f(a)$$

Because all $f(x_i)$ have the same sign and all $f(X^i)$ the same but opposite sign, it is obvious that f(a) must be 0.

It is quite possible that Cauchy's definition of continuity was born when he found this proof. Cauchy's point of departure was the method well-known in his time of finding a good approximation to a root by means of repeated subdivisions of an interval. I suggest that Cauchy applied that method and wondered what property of the function f guaranteed the existence of the root. The required property appeared to be that for all sequences $\{x_i\}$ with lim $x_i=a$, we must have lim $f(x_i)=f(a)$. A modern definition of continuity was born.[20] If this is what happened, Cauchy's definition of continuity was proof-generated. But even if it did not happen in precisely this way, because of the relatively sophisticated nature of the definition it is highly probable that the definition was proof-generated in another context. In that case the above proof shows how the definition was "proof-confirmed".

VIII.5 The transition from the Formalist Tradition to the Conceptual Tradition

VIII.5.1 Cauchy's rejection of the generality of algebra

The fact that the problem of the foundation of the differential and integral calculus could not be solved satisfactorily in the course of the eighteenth century undoubtedly played a role in the shift from eighteenth century to the nineteenth century analysis. Alongside the problem of the foundations of the differential and integral calculus, however, several other factors are involved in the transition from the eighteenth to the nineteenth century tradition.

---

[20]In 1945 Daval and Guilbaud wrote, after describing Cauchy's proof: "En possession d'un procédé effectif de résolution des équations nous avons voulu en juger la validité. La réponse a justement consisté a créer la notion [i.e. the notion of continuity - T.K.] ad hoc: à définir une propriété de la fonction f qui assure le succès du procédé" ([Daval&Guilbaud, 1945],p.117).

Fourier's work, for example, also contributed to the breakdown of the eighteenth century concept of function. The fact that *one* analytical expression like

$$y = \cos x - \frac{1}{3}\cos 3x + \frac{1}{5}\cos 5x - \ldots + \frac{1}{2m-1}\cos(2m-1)x + \ldots$$

is (depending on the value of x) at the same time equal to *three different* analytical expressions $\pi/4$, 0,$-\pi/4$ immediately shows the inconsistency of the notion of a function as an analytical expression.

Also remarkable is the occurrence of certain paradoxes in actual mathematical practice involving divergent series. These paradoxes appeared when attempts were made to develop the functions $(\sin x)^n$ and $(\cos x)^n$ into series. I shall illustrate the problems by means of what Lacroix wrote in his *Traité du calcul différential et du calcul intégral.* In the first volume of this work Lacroix reproduces a derivation given earlier by Lagrange ([Lacroix,1810], pp.274-275). He shows by means of differentiation that $y=(\cos x)^n$ (n arbitrary real number) satisfies the differential equation

$$(7) \qquad \cos x \frac{dy}{dx} + (n \sin x)\, y = 0$$

He then substitutes the following series into the differential equation

$$(8) \qquad y = A \cos mx + B \cos(m-1)x + C \cos(m-2)x + D\cos(m-3)x + \ldots\ldots$$

and he obtains a series of sines

$$(9) \qquad C_1 \sin(m+1)x + C_0 \sin mx + C_{-1} \sin(m-1)x + C_{-2} \sin(m-2)x + \ldots = 0$$

where the C_i are expressions in n, m, A, B, C, etc. By putting $C_i=0$ for all i, Lacroix derives

$$(10) \qquad (\cos x)^n = A\{\cos nx + \begin{bmatrix} n \\ 1 \end{bmatrix} \cos(n-2)x + \begin{bmatrix} n \\ 2 \end{bmatrix} \cos(n-4)x + \begin{bmatrix} n \\ 3 \end{bmatrix} \cos(n-6)x + \ldots\}$$

Putting x=0 in (10) then yields $A=2^{-n}$, a result which Lacroix at that time did not consider as problematic. However, at the end of the third volume of his textbook he added some "Correction et Additions". There he wrote about some difficulties that had arisen ([Lacroix,1819], pp. 616 - 623). By substituting a series of sines into the differential equation instead of a series of cosines, Lacroix can also derive in a similar way

$$(\cos x)^n = A'\{\sin nx + \begin{bmatrix} n \\ 1 \end{bmatrix} \sin (n-2)x + \begin{bmatrix} n \\ 2 \end{bmatrix} \sin (n-4)x + \begin{bmatrix} n \\ 3 \end{bmatrix} \sin (n-6)x + \dots\} \tag{11}$$

Substitution of x=0 yields: $A' = 1/0 = \infty$. Because the series of sines in (11) is not identical to zero for all values of n, Lacroix now runs into problems. An acceptable argument, at least from eighteenth century point of view, leads to a contradiction! I shall not discuss these problems any further here.[21] I will restrict myself to the remark that Lacroix does not succeed in solving them. He sees no solution

> "unless one says that the series [(10) and (11)] do not satisfy equation [(7)] when n is a broken number".[22]

Because

> "No matter at which term one truncates these series, they introduce into the equation [i.e. the differential equation - T.K.] a final term which is not destroyed by any other term, but which disappears because of its coefficient, when n is an integer, in which case the equation is completely satisfied".[23]

Lacroix toys with the idea that the convergence of the series involved might be important and he even writes:

> "If this consideration is well founded, it is not permitted to use other series [...] than those whose form is such that they are convergent in at least one case[viz. for certain values of the variable - T.K.]".[24]

It is clear that Lacroix is very close to the conclusion: "one should not calcu-

---

[21]For an extensive discussion I refer to [Jahnke,1987].

[22]"à moins de dire que les séries [...] ne vérifient pas l'équation [...] quand n est un nombre fractionnaire" ([Lacroix,1819], p.618).

[23]"Quel que soit le terme où l'on arrête ces séries, elles introduisent dans léquation un dernier terme qui n'est détruit par aucun autre mais qui disparaît par son coefficient, lorsque n est un nombre entier, auquel cas l'équation [...] est complètement vérifiée" ([Lacroix, 1819], pp. 618-619).

[24]"Si cette réflexion était bien fondée, il ne serais permis d'employer, même comme développemens, que des séries dont la forme serait telle qu'elles eussent été convergentes au moins dans un cas" ([Lacroix,1819], p.619).

late with divergent series" (Cf. [Burkhardt,1911], p.177). Yet, unlike Cauchy and Abel a few years later, he does not draw this conclusion.
The paradoxes described were considered to be important and several authors published papers on them in the period 1811-1835 (Cf. [Burkhardt, 1911], p.177 and [Jahnke,1987]). It is tempting to assume with Burkhardt that these discussions influenced Cauchy on the subject of divergent series.

VIII.5.2 The competition between the two research traditions.
Although problems with divergent series within the formalist tradition may have influenced Cauchy, the major factor bringing about the transition from the old to the new tradition, from the point of view of the methodology of mathematical research traditions, must have been the degree of progress that Cauchy's foundational research project exhibited. Let us briefly consider the way the transition took place.
In France Cauchy was very influential, and after the publication of his textbooks the view that the use of divergent series should be rejected seems to have been generally accepted ([Burkhardt,1911], p.183). Yet not everybody stopped using divergent series. Poisson, for example, explicitly rejected the use of divergent series but continued using them ([Burkhardt,1911], pp.178-179).
In Germany and England divergent series were not so suddenly rejected. This is understandable. The work of Cauchy in particular offered a very promising new foundation of analysis, but the dogmatic rejection of the use of divergent series was not founded upon conclusive evidence in every aspect. Examples of arguments using divergent series which led to false results could also often be explained in a different way ([Burkhardt,1911], p.178). An illustration of this is De Morgan's point of view in the 1840's. De Morgan hesitated. In the 1842 edition of his textbook on differential and integral calculus, he writes:

> "The history of algebra shows us that nothing is more unsound than the rejection of any method which naturally arises, on account of one or more apparently valid cases in which such method leads to erroneous results. Such cases should indeed teach caution, but not rejection: if the latter had been preferred to the former, negative quantities, and still more their square roots, would have been an effectual bar to the progress of algebra, which would have been confined to the universal arithmetic of which Newton wished it to bear the name: and those immense fields of analysis over which even the rejectors of divergent series now range without fear,

would have been not so much as discovered, much less cultivated and settled" ([De Morgan,1842], p.566).

De Morgan was opposed to a dogmatic rejection of divergent series. Yet he advocates caution in dealing with them. In an 1844 paper De Morgan writes:

"We must admit that many series are such as we cannot at present safely use, except as means of discovery, the results of which are to be subsequently verified: and the most determined rejector of all divergent series doubtless makes this use of them in his closet" ([De Morgan,1844], p.183).

He wrote that integration is not an algebraic operation and that it should not be applied to divergent series without precautions. Yet in the same paper he does not exclude the possibility that the use of divergent series should be abolished altogether:

"Let 1+2+4+... be shown to be any thing but a root of either 1+2z=z, or of another equation which has degenerated into 1+2z=z; that is, let it come out anything but -1 or ∞, and as a result of any process which does not involve *integration performed on a divergent series* - and I shall then be obliged to confess that divergent series must be abandoned, or rather, that the generalizations frequently made on the subject must be curtailed" ([De Morgan,1844], p.187).

One can roughly say that after the appearance of Cauchy's textbooks the mathematical community split up into two groups concerning divergent series. On the one hand there were those who agreed with Cauchy and who, impressed by the apparent possibilities of the alternative approach, dogmatically rejected divergent series. On the other hand there were those who refused to reject divergent series because their use had been, and to a certain extent continued to be, fertile. Burkhardt has carefully described the different points of view ([Burkhardt,1911]). However, the new approach continued to be successful and the defenders of divergent series either did not succeed in defining a clear borderline between acceptable and unacceptable use of such series, or else their research projects progressed less than did Cauchy's. A representative of a foundational project competing with Cauchy's was the German mathematician Ohm. Ohm's work has been studied by Jahnke, who describes Ohm as a mathematician who broke with the 18th century in the sense that he attempted to distinguish clearly between heuristics and justification. Yet, Jahnke writes:

"Ohm turned out to be an author who broke with the style of the eighteenth century in his views and who went along with the nineteenth century inclinations to theorize in mathematics, although to a large extent he used the means of the eighteenth century as far as the 'material side' of mathematics is concerned".[25]

Ohm attempted to solve the foundational problems with 18th century means. In other words, Ohm's project represented an attempt to save the 18th century tradition.

Jahnke pays special attention to the role of the paradoxes that we described above. He shows that the paradoxes were not only solved in 1826 by Abel within the framework of Cauchy's foundational approach, but also by Ohm in 1823 on the basis of a much more formalist approach.[26] Yet the discussion concerning these solutions did not have much effect upon the competition between the projects. Jahnke writes:

"As to the triumph of the positions of Cauchy and Abel it [this discussion] no longer played any role; the triumph was rather brought about by the fact that the problems of Fourier series and the phenomena of divergence, absolute and conditional convergence of number series were increasingly better comprehended, that is to say, because Cauchy's approach turned out to be feasible and in some areas superior".[27]

In terms of the MMRT, Cauchy's approach represented a research project that progressed more than its competitors.

In the second half of the nineteenth century the mathematical community as a whole had rejected the use of divergent series and had switched to a new research tradition. In retrospect one can say that the mathematical community of Europe as a whole behaved quite rationally in doing so, in the sense of the methodology of mathematical research traditions. At the individual level accidental factors undoubtedly played a role. Gradually the mathematical community of Europe as a whole went over to the new tradition. This hap-

---

[25]"Ohm erwies sich[...] als ein Autor, der in seinen Auffassungen mit dem Stil des 18. Jahrhunderts brach und sich in die Theoretisierungstendenzen der Mathematik des 19. Jahrhunderts einordnete, wenn er auch, was die Mathematik nach ihrer 'materiellen Seite' betraf, sich weitgehend der Mittel des 18. Jahrhunderts bediente" ([Jahnke,1987], p.136).

[26]In 1825 Poinsot also came up with the correct solution of the paradoxes ([Jahnke,1987], p.117).

[27]"Für die Durchsetzung der Positionen von Cauchy und Abel spielte sie keine Rolle mehr, vielmehr geschah dies, indem zunehmend die Probleme der Fourierreihen und die Phänomene der Divergenz, absoluten und bedingten Konvergenz numerischer Reihen verstanden wurden, d.h. indem sich Cauchy's Ansatz als machbar und in einigen Bereichen als überlegen erwies"([Jahnke,1987], p.142).

pened in accordance with the gradual way in which the new tradition showed its superiority in the form of highly progressive research projects.[28] The fact that Cauchy's notion of function led to an enormous generalization, creating important new possibilities particularly with respect to trigonometric series for example, played a crucial role. Cauchy must already have been aware of this. Jahnke writes:

> "[...] the development of analysis immediately after Cauchy, which was already very well understood by Abel, and which culminated in Dirichlet's well known 1837 paper, shows that people were aware of the mathematically decisive point: to find a conceptual framework for the treatment of non-real analytic functions and, in particular, of trigonometric series".[29]

It is worth mentioning in this context the phenomenon that several decennia after the eighteenth century tradition had disappeared from sight, the nineteenth century tradition - in its Weierstrassian form - proved its superiority because it appeared to be able to absorb eighteenth century results on divergent series that had disappeared temporarily from established mathematics. In 1880 G. Frobenius launched the renaissance of the theory of divergent series within the context of the new research tradition with the following theorem:

$$\text{If } S_n = \sum_{i=0}^{n} a_i \quad \text{and} \quad \lim_{n \to \infty} \frac{s_0 + s_1 + s_2 + \ldots + s_n}{n} = M$$

---

[28]There is more to the complete story of the reception of Cauchy's foundational work. Other elements played a role as well. For example, Jahnke also distinguishes an "elementarizing reception" in Germany. Jahnke: "In der Rezeption durch Grunert und Schlömilch wurde Cauchys Ansatz also ins *Elementare* umgebogen und auf den Aspekt, eine Methode des sicheren Rechnens zu sein, reduziert" ([Jahnke,1987], p. 170). Schlömilch's "Algebraische Analysis" from the year 1845 was one of the first textbooks that popularized Cauchy's ideas in Germany. Grunert and Schlömilch, when comparing the 18th century ways with Cauchy's approach very much emphasized the rigor of the latter approach. Jahnke connects their appreciation of Cauchy's rigor with "dem sich herausbildenden mathematischen Selbstverständnis der Gymnasien"([Jahnke,1987], p.171). Some apparently admired the rigor. Others like De Morgan, however, refused to sacrifice heuristically interesting methods for lifeless rigor in the presentation (Cf. also [Richards,1987]).

[29]"[...] die Entwicklung der Analysis unmittelbar nach Cauchy, die von Abel bereits sehr gut gesehen wurde und die in Dirichlets bekannter Arbeit von 1837 gipfelte, zeigt, daß man sich über den mathematisch entscheidenden Punkt, einen konzeptionellen Rahmen für die Behandlung nicht-analytischer reeller Funktionen und speziel der trigonometrischen Reihen zu finden, wohl bewußt war" ([Jahnke,1987], p.154).

then $\sum_{n=0}^{\infty} a_n x^n$ converges for $|x| < 1$ and $\lim_{x \to 1} \sum_{n=0}^{\infty} a_n x^n = M$

([Frobenius,1880], p.8).
The Bossut-Lagrange theorem that we discussed above can be proved easily by means of this theorem. Frobenius was followed by Cesaro, Borel, Hardy and others who turned the theory of divergent series into a respectable branch of mathematics ([Tucciarone,1973]).
In terms of the methodology of scientific research traditions, between 1870 and 1880 the nineteenth century research tradition in analysis had progressed (in the sense of the MMRT) so much that it was far superior to the formalist tradition. Yet a considerable number of results concerning divergent series, buried, for example, in Euler's work, still represented an advantage for the formalist tradition that the nineteenth tradition had not matched. However, the theory of divergent series developed by Frobenius and others also cancelled that advantage of the formalist tradition.

VIII.6 Conclusion

On the basis of the existing historiography the transition from eighteenth century analysis to nineteenth century analysis can be satisfactorily described by means of the MMRT in the following sense. Eighteenth century analysis was dominated by the Formalist Tradition. The fundamental domain consisted of formulae. Deduction (for example in the sense of formal derivation) played an important role. But along with it there was a considerable hypothetical element in the tradition. Contradictions were solved by means of ad hoc adjustments in the tradition. In the nineteenth century the Formalist Tradition was replaced by the Conceptual Tradition. The fundamental domain of the Conceptual Tradition consisted of functions, in the sense of relations between variables. The Conceptual Tradition was primarily deductive. As regards the transition from the Formalist Tradition to the Conceptual Tradition my discussion has been rather limited. I have restricted myself to some remarks on Cauchy's foundational project that indicate that the transition took place in accordance with the MMRT.

Chapter IX The history of the interchangeability theorem for partial differentiation from Nicolaus I. Bernoulli to Schwarz[1]

IX.1 Introduction[2]

This main purpose of this chapter is to provide an illustration of how the eighteenth century Formalist Tradition in analysis developed internally, how it was replaced by the nineteenth century Conceptual Tradition in analysis, and how that tradition developed internally. The account shows that mathematical research traditions are not rigid and that they may be subject to modifications at certain stages of their existence. The sequence of different versions of the theorem that I will now discuss supports and supplements the reconstruction of Chapter VIII.

In his paper, "About a complete system of mutually independent conditions to prove the theorem $\frac{\partial}{\partial y}(\frac{\partial f(x,y)}{\partial x}) = \frac{\partial}{\partial x}(\frac{\partial f(x,y)}{\partial y})$" ([Schwarz, 1873]), Schwarz proved the following theorem:

For a continuous function f, defined on a continuous, two-dimensional manifold (stetige, zweifach ausgedehnte Mannigfaltigkeit) for which f_x, f_y and f_{xy} exist[3] holds

$$f_{xy}(x,y) = f_{yx}(x,y)$$

provided that

i) f_x is continuous on the manifold, ii) f_y is continuous on the manifold, and iii) f_{xy} is continuous on the manifold.

This theorem is one of a long series of versions of the interchangeability theorem for partial differentiation.[4] The first version of the theorem appeared in 1719 in the work of Nicolaus I. Bernoulli. In this chapter I will concentrate on the history of the interchangeability theorem for partial differentiation from the time of Bernoulli to that of Schwarz. In the previous

---

[1] I am grateful to Steven Engelsman for drawing my attention to the history of the interchangeability of partial differentiation several years ago.

[2] This chapter is a modification and extension of [Koetsier,1990].

[3] "f_{xy}" denotes a differentiation with respect to x followed by a differentiation with respect to y.

[4] In the past the theorem has enjoyed limited attention by historiographers. In 1867 Lindelöf critically discussed some of the earlier versions ([Lindelöf,1867]) and in 1940 Higgins listed the versions of the theorem that were known to him ([Higgins,1940]). Finally, Engelsman in 1984 payed considerable attention to the versions of the theorem given by Nicolaus I. Bernoulli and Euler ([Engelsman,1984]).

two chapters, I reconstructed two developments in mathematics on the basis of the methodology of mathematical research traditions. Those reconstructions involved rather global descriptions of the traditions concerned. In the previous chapters, I mainly used secondary sources. In this chapter I will base myself almost exclusively upon primary sources.

IX.2 The interchangeability theorem in the eighteenth century

IX.2.1 Nicolaus I. Bernoulli

The first version of the theorem appeared in a manuscript dating from 1719 by Nicolaus I. Bernoulli[5] in the form of a simple rule of the differential calculus and concerned infinitely small geometrical quantities. In 1719 Bernoulli studied families of curves, dealing in particular with orthogonal trajectories. Bernoulli describes variation within a family of curves (Figure 9.1) by means of the parameter a.

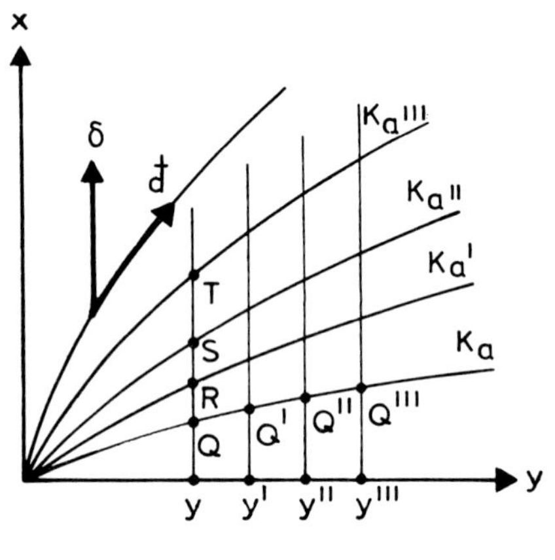

Figure 9.1

To a fixed value of a corresponds a fixed member K_a of the family. For an

[5]The manuscript was first published in 1843 by Fuß ([Engelsman, 1984], p.96) and recently in 1984 by Engelsman together with a translation into English and a thorough analysis of the text ([Engelsman,1984]). My references are to Engelsman's translation.

ordinary differentiation "along the curve K_a" Bernoulli uses the symbol $\mathfrak{d}$. Such a differentiation is based upon the idea of a "progression" of points that are infinitely close to each other.[6] Consider, for example, such a progression of points on the y-axis: y, y', y", y"', y"", etc. (Figure 9.1; N. B. In contrast with our usage Bernoulli draws the y-axis horizontally and the x-axis vertically.) This progression via straight lines parallel to the y-axis induces a progression of points on the curve K_a : Q, Q', Q", etc. In the points of that progression all variables corresponding to the curve assume progressions of values. The first order differentials are by definition the differences of successive values in these latter progressions. In other words, the differentials are the increases of value of the variables corresponding to the curve when one moves with infinitely small steps along the curve. Because differentials are differences, one can apply the normal rules of algebra to them. On the other hand differentials are infinitely small, which means that they can sometimes be omitted. Leibniz's derivation of the product rule for ordinary differentiation illustrates both aspects:

> "d(xy) is the same as the difference between two adjacent xy, of which let one be xy, the other (x+dx)(y+dy). Then d(xy)=(x+dx).(y+dy) - xy or xdy+ydx+dxdy, and this will be equal to xdy+ydx if the quantity dxdy is omitted, which is infinitely small with respect to the remaining quantities, because dx and dy are supposed infinitely small" (quoted in [Bos,1974], p.16).

Partial differentiation can only show up in problem situations with more than one dimension of variability. As soon as one studies a family of curves one has a situation with *two* dimensions of variability. If by *fixing the value of the parameter a* in a family of curves one considers one particular curve, one can differentiate with fixed a "along the curve". Given a family of curves, one can, however, also *fix one particular value of the variable y* and vary the parameter a. Bernoulli uses the symbol δ for this differentiation "from curve to curve" (in contrast with the ordinary differentiation "along a curve"). This differentiation corresponds to a new progression of points (independent of the progressions corresponding to variation along the curves): e.g. the progression Q, R, S, T, etc. (Figure 9.1; N.B. The distances between successive points are infinitely small) and this progression induces also progressions of values for all variables corresponding to these points. The differentials corresponding to this differentiation "from curve to curve" are

---

6 [Bos,1974]

again by definition the differences of the successive values in these progressions.

The first occurrence of the interchangeability property[7] in Bernoulli's work takes the form

đδx = δđx.

In order to prove it Bernoulli considered a mesh of the network of infinitely close lines determined by the two progressions T and P (Figure 9.2).

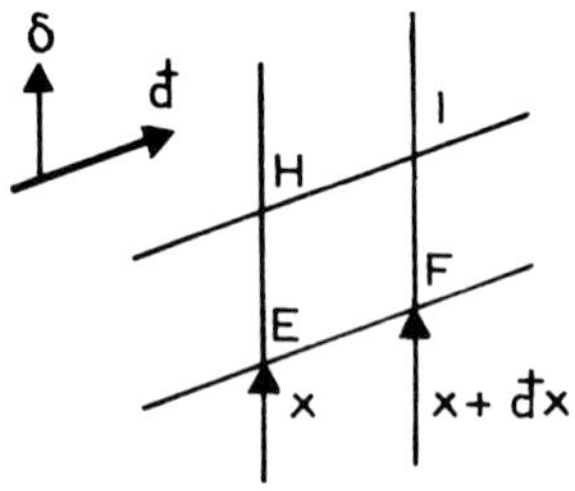

Figure 9.2
(N.B. Bernoulli has a horizontal y-axis and a vertical x-axis.)

Bernoulli wrote:

IF = HE + đHE =δx + đδx

and

---

[7] I will attempt to give a brief indication of the context in which this property appears. Besides đs and δs for an arbitrary variable, Bernoulli also uses the notation ds for a differentiation along an arbitrary curve in the plane. Whenever we move from one point in the infinitely close network of family members and vertical lines to an infinitely close neighbour, the increments of x, y and a are related by the following equation dx=pdy+qda. (=đx+δx). This differential equation determines the family.

Along a family member, i.e. for fixed a, we have dx=pdy. The latter relation implies that we will have along an orthogonal trajectory of the family dy=-pdx. Substitution of dx=pdy+qda yields dy/da=-pq/(1+p^2). In order to determine the orthogonal trajectories from this differential equation, Bernoulli needs q. One of Bernoulli's problems is that sometimes p is given while q is not. In order to determine q from p, Bernoulli uses the interchangeability property as follows. Because δx = qda and đx = pdy, the interchangeability property đδx = δđx implies that đqda = δpdy, and thus we have đq = (δp/da)dy. Assuming p to be independent of x, q can be found by integration (Remember that δ corresponds to fixed y and đ to fixed a) ([Engelsman,1984], pp. 105-107).

$$IF = \delta(x + dx) = \delta x + \delta dx$$

which implies $d\delta x = \delta dx$ ([Engelsman,1984], p.202).

It is clear from the text of the manuscript that Bernoulli knew that the interchangeability property does not only hold for x, but that we also have

$$d\delta s = \delta ds$$

for arbitrary variables.[8]

More than twenty years later Bernoulli was to write that he had considered the 1719 interchangeability property as an axiom "obvious to anybody from the mere notion of differentials, even without inspection of a diagram" (quoted in [Engelsman,1984], p. 106). It is remarkable that the interchangeability theorem developed out of an unconditionally valid triviality.

IX.2.2 Euler and the transition to the Formalist Transition

At the time of Nicolaus I. Bernoulli, the differential calculus was primarily a powerful method to study geometrical problems concerning curves. The version of the interchangeability theorem that Bernoulli gives can be considered as a special case of a more general theorem without any necessary connections with geometry. In a manuscript written about 1730, Euler came up with this more general theorem. This manuscript is one of the earliest sources I am aware of in which the formalist point of view with respect to the differential calculus is clearly present.

The manuscript is called "De differentiatione functionum duas pluresve variabiles quantitates involventium" (About the differentiation of functions of two or more variables).[9] Although Euler developed a theory to solve problems concerning families of curves, this theory is not essentially connected with geometry. It is obvious from this text that already in 1730 it must have been Euler's opinion that analysis, independently of geometry, deals with analytical expressions.

In Euler's text the interchangeability property is called a *principle* and is phrased as follows:

> "When a quantity composed of two variables is differentiated twice, where in the first differentiation one variable, and in the second differentiation the

---

[8]The same proof as for the special case can be used if one replaces x everywhere by s and one reads IF and HE not as x_I-x_F and x_H-x_E respectively but as s_I-s_F and s_H-s_E respectively.

[9]The manuscript was published and translated into English by Engelsman. The translations below are all Engelsman's ([Engelsman,1984], pp. 204-222).

> other variable is treated as a constant - as can be done in two ways - then the results in both cases will be equal" ([Engelsman, 1984], p.214).

According to Euler the correctness of the principle is obvious when one considers a few examples. Euler takes the quantity $x^3y-y^2x^2$ and differentiates it first with fixed x. That yields x^3dy-2x^2ydy. This expression he differentiates with fixed y and he gets $3x^2dxdy-4xydxdy$. (N.B. Euler takes dxdy=dydx for granted because the rules of algebra apply to differentials). Then he differentiates the original quantity $x^3y-y^2x^2$ with fixed y, which yields $3x^2ydx-2y^2xdx$, followed by a differentiation with fixed x, which also yields the same expression $3x^2dxdy-4yxdxdy$. Euler writes:

> "One who wishes to apply these operations in several examples will not only always notice this agreement, but he will also understand that the truth of this principle must necessarily hold in all cases. But for the sake of those who are less able to understand this necessity, the following geometrical demonstration can be produced" ([Engelsman,1984], p.214).

The geometrical demonstration mentioned by Euler uses the fact that any quantity q=q(x,y) that is being differentiated can be interpreted geometrically as a family of curves (e.g. for fixed $y=y_0$, $q=q(x,y_0)$ is then the equation of a curve in the family). I will not further discuss this proof here. The geometrical proof, in Euler's words, is "drawn from an alien source" and he proceeds to give a second proof "from the nature of such differentiation itself" ([Engelsman,1984], p.215). Here Euler obviously meant that the interchangeability principle is an analytical and not a geometrical principle which should preferably be proved in an analytical way. Euler's analytical proof runs as follows.

> "Let P again be a quantity which is made up from a, x and constants; by substituting x + dx in the place of x it will turn into Q. And if a+da is written instead of a, P changes into R. But if both x+dx is substituted for x and a+da for a, then S will emerge instead of P. From this it is clear that if in Q a+da is substituted for a then S will appear, and in the same way one will also have S if in R x+dx is substituted for x.
>
> Now differentiating P while keeping a constant will yield Q-P. Furthermore, taking a to be variable and x to be constant, the differential of this Q-P will be S-Q-R+P, since Q transforms into S and P into R. And the very same will appear if x is kept constant in the first differentiation and a in the second; from the first one will have R-P, and from the second S-R-Q+P. The identity of this expression and the former one can be most clearly perceived" ([Engelsman,1984], pp.215-216).

Phrased in a slightly more modern fashion, Euler's principle looks like this:

$$d_a d_x P(a,x) = d_x d_a P(a,x)$$

And his proof runs as follows. On the one hand we have

$$d_a d_x P(a,x)= d_a(P(a,x+dx) - P(a,x)) =$$
$$=\{P(a+da,x+dx) - P(a+da,x)\} - \{P(a,x+dx) - P(a,x)\}$$
$$= \text{(in Euler's notation) } S-Q-R+P$$

On the other hand we have

$$d_x d_a P(a,x)=d_x(P(a+da,x)-P(a,x))$$
$$=\{P(a+da,x+dx) - P(a,x+dx)\} - \{P(a+da,x) - P(a,x)\}$$
$$=\text{(in Euler's notation) } S-R-Q+P$$

The interchangeability theorem in this form is a theorem about differences, and the proof is faultless. The fact that in Euler's conception differentials are infinitely small differences or "differences equal to zero", is irrelevant for the proof.[10]

In about 1730, Euler still phrased the interchangeability property as a property of differentials. In the course of the further development of Euler's work an important shift would take place. Bos has shown that in the course of time Euler eliminated (in particular) the use of higher order differentials from analysis and started using *differential coefficients* systematically. If $dP=P_x dx+P_y dy$ is the total differential of a function P(x,y) the two expressions P_x and P_y are differential coefficients. Because dy=0 yields $P_x=dP/dx$ and dx=0 yields $P_y=dP/dy$, it is obvious that the differential coefficients are equal to (partial) *differential quotients*. One of the reasons for avoiding higher order differentials was the fact that certain properties expressable in terms of higher order differentials depended in an unfortunate way upon the particular choice of the infinitesimal progressions.[11]

It is interesting that in the manuscript from c. 1730, as soon as he starts using the interchangeability principle Euler, without explicitly saying so, in fact already derives the theorem in its *differential coefficient form*. Euler wrote:

> "Let P be a function of those x,y and constants, and let its differential be equal to Qdx+Rdy, where Q and R are also functions of x and y. Therefore, take dQ=Kdx+Ldy and dR=Mdx+Ndy. By means of our principle we shall be able to find a certain relation between these quantities K, L, M and N. Namely, differentiating P first with constant y yields Qdx. In turn, its

---

[10] Although Euler later declared that differentials are equal to zero, he never stopped using infinitely small differences.

[11] [Bos,1974]

differential while keeping x constant will be Ldxdy. Thereupon one will have Rdy by differentiating P while keeping x constant, the differential of which with constant y will be Mdxdy. By force of the principle, therefore: Ldxdy=Mdxdy, and hence: M=L" (quoted in [Engelsman,1984], p.216).

The argument is based upon the implicit assumption that the total differential of a function P(x,y) always possesses the form $dP=P_x dx+P_y dy$ where the differential coefficients P_x and P_y are expressions in x and y. Euler's argument also uses the total differentials $dP_x=P_{xx}dx+P_{xy}dy$ and $dP_y=P_{yx}dx+P_{yy}dy$. Because $M=P_{yx}$ and $L=P_{xy}$, Euler shows that $P_{yx}=P_{xy}$, i.e. the interchangeability theorem in its differential coefficient form.

Euler would repeat this proof of the interchangeability theorem again and again, for example in his *Institutiones calculi differentialis* ([Euler,1755], Pars 226-231).

In Chapter VIII, section VIII.2.2 we saw that Euler later gave a proof of the implicit assumption concerning the form of the total differential. In Lakatosian terms one could say that Euler had apparently become aware of a hidden lemma and had proved it. The example shows that Lakatos exaggerated when he stated that critical proof analysis was an invention of the nineteenth century.[12]

IX.2.3 Fontaine's fluxio-differential calculus

It is interesting to pay some attention in this context to Alexis Fontaine (1707-1771) in whose work we can also observe the transition from the differential calculus - as a tool to solve geometrical problems - to a formalist calculus, largely independent of geometry. According to Clairaut, Fontaine must be considered as an independent discoverer of the interchangeability theorem.[13]

In a 1732 paper, "Nouvelle Méthode pour la Solution des Problèmes de Maximis et de Minimis" ([Fontaine, 1732])[14], Fontaine introduced a new calculus: the *fluxio-differential calculus*. By means of this fluxio-differential calculus, Fontaine succeeded in solving several geometrical and mechanical problems.

Because this calculus is not very well known[15], I shall illustrate it by means of

---

[12][Lakatos,1976], pp.136-141

[13]Clairaut claims to have discovered the theorem himself and mentions Fontaine and Euler as other independent discoverers ([Clairaut,1740], p. 294, footnote).

[14]The paper was published in 1764 ([Fontaine,1764], pp. 1-14).

[15]Only Greenberg has written in the recent past on Fontaine ([Greenberg,1981],

Fontaine's solution of the classical *isoperimetric problem* ([Fontaine,1732], pp. 4-5). Fontaine's goal is the derivation of a differential equation for the curve k of fixed length, which is such that the area enclosed by k and the x-axis is as large as possible. The points A and B are fixed (Figure 9.3). Fontaine considers a progression ....x, x', x'', x''', ... which induces via the curve progressions ...y, y', y'', y''', ... and ...s, s', s'', s''', ..., etc. Here s denotes the curve length, measured from, for example, A. To denote the differentials, Fontaine uses the English fluxional notation $\dot{x}$ = x'-x, $\dot{s}'$ = s''-s' etc. Instead of the fluxional dot I will use the letter δ and write δ(x)=x'-x, δ(s')=s''-s', etc.

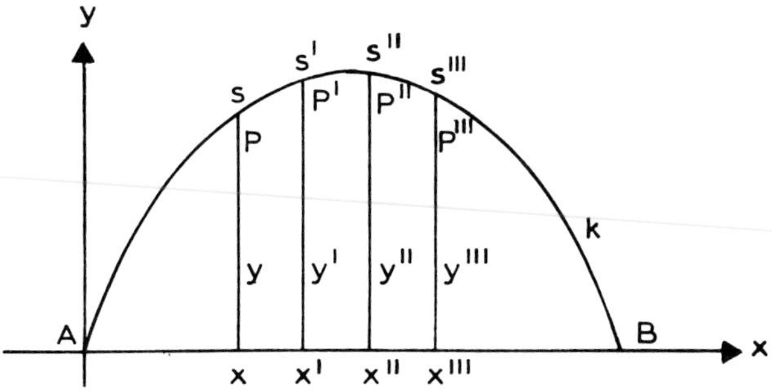

Figure 9.3

In this particular case Fontaine concentrates on the section PP'P''P''' of the curve and remarks that if curve k is the solution of the isoperimetric problem, then the extremum property should also locally hold: i.e. if one varies P' and P'' such that the length of PP'P''P''' remains unchanged, the area underneath PP'P''P''' has its maximum value for P' and P'' on k. This means that the area underneath PP'P''P''', which equals

(9.1) y.δ(x) + y'.δ(x') + y''.δ(x'')

(without comment Fontaine neglects higher order differentials), has a

[Greenberg,1984]).

maximum value under the condition that the length

(9.2) $\delta(s) + \delta(s') + \delta(s'') = \text{constant}.$

Treating the "fluxions" $\delta(x)$, $\delta(s)$, etc. as variables, Fontaine applies the ordinary differential calculus to the extremum problem defined by (9.1) and (9.2). He obtains the following two necessary conditions for an extremum

(9.3) $d(y.\delta(x) + y'.\delta(x') + y''.\delta(x''))=0$

(9.4) $d(\delta(s) + \delta(s') + \delta(s''))=0$

Fontaine now writes

(9.5) $\delta(s)^2=(x'-x)^2 + (y'-y)^2$ and
$\delta(s')^2=(x''-x')^2 + (y''-y')^2$ and
$\delta(s'')^2=(x'''-x'')^2 + (y'''-y'')^2$,

and differentiates these equations, which yields (because y, y''' and x, x', x'', x''' are fixed!)

(9.6) $$d\delta(s) = \frac{\delta(y)d(y')}{\delta(s)} \; ; \; d\delta(s') = \frac{\delta(y')dy''}{\delta(s')} - \frac{\delta(y')dy'}{\delta(s')} \text{ and}$$

$$d\delta(s'') = -\frac{\delta(y'')dy''}{\delta(s'')}$$

and substitution of (9.6) in (9.4) gives

(9.7) $$-\{\frac{\delta(y')}{\delta(s')} - \frac{\delta(y)}{\delta(s)}\} dy' - \{\frac{\delta(y'')}{\delta(s'')} - \frac{\delta(y')}{\delta(s')}\} dy'' = 0$$

or[16]

(9.8) $$\delta\{\frac{\delta(y)}{\delta(s)}\} dy' + \delta\{\frac{\delta(y')}{\delta(s')}\} dy'' = 0$$

Equation (9.3) yields

$$\delta(x').dy' + \delta(x'').dy''=0$$

Elimination of dy' and dy'' from this equation and (9.8) yields

(9.9) $$\frac{\delta\{\frac{\delta(y')}{\delta(s')}\}}{\delta(x'')} - \frac{\delta\{\frac{\delta(y)}{\delta(s)}\}}{\delta(x')} = 0$$

or

---

[16]Remember that the definition of the "prime operation" is such that F(x',y',z', etc.)-F(x,y,z, etc.)=δ(F(x,y,z,etc.)) for any expression F.

(9.10) $$\delta\left\{\frac{\delta\frac{\delta(y)}{\delta(s)}}{\delta(x')}\right\}=0$$

We can choose the progression ... x, x', x", x"', ... in such a way that it is equidistant, i.e. such that $x^{i+1}-x^i$ is a constant. This implies that $\delta(x)=\delta(x')=\delta(x'')$ and (9.10) can be written as follows

(9.11) $$\delta\left\{\frac{\delta\frac{\delta(y)}{\delta(s)}}{\delta(x)}\right\}=0$$

And this implies that

(9.12) $$\delta\left\{\frac{\delta(y)}{\delta(s)}\right\}=c_1\delta(x)$$

or

(9.13) $$\frac{\delta(y)}{\delta(s)}=c_1x+c_2$$

wherein c_1 and c_2 are constants.
It is not difficult to show that differential equation (9.13) is satisfied only by a circle (if $c_1\neq 0$).
The above example shows Fontaine's fluxio-differential calculus in action.
Fontaine gave the most spectacular application of his calculus in "Sur les Courbes Tautochrones", presented to the Académie in 1734, published in 1736 and again in 1764 ([Fontaine, 1734]). In this paper, Fontaine solved the problem of the tautochrones[17] for the case in which the resistance of the medium is a quadratic function of the speed of the moving body. Implicitly he twice used the following interchangeability property:

(9.14) $$\delta(ds)=d\delta(s)$$

In 1767, more than thirty years later, Fontaine wrote that he had already proved (9.14) in 1734. Fontaine wrote:

> "This [...] theorem can be found [in the calculations to determine the function that solves the problem of the tautochrones, which I gave in 1734],

---

[17] A tautochrone or isochrone is a curve along which an oscillation of a mass point under the influence of gravity takes the same time whether it swings through a large or small arc.

and I also proved it : $d\delta x=\delta x'-\delta x=\delta(x'-x)=\delta dx$ [...]".[18]
Fontaine's proof has the following structure:

$$d\delta(x)=\delta(x)'-\delta(x)=(!)\ \delta(x')-\delta(x)=\delta(x'-x)=\delta(dx)$$

(N.B. This time the primes do correspond to the differentiation and not to the fluxation!). The proof is based on the following implicit assumption

$$\delta(x)' = \delta(x')$$

This rule holds, but not obviously. If we associate independent variables u and v with respectively "d" and "δ", the truth of the rule can be seen as follows

$$\begin{aligned}(9.15)\qquad \delta(x)'&=\delta(x(u,v))'=(x(u,v+\delta(v))-x(u,v))'=\\ &=x(u+du,v+\delta(v))-x(u+du,v)=\\ &=\delta(x(u+du,v))=\delta(x(u',v))=\delta(x')\end{aligned}$$

Fontaine's fluxio-differential calculus is both an early partial differential calculus and an ancestor of the calculus of variations. I will not elaborate on its relationship with the calculus of variations but I will explain in which way the fluxio-differential calculus can be considered as a partial differential calculus. In section IX.2.1, I wrote that partial differentiation can only show up in problem situations with more than one dimension of variability. The families of curves that were studied by Nicolaus I. Bernoulli represent such situations with two dimensions of variability. In Fontaine's fluxio-differential calculus we also have two dimensions of variability. The fluxions correspond to one dimension of variability, the differentials to another. The variability corresponding to the differentiation is one-dimensional. Fontaine here leaves no doubt: he says that in order to have a "real differentiation" ("pour différencier réellement"), the number of ordinates that is varied should be equal to the number of conditions ([Fontaine,1732], p.3 and p.4). In the above example we have two conditions (9.3 and 9.4), and Fontaine varies the two ordinates y' and y". It is remarkable in this context that in 1755 in the letter to Euler in which he wrote that he had discovered what would later be called the calculus of variations Lagrange used Euler's version of the interchangeability

---

[18]"Ce[...]théorème ce trouve [dans les calculs de la fonction du problème des courbes tautochrones, que je donnai en 1734], et je l'ai démontré de même, $dfx=fx'-fx=f(x'-x)=$ $=fdx$ [...]" ([Fontaine,1767], p.597, footnote). At that time Fontaine used the letter f instead of the fluxional point.

theorem to change the order of a differentiation and a variation. In his reply to Lagrange, Euler pointed out that this application of the interchangeability theorem for partial differentiation is problematic because in Lagrange's calculus the variation δy is a function of x and no differentiation, at least no "real differentiation" (as Fontaine would say).[19]

Fontaine's proof of the interchangeability theorem $\delta(ds) = d\delta(s)$ is rather superficial. We do not even know whether Fontaine's 1767 claim - that he consciously used the theorem thirty years earlier - is trustworthy. Yet the fluxio-differential calculus is a differential calculus based on two dimensions of variability and naturally leads to that interchangeability theorem. Moreover, the further development of Fontaine's work very well illustrates the transition from the 17th Century Geometric Tradition to the Formalist Tradition.

IX.2.4 Fontaine's formalist calculus

The fluxio-differential calculus does not belong to the formalist tradition. It is an analytical method but is used as a tool to solve geometrical problems. However, in a memoir, "Le Calcul Intégral, Première Méthode", dated November 19th, 1738 ([Fontaine,1738]) Fontaine developed a calculus of several variables which falls entirely within the formalist tradition. In this calculus Fontaine proves another version of the interchangeability theorem for partial differentiation.

Fontaine must have considered it as obvious that the total differential of a function F(x,y,...) is a linear expression in the differentials of the variables involved:

$$dF=Adx +Bdy + \ldots$$

He denotes the differential coefficients A, B, etc. respectively by dF/dx, dF/dy, etc. So we have

$$dF = \frac{dF}{dx}dx + \frac{dF}{dy}dy + \ldots$$

Obviously dF/dx, etc. are initially no quotients but only notations for differential coefficients. Because one can easily prove that the differential coefficients are equal to partial differential quotients, the notation is well chosen. Similarly Fontaine uses the notation d^2F/dx^2, $d^2F/dxdy$ to denote the coefficients in

---

[19][Koetsier,1986], pp. 195-197

$$d\frac{dF}{dx} = \frac{d^2F}{dx^2}dx + \frac{d^2F}{dxdy}dy + \ldots$$

Fontaine first proves an interchangeability theorem for differentiation and integration. Fontaine writes:

"Let μ still be a function of p, of x, of y, of z etc. The coefficient of dx in the difference [= total differential - T.K.] of ∫μdx is μ; but what, for example, is the coefficient of dy? I say that

$$\frac{d\int\mu dx}{dy} = \int\frac{d\mu}{dy}dx\text{''}.$$ [20]

(The expression ∫μdx denotes a primitive function of μ with respect to x.) Fontaine's proof runs as follows.[21] We have

$$d_y\int\mu(p,x,y,z,..)dx = \int\mu(p,x,y+dy,z,..)dx - \int\mu(p,x,y,z,..)dx =$$

$$\int(\mu(p,x,y+dy,z,..)dx - \mu(p,x,y,z,..)dx) = \int(d_y\,\mu/dy)dy\,dx,$$

which yields, after division by dy (because the integration concerns x, dy can be treated as a constant with respect to ∫)

$$\frac{d\int\mu dx}{dy} = \int\frac{d\mu}{dy}dx.$$

Fontaine then proceeds to prove the interchangeability theorem for partial differentiation.[22] Fontaine:

---

[20]"Soit toujours μ une fonction de p, de x, de y, de z etc. Le coefficient de dx dans la différence de ∫μdx est μ; mais quel est, par example, celui de dy! Je dis que (d∫μdx)/dy = ∫(dμ/dy)dx" ([Fontaine,1738], p.26).

[21]"Démonstration. Car en général d∫μdx = ∫μ'dx' - ∫μdx = ∫(μ'dx' - μdx)= ∫d(μdx) = (en ne faisant varier que y) ∫(dμ/dy)dydx; donc(d∫μdx)/dy = ∫(dμ/dy)dx C.Q.F.D." ([Fontaine,1764], p.26). Fontaine writes in the first part of his proof dx' instead of dx and later explicitly restricts himself to a variation of y only. Apparently, he thought of an arbitrary variation in the first part of his proof, not only of y but of all the variables p, x, y, z, etc. and only later restricted himself. In my paraphrase of Fontaine's proof I restrict myself right from the beginning to a variation of y only.

[22]"Je suppose que Adx et Bdy soient deux termes de la différence d'une meme fonction ϕ de p,de x, de y, de z, etc. et je dis que dA/dy =dB/dx. Démonstration. On aura ∫Adx+a=ϕ, ∫Bdy+=ϕ (a est une fonction de p, y, z, etc. et b est une fonction de p, x, z, etc.); donc ∫Adx + a = ∫ Bdy + b. Je différencie cette dernière équation en faisant varier x; j'aurai A = ∫ (dB/dx)dy + db/dx. Je différencie derechef en faisant varier y; j'aurai dA/dy = dB/dx

"I suppose that Adx and Bdy are two terms of the difference [= total differential - T.K.] of the same function ϕ of p, of x, of y, of z, etc. and I say that $\frac{dA}{dy}=\frac{dB}{dx}$".

And his proof runs as follows. If

$$d\phi(p,x,y,z,...) = \frac{d\phi}{dp}dp + \frac{d\phi}{dx}dx + \frac{d\phi}{dy}dy + ...$$

then

$$\phi = \int\frac{d\phi}{dx}dx + a(p,y,z,...) = \int\frac{d\phi}{dy}dy + b(p,x,z,...)$$

Differentiation with respect to x and application of the first theorem yields

$$\frac{d\phi}{dx} = \int\frac{d^2\phi}{dxdy}dy + \frac{db(p,x,z,...)}{dx}$$

Differentiation with respect to y yields then

$$\frac{d^2\phi}{dxdy} = \frac{d^2\phi}{dydx}$$

Q.E.D.

Fontaine's reasoning illustrates the formalist tradition very well. The geometrical elements, so clearly present in his work on the fluxio-differential calculus, have almost totally disappeared and the discussion exclusively concerns analytical expressions.[23]

C.Q.F.D." ([Fontaine,1738], p. 26)

[23]In two papers ([Greenberg,1981],[Greenberg,1984]) Greenberg has investigated the problem of the relationship between Fontaine's fluxio-differential calculus and his calculus of several variables. A critical discussion of Greenberg's ideas is out of place here.Yet some remarks are relevant. Greenberg seems to defend the thesis that Fontaine's calculus of several variables developed out of the fluxio-differential calculus as follows. At some time Fontaine must have discovered that the two dimensions of variability of the fluxio-differential calculus could be handled by means of one single differential operator "d" ([Greenberg,1984], pp.32-33). In other words, Fontaine first turned the fluxio-differential calculus into a calculus of two variables, which he then generalized into a calculus of more variables. Greenberg suggests that the interchangeability theorem for partial differentiation in its differential coefficient form developed out of the interchangeability theorem of the fluxio-differential calculus, which we discussed above. In the notation of Greenberg, who also replaces the dot-operator by "δ", the theorem $d(\delta F(x,X)) = \delta(dF(x,X))$ turned into $d(\delta F/\delta X)dx= =\delta(dF/dx)\delta X$ ([Greenberg,1981], pp. 265-266), which then became

IX.2.5 Clairaut on the interchangeability theorem

Clairaut, whose interests at that time parallelled Fontaine's in several respects, gave an entirely different proof of the theorem in 1739. He wrote:

> "One can immediately see the truth of this proposition by supposing that the given function is simply $rx^m y^n p^q$ because every function will consists of terms of this form if it is expanded in a series, and if the theorem is true for one term it will be true for infinitely many terms. Well then, it is so easy to assure oneself of the truth of my theorem for one term so that it is not necessary to digress more on this point".[24]

One year later in another paper, after having repeated the above argument, Clairaut wrote:

> "But because a proof of this kind is only an induction, we are going to look a priori for the proof of our theorem".[25]

The quotation shows that eighteenth century mathematicians were aware of the inductive elements in their reasoning. From a modern point of view there are at least two weak points in Clairaut's inductive proof - the assumption that

---

d(dF/dy)dx= =d(dF/dx)dy for a function F of an arbitrary number of variables. Although Greenberg may be right, I think that his thesis is quite speculative. When Fontaine in 1767 returned to the fluxio-differential calculus, he apparently considered it much more as a chapter in the history of the calculus of variations than as a theory that had preceded his own calculus of several variables. Just as speculative and, from my point of view, just as acceptable is the following possible line of development. In the fluxio-differential calculus, Fontaine repeatedly differentiated expressions containing several variables and the results are always of the form

$$dF(x,y,z,..) = A(x,y,z,..)dx + B(x,y,z,..)dy + C(x,y,z,..)dz + \ldots$$

As we have seen above, Fontaine was keenly aware of the fact that in order to have a "real differentiation", i.e. a differentiation corresponding to one dimension of variability, one should have a number of conditions restricting the possible values of the variables. It is quite possible that Fontaine realized that total differentials can be studied independently of those conditions. If that insight lies at the basis of Fontaine's calculus of several variables, it would have developed more as a by-product of the fluxio-differential calculus than as a generalization of it.

[24]"On peut voire tout d'un coup la vérité de cette proposition en supposant que la fonction donnée soit simplement $rx^m y^n p^q$; car toute fonction étant réduite en suite, sera composée de termes qui auront cette forme, et si le Théorème est vrai pour une terme, il sera vrai pour une infinité de termes. Or il est si facile de s'assurer de la vérité de mon Théorème pour un terme, qu'il n'est pas nécessaire de s'étendre davantage sur cette article" ([Clairaut,1739], pp. 427-428).

[25]"Mais comme une démonstration de cette nature n'est qu'une induction, nous allons chercher *a priori* la démonstration de notre Théorème" ([Clairaut,1740], pp.295-296).

all functions can be expressed as a power series, and the assumption that an infinite sum can be handled as a finite sum. However, since handling infinite sums as if they were finite was such common practice in the eighteenth century, it must have been the former assumption that bothered Clairaut. Clairaut knew that it is sometimes difficult to find the power series corresponding to a given function, and he must have realized that it was not a priori clear that in such cases the solution always exists.

I will not discuss Clairaut's first a priori proof, because it is similar to Fontaine's 1738 proof. In the same 1740 paper, however, there is yet another proof of the interchangeability theorem for partial differentiation in its differential coefficient or differential quotient form ([Clairaut, 1740], pp.312-313).

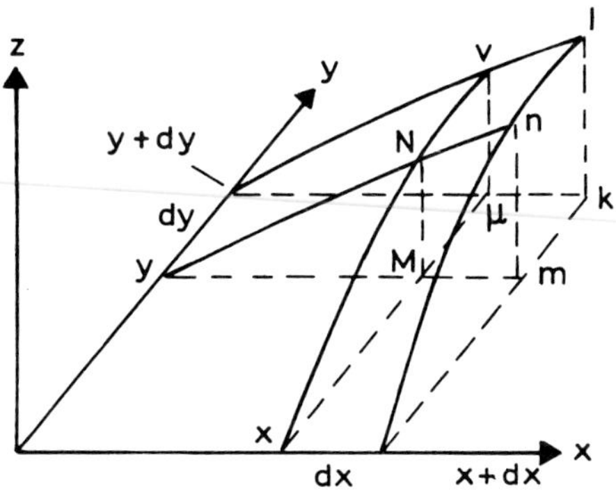

Figure 9.4

The proof is mainly interesting because Clairaut considers a function of two variables z = z(x,y) and interprets it as representing a surface (Figure 9.4). Such an interpretation, which was to become standard in the differential calculus of two variables, does not occur in the earlier texts that I have discussed so far. Clairaut's proof runs as follows. Let dz = Adx + Bdy. Clairaut considers an arbitrary (x,y) and the corresponding value z (MN in Figure 9.4). He then determines the value of z corresponding to (x+dx,y+dy) (kl in Figure 9.4) in two different ways.

First y grows with dy and x remains constant. Then z becomes

$$\nu\mu = z + Bdy.$$

Subsequently x increases with dx and $\nu\mu$ becomes

$$kl = z + Adx + Bdy + \frac{dB}{dy} dydx$$

Secondly x increases with dx and y remains constant. Then z becomes

$$nm = z + Adx.$$

Subsequently y grows with dy and nm becomes

$$kl = z + Bdy + Adx + \frac{dA}{dy} dydx$$

Equation of the two expressions for kl immediately yields

$$\frac{dB}{dx} = \frac{dA}{dy}$$

([Clairaut,1740], pp. 312-313).
Apart from the geometrical interpretation, this proof is not much different from Euler's 1730 proof: z(x+dx,y+dy) is determined in two ways and the two expressions that one obtains are equated.

IX.2.6 Lagrange

From the point of view of the methodology of mathematical research traditions, the formalist tradition was born when highly progressive formalist research projects appeared. All versions of the interchangeability theorem that we have discussed so far were given in the context of more or less clearly definable mathematical research projects. It would lead us astray if we attempted here to define these projects and consider their rates of progress. The next version of the interchangeability theorem for partial differentiation which I want to discuss was, however, given within the context of a formalist mathematical research project; this deserves a little more attention. Although mathematical practice was hardly influenced by it, most of the eighteenth century mathematicians were aware of the existence of some unsolved foundational problems concerning the status of differentials. In this respect, Lagrange made an important attempt to rigorize analysis. His *Théorie de fonctions analytiques* ([Lagrange,1797]) and his *Leçons sur le calcul des fonctions* ([Lagrange,1806]) contain an attempt to give a foundation of analysis without the use of infinitesimals. Lagrange's notion of a function is formalist. As we have seen in Chapter VIII, there is no difference between Lagrange and Euler as regards the notion of function they use. Lagrange's starting point however, is his conviction that every function can be developed into a power series. This means that a substitution of the expression "x+i" for "x" in a function f(x) in principle leads to an expansion

$$f(x+i) = f(x) + p(x).i + \frac{q(x)}{2!}i^2 + \frac{r(x)}{3!}i^3 + \ldots$$

By definition, Lagrange calls p(x), q(x), r(x), etc. respectively the first derivative of f(x), the second derivative of f(x), the third derivative of f(x), etc. and he denotes them by "f'(x)", "f''(x)", "f'''(x)", etc. This definition enables Lagrange to derive all kinds of properties of the derivative by means of formal manipulations, without the use of infinitesimals. This is an important element of the heuristic of Lagrange's foundational research project.

Actually the notion of derivative already shows up in Lagrange's work in a paper dating from 1772 ([Lagrange,1772]), which also contains a proof of the interchangeability theorem in differential quotient form. I will first discuss this proof. In the paper he defines the notion of derivative (or derived function). He also shows (by substitution of $x+\xi+\omega$ for x in a function of x) that one can determine higher derivatives by repeatedly determining a first derivative (the first derivative of the first derivative equals the second derivative, etc.). Yet in 1772 the derivative still plays a secondary role in Lagrange's work. It is merely a means to obtain other results. In the 1772 paper Lagrange derives the interchangeability theorem as follows. First he shows that the first derivative equals a differential quotient. After substitution of $x+\xi$ for x in a function u, this function becomes

$$u + u'\xi + \frac{u''\xi}{2} + \frac{u'''\xi^2}{2.3} + \text{etc.}$$

(Here u' is the first derivative of u, u'' is the second, etc.).

Lagrange writes:[26]

---

[26]"si l'on regarde ξ comme infiniment petit et qu'on néglige les puissances ξ^2, ξ^3, ..., on aura simplement $u'\xi$ pour l'accroissement de u; de sorte que, désignant cet accroissement par du, et l'accroissement x de u par dx, on aura du=du' and u'=du/dx " ([Lagrange,1772], p.446).

"if one considers ξ as infinitely small and one neglects the powers ξ^2, ξ^3,..., the increment of u will simply be u'ξ, so that if we denote this increment by du and the increment of x by dx, one will simply have du=u'dx and u'=du/dx".

In order to obtain the interchangeability theorem, Lagrange substitutes x+ξ for x in a function of two variables x and y, and gets the above given expansion. Then he substitutes y+ψ for y in that expansion and expands further ([Lagrange,1772], p. 446):

$$u + u^{1,}\xi + u^{,1}\psi + \frac{u^{2,}}{2}\xi^2 + \frac{u^{1,1}}{1.1}\xi\psi + \frac{u^{,2}}{2}\psi^2 + \text{etc.}$$

(In "$u^{n,m}$" the n refers to the n-th partial derivative with respect to x and the m refers to the m-th partial derivative with respect to y). Using the fact that a derivative equals a differential quotient Lagrange writes

$$u^{,1} = \frac{du}{dy} \text{ and } u^{1,1} = \frac{du^{,1}}{dx} = \frac{d\frac{du}{dy}}{dx} = \frac{d^2u}{dxdy}$$

Similarly he obtains

$$u^{1,} = \frac{du}{dx} \text{ and } u^{1,1} = \frac{du^{1,}}{dy} = \frac{d\frac{du}{dx}}{dy} = \frac{d^2u}{dxdy}$$

And obviously one has

$$\frac{d^2u}{dxdy} = \frac{d^2u}{dydx}$$

([Lagrange,1772], p.448).

In 1772 Lagrange does not pay any attention to the fact that the expansion of u(x+ξ,y+ψ) can be obtained in two ways: one can first substitute x+ξ for x in u and then y+ψ for y in the resulting expansion, but one can also turn things around. The question is: does one get the same coefficient $u^{1,1}$? It is remarkable that in 1797 when the derivative has become a central notion in Lagrange's foundational work, the interchangeability theorem that he proves concerns exactly this point: he proves the equality of the two mixed derivatives. One could say that Lagrange's interchangeability theorem of 1797 is exactly the hidden interchangeability lemma in the proof of the interchangeability theorem of 1772.

I will reproduce Lagrange's 1797 proof of the interchangeability theorem faithfully, except for the fact that I will write "$f_x(x,y)$" where Lagrange writes "f'(x,y)", and "$f_y(x,y)$" where Lagrange writes "$f_{,}(x,y)$". The disadvantage of Lagrange's notation is that the notation does not distinguish between "$f_{yx}(x,y)$" and "$f_{xy}(x,y)$" in advance, i.e. before the equality has been proved, because both are denoted by means of an "f" followed by a lower and an upper prime. Lagrange's proof runs as follows. If f(x,y) is a function of two variables x and y, then we can substitute x+i for x in f(x,y) and develop the resulting expression in a series

(9.16) $$f(x+i,y) = f(x,y) + i.f_x(x,y) + \frac{i^2}{2!}f_{xx}(x,y) + \text{etc.}$$

Similarly substitution of y+o for y in f(x,y) yields

(9.17) $$f(x,y+o) = f(x,y) + o.f_y(x,y) + \frac{o^2}{2!}f_{yy}(x,y) + \text{etc.}$$

Lagrange now substitutes in (9.16) the expression y+o for y and develops the terms of the right hand side with respect to y. After rearrangement he obtains

(9.18) $$f(x+i,y+o)=f(x,y) + i.f_x(x,y) + o.f_y(x,y) +i.o.f_{xy}(x,y) + \text{etc.}$$

Substitution of x+i for x in (9.17) yields analogously

(9.19) $$f(x+i,y+o)=f(x,y) + i.f_x(x,y) + o.f_y(x,y) +i.o.f_{yx}(x,y) + \text{etc.}$$

And the unicity of the development now implies that $f_{xy}(x,y) = f_{yx}(x,y)$. With reference to (9.19) Lagrange writes that one obtains:

> "[...] in this way the same formula as above, which is as it should be [...]. From which it follows that it does not matter in which order the double operation, which is necessary in order to go from the primitive function f(x,y) to the derived function, is executed [...]".[27]

Because Lagrange writes $f'_{,}(x,y)$ for both $f_{xy}(x,y)$ and $f_{yx}(x,y)$ the formulae (9.18) and (9.19) are in his text completely identical.

---

[27]"[...] ainsi la même formule que ci-dessus, comme cela doit être [...] D'où il suit qu'il est indifférent dans quel ordre se fasse la double opération nécessaire pour passer de la fonction primitive f(x,y) à la fonction dérivée [..]" ([Lagrange,1797], p.93).

Nicolaus I. Bernoulli (c. 1719 manuscript) The theorem concerns geometrically interpreted differentials; it expresses the equality of mixed differentials; it holds generally.

Fontaine (1734 manuscript presented to the Académie) The theorem concerns geometrically interpreted fluxions and differentials; it expresses the interchangeability of fluxionation and differentiation; it holds generally.

Euler (c. 1730 manuscript) The theorem concerns analytical expressions; it expresses the equality of mixed differentials; the geometrical interpretation has become inessential; the theorem holds generally.

Euler (c.1730 manuscript) Euler in fact already derives the theorem in its differential coefficient form (or differential quotient form). Euler's analytical proof of c.1730 appeared again in Euler's later work (e.g. [Euler,1755]). The theorem in this form concerns analytical expressions; it expresses the equality of mixed differential coefficients; it holds generally

Fontaine (1738 manuscript) The theorem concerns analytical expressions; it expresses the equality of mixed differential coefficients; it holds generally.

Clairaut (1739 paper) The theorem concerns analytical expressions; it expresses the equality of mixed differential coefficients; it holds generally.

Clairaut (1740 paper) The theorem concerns analytical expressions; it expresses the equality of mixed differential coefficients; it holds generally.

Clairaut (1740 paper) The theorem concerns analytical expressions; it expresses the equality of mixed differential coefficients; it holds generally. Clairaut interprets expressions as $z = f(x,y)$ as the equation of a surface in (Euclidean) space.

Lagrange (1772 paper) The theorem concerns analytical expressions; it expresses the equality of mixed partial differential quotients; it holds generally.

Lagrange (1798 textbook) The theorem concerns analytical expressions; it expresses the equality of mixed partial derivatives; it holds generally.

Table I The eighteenth century versions of the theorem discussed in the text.

IX.2.7 Concluding remarks on the eighteenth century versions of the theorem

Table I contains a list of the various eighteenth century versions of the interchangeability theorem that I discussed. In spite of considerable differences between the different versions, they also have very much in common. In the eighteenth century the theorem is always presented as an *unconditional truth*. If we interpret the reasoning from a modern point of view there are hidden assumptions, for example concerning the existence of differential coefficients. Such a point worries us because we think "locally".

For a modern mathematician, existence of a differential coefficient means existence of a value in particular points of a domain. Eighteenth century mathematicians, however, thought "globally" when it came to the existence of functions. For them all differential coefficients obviously existed, namely as *expressions*. And although they must have been aware of the fact that expressions sometimes "misbehave" locally for particular values of the variables, this evidently did not bother them.
One element the various versions also obviously have in common lies in the concrete practice of the calculus. Eighteenth century mathematicians noticed that in all concrete examples the interchangeability property holds. Euler's example illustrates it. Mixed partial differentiation of the function $x^3y-y^2x^2$ yields $(3x^2-4yx)dxdy$ independently of the order of differentiation. Moreover, it is evident that this example is representative for all polynomials. Clairaut's "inductive" 1739 proof - consisting in the remark that the theorem clearly holds for all finite and infinite linear combinations of terms of the form x^my^n because it obviously holds for the individual terms of that form - is probably just a more explicit way of saying what Euler meant when he wrote that the consideration of a few examples were sufficient to see the necessary truth of the theorem. In spite of the great differences that exist between them, the proofs of the theorem by Bernoulli in 1719, by Euler in 1730, by Fontaine in 1734 and 1738, by Clairaut in 1739 and 1740 and finally by Lagrange in 1772 and 1797, have in common the fact that they all explained the interchangeability phenomenon which was present at the level of formulae manipulation. For Bernoulli in 1719 and Fontaine in 1734 the theorem was still very much connected with geometry, though the others interpreted the theorem in formalist terms. To a certain extent, Clairaut, who was rather geometrically inclined, might be an exception here. Yet two of the three proofs that Clairaut gave are completely independent of geometrical interpretation, and in the case of his third proof, the geometrical interpretation can be viewed as a mere illustration.
The different eighteenth century versions of the interchangeability theorem that I have discussed so far support the rational reconstruction on the basis of the MMRT given in the previous chapter in the sense that they show the unity of the formalist tradition. Excluding Nicolaus I. Bernoulli's version and the version belonging to Fontaine's fluxio-differential calculus, all versions have the following in common - *they concern formulae, they are believed to hold unconditionally and they are proved by formal means*. Lagrange's second version of the theorem, moreover, nicely illustrates how foundational

problems in a tradition can lead to proposals for modifying the tradition. The theorem concerns the same interchangeability property viewed in formalist terms, but his proof represents a modification of primarily the methodology of proof of the Formalist Tradition.

IX.3 The nineteenth century Conceptual Tradition in analysis

IX.3.1 Cauchy

The first proof belonging to the nineteenth century Conceptual Tradition was given by Cauchy. He gave a proof of the theorem in his *Résumé des leçons sur le calcul infinitésimal* ([Cauchy,1823]).[28] The theorem, which is not explicitly presented as a theorem, has precisely the same form that it had with Nicolaus I. Bernoulli,

$$d_x d_y u = d_y d_x u$$

([Cauchy,1823], p.77), but quite a different meaning, as we shall see below. We saw in Chapter VIII that Cauchy's *Cours d'analyse* represents a completely new approach to analysis. The generality of algebra is rejected and real functions are no longer just analytical expressions, but relations between real quantities expressed by formulae. Moreover, Cauchy defines the fundamental notions - such as convergence of a series, continuity, the derivative and the integral - all by means of the limit concept. In the course of the nineteenth century, Cauchy's approach would be refined considerably, in particular by Weierstraß and the mathematicians influenced by him. To a certain extent this process can be described in Lakatosian terms: after repeated proof analysis and the discovery of hidden lemmas etc., the nineteenth century tradition turned into a firm deductive tradition. The different versions of the interchangeability theorem that appeared during the nineteenth century reflect this development.

---

[28] He gives the same proof in his *Leçons sur le calcul différentiel* ([Cauchy, 1829], p. 525).

Continuity of functions of several variables.

I will now first consider the conceptual background of the theorem in Cauchy's work. In Chapter III we saw how Cauchy defined the continuity of a function of one variable in his *Cours d'analyse*: a function f(x) is continuous on an interval if for all x the value of $f(x+\alpha)-f(x)$ decreases indefinitely with the value of α or, in other words, if for all x the increment $f(x+\alpha)-f(x)$ is infinitely small for infinitely small α. It is remarkable that Cauchy does not separately define the notion of continuity in the case of a function of more than one variable. Apparently Cauchy thought that everything about the continuity of a function could be said in terms of the continuity of the function in its individual variables separately. In the *Cours d'analyse* we have the following argument. Cauchy considers a function that is continuous in all variables in a neighbourhood of (X,Y,Z,..), and he argues that in this neighbourhood the difference

$$F(x+\alpha,y+\beta,z+\gamma,..) - f(x,y,z,..)$$

is infinitely small, if α , β, γ, etc. are infinitely small ([Cauchy,1821], p.46). In order to prove that statement, Cauchy considers the equation:

$$\begin{aligned} F(x+\alpha,y+\beta,z+\gamma,...)-f(x,y,z,...)= & \\ & F(x+\alpha,y,z)-f(x,y,z,...) + \\ & F(x+\alpha,y+\beta,z,...)-F(x+\alpha,y,z,...)+ \\ & F(x+\alpha,y+\beta,z+\gamma,...)-F(x+\alpha,y+\beta,z,...)+ \\ & .... \end{aligned}$$

with α, β, γ converging to zero.

In Cauchy's view the whole right hand side of the equation then converges to zero and so, obviously, does the left hand side. ([Cauchy,1821], p.46). He added[29]:

> "The proposition [...] obviously also holds if one were to establish certain relations between the new variables α, β, γ. It is sufficient that these relations allow the new variables to converge all at the same time to the limit zero".

---

[29]"La proposition [...] subsiste évidemment dans le cas même où l'on établirait entre les nouvelles variables α, β, γ, ...certaines relations. Il suffit que ces relations permettent aux nouvelles variables de converger toutes en même temps vers la limite zéro" ([Cauchy,1821], p.46).

From a modern point of view Cauchy underestimated the situation regarding the simultaneous convergence of several variables.[30] The second half of the nineteenth century in particular was to show considerable conceptual refinement concerning the continuity of functions of several variables.

Differentiability of functions of several variables

Before I can turn to the interchangeability theorem for partial differentiation, it is necessary to have a look at Cauchy's theory of differentiability. Cauchy does not explicitly say so, but in fact he proves the interchangeability theorem in a context where he considers respectable functions, i.e. functions of several variables that are differentiable, an unspecified number of times. First Cauchy defines the total differential of a function u = f(x,y) ([Cauchy,1823], p.48).[31] He writes

---

[30]Laugwitz gives a non-standard interpretation of Cauchy's argument. Cauchy's proof runs as follows: "It is indeed clear that [....] the numerical values of the differences $f(x+\alpha,y,z,...)-f(x,y,z,...)$, $f(x+\alpha,y+b,z,...)-f(x+\alpha,y,z,...)$, $f(x+\alpha,y+\beta,z+\gamma,...)-f(x+\alpha,y+\beta,z,...)$, ... will decrease indefinitely with the values of the variable quantities $\alpha,\beta,\gamma,...$, i.e. the numerical value of the first difference with the numerical value of α, the numerical value of the second difference with the numerical value of β [...] and so on. One has to conclude that the sum of all those differences $f(x+\alpha,y+\beta,z+\gamma,...)-f(x,y,z,...)$ will converge to the limit zero, if $\alpha,\beta,\gamma,...$ converge to this limit" ([Cauchy,1821], p.46).
From the point of view of Weierstrassian analysis the theorem is simply wrong. The example $f(x,y) = \frac{2xy}{x^2+y^2}$ for $(x,y) \neq (0,0)$, $f(x,y) = 0$ shows, from a standard point of view, that Cauchy made a mistake. A Weierstrassian analysis of such proof reveals hidden lemmas concerning uniform convergence. Laugwitz reads the text as follows: "The simple way out is to consider infinitesimals α, β as mathematical entities of their own specific rank ..." ([Laugwitz,1987],p.263). And indeed, interpreted as a non-standard theorem, it could be proved easily as follows: All the following differences: $f(x+\alpha,y,z,...)-f(x,y,z,...)$, $f(x+\alpha,y+\beta,z,..)-f(x+\alpha,y,z,...)$, $f(x+\alpha,y+\beta,z+\gamma,...)-f(x+\alpha,y+\beta,z,...)$, ...are infinitesimal. Hence, their sum is obviously also infinitesimal. However, Cauchy did not give this proof! In his proof, indefinitely decreasing values of variables are all over the place. Cauchy even added:"The proposition [...] obviously holds also if one were to establish certain relations between the new variables α, β, γ. It is sufficient that those relations allow the new variables to converge all at the same time to the limit zero" ([Cauchy, 1821],p.46). Certainly when he wrote down this sentence Cauchy did not think of fixed infinitely small quantities at all, but thought completely in terms of variables converging to zero in some order in time: first α, then β, etc. or all at the same time in accordance with a relation existing between them.

[31] Cauchy develops the theory for functions of arbitrarily many variables. Without loss of

(9.20) $$\Delta u = f(x+\alpha h, y+\alpha k) - f(x,y)$$

with finite h and k and infinitely small α. Because of the continuity of f(x,y), the value of (9.20) will be infinitely small. Cauchy divides by the infinitely small α, obtaining:

(9.21) $$\frac{\Delta u}{\alpha} = \frac{f(x+\alpha h, y+\alpha k) - f(x,y)}{\alpha}$$

He remarks that "in general" the ratio converges to a finite limit different from zero. Cauchy calls this limit by definition the *total differential* du. The total differential thus defined depends on h and k.

The modern theory of differentiability of functions of several variables is conceptually more refined. Cauchy actually defines what is nowadays called the Gâteaux-differential, since (9.21) implies that when α converges to zero, the point (x,y) is approached along an arbitrary straight line. A function such as, for example,

$$f: R^2 \to R, \text{ with } f(x,y) = 1 \text{ if } x=y^2 \neq 0 \text{ and } f(x,y) = 0 \text{ elsewhere,}$$

is not continuous in (0,0), but possesses a Gâteaux-differential in (0,0). Such functions show that one should distinguish Gâteaux-differentiability from Fréchet-differentiability. Fréchet-differentiability corresponds by definition to the fact that the differential quotient always yields the same limit for a *completely arbitrary* approach of the point concerned.[32]

The value of the total differential for k=0 is denoted by $d_x u$ and for h=0 by $d_y u$. By taking u(x,y) = x or u(x,y) =y one finds h=dx and k=dy. Apparently one can write dx instead of h and dy instead of k.

It is clear that if ϕ and χ denote the partial derivatives of u, we can write

$$\phi(x,y) = \frac{d_y u}{dx} \text{ and } \chi(x,y) = \frac{d_y u}{dy}$$

By means of the mean-value theorem, Cauchy proves that

generality, I will restrict myself to functions of two variables.

[32]An extensive modern treatment of Fréchet- and Gâteaux-differentiability can be found in [Flett,1980].

$$du = \phi(x,y)dx + \chi(x,y)dy.$$[33]

Because Cauchy explicitly uses the continuity of the derivative in his proof of the mean-value theorem, he also implicitly uses the continuity of the partial derivatives here.

Cauchy proves the interchangeability theorem for partial differentiation for functions of several variables that are repeatedly totally differentiable. Cauchy proves

$$(1) \quad d_x d_y u = d_y d_x u$$

saying:

"Suppose that when one lets only x increase by an infinitely small quantity αdx, one indicates the increment of a function of x, y, z, ... by means of the letter x placed at the foot of the symbol Δ. One will then find

$$(2) \quad \Delta_x u = f(x+\alpha dx, y, z, \ldots) - f(x, y, z, \ldots),$$

$$d_x u = \lim \frac{\Delta_x u}{\alpha}$$

$$(3) \quad \Delta_x d_y u = d_y(u + \Delta_x u) - d_y u = d_y \Delta_x u$$

and consequently

$$\frac{\Delta_x d_y u}{\alpha} = \frac{d_y \Delta_x u}{\alpha} = d_y \frac{\Delta_x u}{\alpha}\text{"}$$

Then by letting α converge to zero and by taking the second of the formulas

---

[33]This is what the proof looks like. If ϕ, χ are the partial derivatives of u, then we have
$u(x+ \Delta x, y) - u(x,y) = \Delta x\, \phi(x + \theta_1 \Delta x, y)$ and
$u(x+ \Delta x, y+ \Delta y) - u(x+ \Delta x, y) = \Delta y\, \chi(x + \Delta x, y + \theta_2 \Delta y)$ and also
$\Delta u = u(x+ \Delta x, y+ \Delta y) - u(x, y) = \Delta x\, \phi(x + \theta_1 \Delta x, y) + \Delta y\, \chi(x + \Delta x, y + \theta_2 \Delta y)$
If we write $\Delta x = \alpha h$ and $\Delta y = \alpha k$, then
$\Delta u/\alpha = h\phi(x + \theta_1 \alpha h, y) + k\chi(x + \alpha h, y + \theta_2 \alpha k)$
By taking the limit Cauchy obtains
$du = \phi(x,y)dx + \chi(x,y)dy$
Although he does not say so, Cauchy needs the continuity of the partial derivatives here ([Cauchy,1823], pp. 53 - 54).
In Laugwitz' interpretation of Cauchy's analysis, a derivative is always continuous ([Laugwitz 1987], p.268). I disagree. In Cauchy's proof of the mean-value theorem $f(x+h)-f(x)=h.f'(x+\theta h)$, the condition of the continuity of $f'(x)$ is used independently of the condition of the existence of $f'(x)$ ([Cauchy,1823], p.45).

(2) into account, one will obtain equation (1)".[34]
(2) is a repetition of the definition of $d_x u$. Because differentiation is a distributive operation, (3) is also correct. However, in the end, Cauchy interchanges lim and d_y in

$$\lim d_y \frac{u(x+\alpha dx,y) - u(x,y)}{\alpha}$$

Even if one restricts oneself like Cauchy to respectable, repeatedly differentiable functions, there is an important gap here in the proof. This gap was noticed by Blanchet.

IX.3.2 Blanchet fills the gap in Cauchy's proof

Blanchet, who undoubtedly had read Cauchy's proof, first proves a lemma and then the interchangeability theorem in an 1841 letter to Liouville. The lemma concerns precisely the problematic part of Cauchy's proof. Although from a modern point of view Blanchet's proof is not very precise, it was a considerable improvement on Cauchy's proof. That is why I will reproduce Blanchet's proof entirely. The lemma runs as follows:

"If one has $\lim_{\alpha \to 0} \phi(x,\alpha) = \psi(x)$,
one will also have $\lim_{\alpha \to 0} \phi_x(x,\alpha) = \psi'(x)$
under the normal conditions of continuity in the neighbourhood of the values that one will consider".[35]

And this is the proof.
One has $\lim_{\alpha \to 0} \phi(x+h,\alpha) = y(x+h)$,

and also $\lim_{\alpha \to 0} [\phi(x+h,\alpha)-\phi(x,a)] = \psi(x+h)-\psi(x)$.

Which implies $\lim_{\alpha \to 0} h.\phi_x(x+\theta h,\alpha) = h.\psi'(x+\tau h)$

---

[34]"Concevons que l'on indique par la lettre x, placée au bas de la caractéristique Δ, l'accroissement que reçoit une fonction de x,y,z,... lorsqu' on fait croître x seule d'une quantité infiniment petite αdx. On trouvera (2) $\Delta_x u = f(x+\alpha dx,y,z,...)-f(x,y,z,...)$, $d_x u=\lim(\Delta_x u/\alpha)$ (3) $\Delta_x d_y u=d_y(u+\Delta_x u)-d_y u=d_y \Delta_x u$ et, par suite, $(\Delta_x d_y u/\alpha)= =(d_y \Delta_x u/\alpha)=d_y(\Delta_x u/\alpha)$ puis, en faisant converger α vers zéro, et ayant égard à la seconde des formules (2), on obtiendra l'équation (1)" ([Cauchy,1823], p. 77).

[35]"Si l'on a $\lim._\alpha \phi(x,\alpha)=\psi(x)$ on aura aussi $\lim._\alpha \phi_x(x,\alpha)=\psi'(x)$ sous les conditions ordinaires de continuité dans le voisinage des valeurs que l'on considérera" ([Blanchet,1841], p.67; $\mathrm{Lim}._\alpha$ denotes the limit for $\alpha \to 0$)).

(existence and continuity in x of $\phi_x(x,\alpha)$, at least for $\alpha \neq 0$, and existence and continuity of $\psi'(x)$ are implicitly used)

or $$h. \lim_{\alpha \to 0} \phi_x(x+\theta h,\alpha) = h\psi'(x+\tau h).$$

Because h is infinitely small but not equal to zero, we have

$$\lim_{\alpha \to 0} \phi_x(x+\theta h,\alpha) = \psi'(x+\tau h).$$

The left hand side differs infinitely little from $\lim_{\alpha \to 0} \phi_x(x,\alpha)$ (Blanchet also implicitly assumes the existence and continuity in x of $\lim_{\alpha \to 0} \phi_x(x,\alpha)$ and the right hand side differs infinitely little from $\psi'(x)$.

This implies that $\lim_{\alpha \to 0} \phi_x(x,\alpha)-\psi'(x)$ is infinitely small and "because it [the difference - T.K.] is constant it is zero".

In his proof Blanchet uses the existence and continuity in x of $\phi_x(x,\alpha)$ (at least for $\alpha \neq 0$), of $\psi'(x)$ and of $\lim_{\alpha \to 0} \phi_x(x,\alpha)$.

After having proved the lemma, Blanchet turns to the interchangeability *theorem*:

$$f_{yx}(x,y) = f_{xy}(x,y)$$

Proof: We have

$$\lim_{h \to 0} \frac{f(x+h,y)-f(x,y)}{h} = f_x(x,y)$$

Application of the lemma yields

$$\lim_{h \to 0} \frac{f_y(x+h,y)-f_y(x,y)}{h} = f_{xy}(x,y).$$

According to Blanchet, only the continuity of "the second derivative" and of "the first derivatives" is needed. Presumably he means that he needs the continuity in y of $\{f_y(x+h,y)-f_y(x,y)\}/h$ (at least for $h \neq 0$) and the continuity in y of f_{xy} and f_{yx}.

This proof is impressive because it shows a very critical attitude.[36] It is quite

---

[36]In a footnote in Blanchet's paper there is a statement that could be interpreted as meaning that Duhamel had told him [Blanchet] that he [Duhamel] had already given the same proof of the interchangeability property several years earlier in his *Cours lithographié* ([Blanchet,1841], p.65). The only text mentioned in the catalogues of the Bibliothèque

possible that Blanchet subjected Cauchy's proof to a critical analysis, and that he discovered a hidden lemma and proceeded to prove it. Blanchet also clearly states his awareness of the fact that the theorem does not hold unconditionally.

IX.3.3 Two other nineteenth century French proofs of the theorem.

Cauchy-Moigno

I will briefly discuss two other nineteenth century proofs of the interchangeability theorem. In 1840 Moigno published a textbook based on Cauchy's published and unpublished papers, the *Leçons de calcul différentiel rédigées d'après les méthodes et les ouvrages publiés ou inédits par M. l'abbé Moigno* ([Moigno,1840]). Moigno included Cauchy's 1823 proof of the interchangeability theorem, but also another proof. That second proof is based upon the property that in general

(9.22) $$f(x+\Delta x) = f(x) + f_x(x)\,\Delta x + \varepsilon\Delta x$$

with $\lim \varepsilon(x,\Delta x) = 0$ for $\Delta x \to 0$.

Moigno takes f(x,y), substitutes $x + \Delta x$ for x and applies (9.22), which yields

(9.23) $$f(x+\Delta x,y) = f(x,y) + f_x(x,y)\,\Delta x + \varepsilon_1\Delta x$$

Moigno now substitutes in (9.23) $y+\Delta y$ for y. With respect to the terms on the right-hand side we can say the following. We have

(9.24) $$f(x,y+\Delta y) = f(x,y) + f_y(x,y)\,\Delta y + \varepsilon_3\Delta y$$

and

(9.25) $$f_x(x, y+\Delta y) = f_x(x,y)) + f_{xy}(x,y)\,\Delta y + \varepsilon_4\Delta y,$$

Moreover, because ε_1 turns into $\varepsilon_1 + \varepsilon_2\Delta y$, we find

(9.26) $$f(x+\Delta x, y+\Delta y) = f(x,y) + f_x(x,y)\Delta x$$
$$+f_y(x,y)\Delta y + f_{xy}(x,y)\Delta x\,\Delta y +\varepsilon_1\Delta x + \varepsilon_3\Delta y + \varepsilon'\Delta x\Delta y$$

Nationale and the École polytechnique, which could be this above-mentioned *Cours lithographié* is "Duhamel, J.M.C., Suite du cours d'analyse, Leçons et cours autographiés École polytechnique 1837-1838". However, this text does not contain a proof of the interchangeability property. It is moreover improbable that Duhamel would have given a similar proof, because in [Duhamel,1847] he gives a proof of the interchangeability theorem that shows no sophistication whatsoever. Duhamel simply interchanges the order of two limits in a difference quotient ([Duhamel,1847], p.63).

Here ε' equals $\varepsilon_2 + \varepsilon_4$. By reversing the order of substitution (of $x+\Delta x$ and $y+\Delta y$ for x and y in $f(x,y)$), Moigno also derives

(9.27) $$f(x+\Delta x, y+\Delta y) = f(x,y) + f_x(x,y)\Delta x + f_y(x,y)\Delta y + f_{yx}(x,y)\Delta x\, \Delta y + \varepsilon_1\Delta x + \varepsilon_3\Delta y + \varepsilon''\Delta x\Delta y$$

($\varepsilon''=\varepsilon_1 + \varepsilon_3$). Moigno substracts (9.26) from (9.27) and divides by $\Delta x\ \Delta y$:

(9.28) $$f_{yx}(x,y)+ \varepsilon' = f_{xy}(x,y)+ \varepsilon''$$

According to Moigno, both ε' and ε'' vanish with Δx or with Δy, and that implies the desired equality of the mixed partial derivatives.
The situation, however, is not that simple. We know that ε' equals $\varepsilon_2 + \varepsilon_4$. It follows from (9.25) that $\varepsilon_4 = \varepsilon_4(x,y,\Delta y)$ vanishes with Δy. As regards $\varepsilon_2 = \varepsilon_2(x,y,\Delta x,\Delta y)$ we have

(9.29) $$\varepsilon_1(x,y+\Delta y,\Delta x) = \varepsilon_1(x,y,\Delta x) + \varepsilon_2(x,y,\Delta x,\Delta y)\Delta y.$$

and ε_2 apparently vanishes with Δx together with ε_1. However, it is not clear that ε_2 also vanishes with Δy. Conclusion: ε' vanishes when first Δx goes to zero and then Δy. We can say the opposite about ε'': it vanishes if first Δy goes to zero and then Δx. In order to draw the conclusion from (9.28) that f_{xy} equals f_{yx}, one needs to assume that the order of limits can be interchanged either for ε' or for ε''. Moigno does not prove this hidden lemma.

Bertrand's proof

A similar hidden lemma occurs in a proof which was given by Bertrand in a book published in 1864. The proof, however, may be older.
Bertrand's proof runs as follows ([Bertrand,1864], pp.156-157).
He remarks that

(9.30) $$f_x(x,y) = \frac{f(x+\Delta x,y)-f(x,y)}{\Delta x} + \varepsilon_1(x,y,\Delta x)$$

(9.31) $$f_y(x,y) = \frac{f(x,y+\Delta y)-f(x,y)}{\Delta y} + \varepsilon_2(x,y,\Delta y)$$

with $\lim \varepsilon_1(x,y,\Delta x) =0$ for $\Delta x \to 0$ and $\lim \varepsilon_2(x,y,\Delta y) = 0$ for $\Delta y \to 0$.
He then writes

(9.32) $$\frac{\Delta f_x(x,y)}{\Delta y} = \frac{f(x+\Delta x,y+\Delta y)-f(x+\Delta x,y)-f(x,y+\Delta y)+f(x,y)}{\Delta x\Delta y} + \frac{\Delta\varepsilon_1}{\Delta y}$$

$$(9.33)\qquad \frac{\Delta f_y(x,y)}{\Delta x} = \frac{f(x+\Delta x,y+\Delta y)-f(x+\Delta x,y)-f(x,y+\Delta y)+f(x,y)}{\Delta x\Delta y} + \frac{\Delta\varepsilon_2}{\Delta x}$$

At this point Bertrand refers to a lemma that he has given twenty pages earlier:

> "If a function $\Psi(x,a)$ contains, besides the variable x, a letter a that is independent of it, and if this function is zero, supposing that a is equal to zero, then the same holds for the derivative: $d\Psi(x,a)/dx = 0$".[37]

The lemma is based upon the following argument:

> "If the letter is treated as a constant in the differentiation and the value of this constant remains undetermined, it does not make any difference whether we fix its value before or after the operation [...]. It is consequently evident that if the function $\Psi(x,a)$ is infinitely small for an infinitely small value of a, then whatever x is, the same holds for its derivative $\Psi_x(x,a)$".[38]

According to Bertrand the lemma implies that lim $\Delta\varepsilon_1/\Delta y = 0$ and lim $\Delta\varepsilon_2/\Delta y = 0$ and this is why (9.32) and (9.33) imply:

$$(9.34)\qquad \lim \frac{\Delta f_x(x,y)}{\Delta y} = \lim \frac{\Delta f_y(x,y)}{\Delta x}$$

or

$$(9.35)\qquad f_{xy}(x,y) = f_{yx}(x,y)$$

It is remarkable that, in order to prove the lemma, Bertrand compares the taking of a limit to a substitution. We have lim $\varepsilon_1(x,y,\Delta x) = 0$ for $\Delta x \to 0$. Taking this limit is compared by Bertrand to giving Δx the value "infinitely small". Because Δx is a "constant" as to the differentiation with respect to y, it is irrelevant whether we differentiate ε_1 with respect to y before or after fixing the infinitely small value of Δx.

From lim $\varepsilon_1(x,y,\Delta x)=0$ then follows:

---

[37] "Si une fonction Y(x,a) contient, outre la variable x, une lettre a qui en est indépendante, et que cette fonction soit nulle, quelle que soit x, lorsqu'on y suppose a égal à zéro, il en est de même de la dérivée dY(x,a)/dx =0" ([Bertrand,1864], p. 135).

[38] "La lettre étant traitée comme une constante dans la différentiation et la valeur de cette constante restant indéterminée, il est indifférent de la fixer avant ou après l'opération. [...] Il est évident, d'après cela, que si la fonction Y(x,a) est infiniment petite, quel que soit x, pour une valeur infiniment petite de a, il en est de même de sa dérivée $\Psi_x(x,a)$" ([Bertrand,1864], p.135).

(9.36) $$\lim \frac{\Delta\varepsilon_1}{\Delta y} = \lim_{\Delta x \to 0} \frac{d\varepsilon_1}{dy} = \frac{d(\lim_{\Delta x \to 0} \varepsilon_1)}{dy} = \frac{d0}{dy} = 0$$

In a similar way Bertrand's lemma implies

(9.37) $$\lim \frac{\Delta\varepsilon_2}{\Delta x} = 0$$

In (9.34) Bertrand seems to think of Δx and Δy as approaching 0 "at the same time". The implicit assumption then is that the order of limits is interchangeable.
Below I will describe an 1847 proof by Schlömilch that is similar to Bertrand's proof.

IX.3.4 Schlömilch and Lindelöf

Authors like Cauchy, Moigno, Blanchet and Bertrand clearly felt the need for a proof of the interchangeability theorem. And although they all could have been aware of the fact that the interchangeability theorem does not hold unconditionally, it was only Blanchet who paid some attention to the conditions. The proofs of the others create the impression that they were not given at the cost of any great effort. It is possible that the authors gave a proof because they had to give a proof, but not being really interested in the matter, they simply gave the first proof-like argument they came across and proceeded to do things they were more interested in. Except for Blanchet's proof, the proofs are circular. They are based upon an assumption of the interchangeability of two limits that is just as hard to prove as the theorem itself: the interchangeability of limits is proved by means of the interchangeability of limits.

Schlömilch[39]

In various books Schlömilch also proved the interchangeability theorem. The first proof is in his *Handbuch der Differenzial- und Integralrechnung* ([Schlömilch, 1847], pp. 114-117). There he argues as follows:

(9.39) $$f_x(x,y) = \frac{f(x+\Delta x,y)-f(x,y)}{\Delta x} + \xi \quad \text{with} \quad \lim_{\Delta x \to 0} \xi = 0$$

---

[39]I am grateful to Detlef Spalt for helping me to get access to Schlömilch's early proofs of the interchangeability theorem.

which implies

(9.40) $$f_{xy}(x,y) = \frac{d}{dy}\left(\frac{f(x+\Delta x,y)-f(x,y)}{\Delta x}\right) + \frac{d\xi}{dy},$$

or, in other words,

(9.41) $$f_{xy}(x,y) = \lim \frac{f(x+\Delta x,y+\Delta y)-f(x+\Delta x,y)-f(x,y+\Delta y)+f(x,y)}{\Delta x\Delta y} + \frac{d\xi}{dy},$$

Schlömilch writes:

"Here the Lim sign only refers to the vanishing of Δy while Δx still remains unchanged. But because Δx is arbitrary we can let Δx vanish together with Δy (although always independently of y); as a result ξ and consequently also $d\xi/dy$ will approach the limit zero. We then have

[(9.42)] $$f_{xy}(x,y) = \lim \frac{f(x+\Delta x,y+\Delta y)-f(x+\Delta x,y)-f(x,y+\Delta y)+f(x,y)}{\Delta x\Delta y}$$

and now the limit sign refers to a joint unlimited decreasing of Δx and Δy".[40]

In exactly the same way Schlömilch proves:

(9.43) $$f_{yx}(x,y) = \lim \frac{f(x+\Delta x,y+\Delta y)-f(x+\Delta x,y)-f(x,y+\Delta y)+f(x,y)}{\Delta x\Delta y}$$

in which Lim stands for $\Delta x \to 0$ and und $\Delta y \to 0$ "at the same time". From (9.42) and (9.43) the theorem then follows immediately.

Schlömilch's approach is analogous to Bertrand's, but where Bertrand tries to explain why lim $\xi = 0$ implies lim $d\xi/dy = 0$ (for $\Delta x \to 0$), Schlömilch gives no argument whatsoever. On the other hand Schlömilch explicitly speaks about Δx and Δy approaching zero "at the same time".

In the first edition of his *Compendium der höheren Analysis* ([Schlömilch, 1853]), Schlömilch briefly repeated the same proof. In the 1862 edition of the same *Compendium* Schlömilch wrote:

"It has already been remarked from several sides that the foundation and in

---

[40]"Hier bezieht sich das Lim. Zeichen allein auf die Abnahme des Δy, während Δx noch ungeändert bleibt. Da aber Δx beliebig ist, so können wir auch Δx mit Δy gleichzeitig (wiewohl immer unabhängig davon) abnehmen lassen, wodurch sich ξ, folglich auch $d\xi/dy$ der Gränze Null nähert. Wir haben dann $f_{xy}(x,y)$ = Lim $\{f(x+\Delta x,y+\Delta y) -f(x+\Delta x,y)- f(x,y+\Delta y) +f(x,y)\}/\Delta x\Delta y$ und jetzt geht das Zeichen Lim auf eine gemeinschaftliche unbegränzte Abnahme von Δx und Δy" ([Schlömilch,1847], p.116).

particular the foundation of certain fundamental theorems of the differential and integral calculus is nowhere given with full rigor (not even by Cauchy and Moigno); this is why I made it my duty to be extremely precise and I hope that I wil satisfy the most rigorous demands".[41]

And indeed in 1862 Schlömilch also found it necessary to change the proof of the interchangeability theorem of partial differentiation. His proof runs as follows. Let

(9.44) $\Delta(h,k)=\{f(x+h,y+k) -f(x+h,y)-f(x,y+k) +f(x,y)\}$

(this abbreviation is mine - T.K.). By means of the mean-value theorem, Schlömilch then gets

(9.45) $$f_x(x+\theta h,y) = \frac{f(x+h,y)-f(x,y)}{h}$$

According to Schlömilch, θ is independent of y ([Schlömilch,1862], p. 70) and another application of the mean value theorem yields

(9.46) $$f_{xy}(x+\theta h,y+\eta k) = \frac{\Delta(h,k)}{hk}$$

From (9.46) Schlömilch draws the conclusion that

(9.47) $$f_{xy}(x,y) = \lim \frac{\Delta(h,k)}{hk}$$

where "Lim" refers to h and k vanishing "at the same time" (Ibidem, p.71). By applying the mean value theorem first with respect y and then to x, Schlömilch also derives

(9.48) $$f_{yx}(x,y) = \lim \frac{\Delta(h,k)}{hk}$$

and (9.47) and (9.48) together yield

(9.49) $$f_{yx}(x,y) = f_{xy}(x,y).$$

---

[41]"Ueber die Begründung der einzelnen und namentlich gewisser fundamentaler Sätze der Differenzial- sowie der Integralrechnung ist schon von mehreren Seiten bemerkt worden, dass sie fast nirgends (selbst bei Cauchy und Moigno nicht) in voller Strenge zu finden sei; ich habe mir daher gerade nach dieser Richtung hin die äusserste Genauigkeit zur Pflicht gemacht und hoffe, auch den rigorosesten Anforderungen zu genügen."([Schlömilch,1862], p. vi).

In the 1862 proof, Schlömilch uses the mean-value theorem in the following form (modern notation): If $f:[a,b]\to \mathbb{R}$ is differentiable on [a,b] and f' is continuous on [a,b], then there exists a σ with $a<\sigma<b$ such that $f(b)-f(a)=(b-a)f'(\sigma)$ ([Schlömilch, 1862], pp. 40-48). Explicitly referring to the condition of the continuity of the derivative in the proof of the mean-value theorem, Schlömilch finished his proof with a remark concerning the conditions under which the interchangeability theorem holds:

> "[...] the variables may not assume such values that one of the functions
>
> $f(x,y)$, $\partial f(x,y)/\partial x$, $\partial f(x,y)/\partial y$, $\partial^2 f(x,y)/\partial x\partial y$
>
> could become discontinuous".[42]

Schlömilch gave one more proof of the theorem. However, I will first discuss a paper by Lindelöf.

Lindelöf

In 1867 the Finnish mathematician L. Lindelöf wrote a paper in which he gave a critical survey of the existing proofs of the interchangeability theorem. Concerning the eighteenth century proofs he is a true representative of the nineteenth century tradition and is very critical. Euler's proofs do not hold because " one confuses differentials with finite differences".[43] Lagrange's proof is rejected because it is based upon the Taylor series. In Lindelöf's view, an elementary result is in this way firstly based upon a less elementary result and, secondly, cases where the Taylor series is not applicable are excluded from the start. More interesting is Lindelöf's criticism of the nineteenth century proofs. He criticizes Schlömilch's 1862 proof and the 1864 proof of Bertrand by means of counter-examples. The counter-examples are used to reject "lemmas" that are used in those proofs. Schlömilch's 1862 proof is attacked where Schlömilch assumes that in

$$f_x(x+\theta h,y) = \frac{f(x+h,y)-f(x,y)}{h} \qquad (9.50)$$

the value of θ does not depend on y. In the case of a function $y = \sqrt{(a^2-x^2)}$ on the interval [0,h], it is easy to show that the θ depends on the parameter a. Bertrand's proof is criticised with respect to the lemma: If $\Psi(x,a)$ is such that

---

[42] "[...] die Variabelen dürfen keine solchen Werthe erhalten, wodurch eine der Functionen $f(x,y)$, $\partial f(x,y)/\partial x$, $\partial f(x,y)/\partial y$, $\partial^2 f(x,y)/\partial x\partial y$ discontinuierlich werden könnte" ([Schlömilch, 1862], p.72).

[43] "on confond des différentielles avec des différences finies" ([Lindelöf,1867], p. 208).

for a->0 lim $\Psi(x,a)=0$ for all x, then also lim $\Psi_x(x,a) = 0$ for all x. Lindelöf admits that the lemma is true "in general", but that it nevertheless suffers exceptions. He gives the example y = a sin(x/a). For a->0 we have lim y(x) = 0, but for a->0 lim $y_x(x)$ is undetermined. Without giving counter-examples, Lindelöf also criticizes Bertrand and "all the others", which undoubtedly includes also Schlömilch, for assuming that the value of the limit of a function of two variables is independent of the relation that exists between those variables:[44]

> "There is another principle intimately related to the nature of the question that Monsieur Bertrand, without explicitly saying so, uses like all the others. This is the principle that the limit to which a function of two variables converges when these variables vanish or approach certain determined values, is independent of the relation that exists between the variables. This is in fact the essential character of a monodromic and continuous function. But after admitting this principle any other hypothesis becomes superfluous and our theorem follows almost immediately from it".

One could say that in order to solve these problems, Lindelöf applies (in Lakatosian terms) the method of lemma incorporation. The resulting proof runs as follows. First Lindelöf argues:[45]

---

[44]"Il y a un autre principe intimement lié à la nature de la question et dont M. Bertrand, sans le dire expressement, fait usage aussi bien que tous les autres. C'est que la limite vers laquelle converge une fonction de deux variables, lorsque celles-ci s'évanouissent ou s'approchent de certaines valeurs déterminées, est indépendante de la liaison établie entre les variables. Ceci est en effet le caractère essentiel d'une fonction monodrome et continue. Mais en admettant ce principe, toute autre hypothèse devient superflue et notre théorème en résulte presque immédiatement" ([Lindelöf,1867], p. 212).

[45]"Soit, en effet, z=f(x,y) l'équation d'une surface continue aux environs d'un point P, que l'on considère en particulier et dont les coordonnées sont x,y,z; donnons à x et y des accroissements h,k et considérons l'expression {f(x+h, y+k) - f(x+h,y) - f(x,y+k) +f(x,y)}/hk dans laquelle nous regardons h et k comme variables. Celle-ci est évidemment continue pour des petites valeurs de h et k différentes de zéro et il est facile de voir que la continuité de sera pas interrompue, lorsque l'une ou l'autre de ces variables passe par zéro, pourvu que les dérivées partielles dz/dx, dz/dy soient aussi continues aux environs du point P. Dans cette hypothése l'expression précédente sera donc monodrome et continue tout autour du point P, de sorte qu'en faisant h et k décroître jusqu'à zéro, elle tendra vers une limite qui sera indépendante de la liaison supposée entre ces variables.Or si l'on fait d'abord h=0 et qu'on appelle $\phi(x,y)$ la dérivée partielle de f(x,y) prise par rapport à x, cette expression se réduit à $\{\phi(x,y+k)-\phi(x,y)\}/k$; faisant ensuite k=0, elle devient $d\phi(x,y)/dxdy$ =

"Let, in fact, z = f(x,y) be the equation of a continuous surface in the neighbourhood of a point P that one considers in particular and whose coordinates are x, y, z; let us give x and y increments h and k and let us consider the expression

$$\frac{f(x+h,y+k)-f(x+h,y)-f(x,y+k)+f(x,y)}{hk}$$

in which we consider h and k as variables. This expression is evidently continuous for small values of h and k that differ from zero and it is easy to see that the continuity will not be interrupted if one or the other of those variables passes through zero, provided that the partial derivatives dz/dx, dz/dy are also continuous in the neighbourhood of the point P. Under this hypothesis the preceding expression will consequently be monodromic and continuous all around the point P, so that, if h and k decrease towards zero, the expression will tend towards a limit that will be independent of the supposed relationship between those variables. Well then, if one puts first h = 0 and calls $\phi(x,y)$ the derivative of f(x,y) taken with respect to x, this expression will be reduced to

$$\frac{\phi(x,y+k)-\phi(x,y)}{k};$$

Taking next k = 0, the expression becomes

$$\frac{d\phi(x,y)}{dxdy}=\frac{d^2z}{dxdy}.$$

The same substitutions made in the inverse order would give the result $\frac{d^2z}{dydx}$. Consequently one has $\frac{d^2z}{dxdy}=\frac{d^2z}{dydx}$".

What is happening here? I read Lindelöf as follows. The second quotation shows clearly that Lindelöf all the time interprets the function f(x,y) geometrically. He thinks of a surface with equation z = f(x,y). The hidden lemma that Lindelöf discovered in the work of Bertrand and others consists of the principle that the surface corresponding to z = f(x,y) consists of one sheet and is continuous; or, in other words, in the area considered we have: (i) the value f(x,y) is unique and (ii) no matter how (h,k) approaches (0,0): Lim

---

$d^2z/dxdy$. Les mêmes substitutions faites dans l'ordre inverse donneraient pour résultat $d^2z/dydx$. On aura donc $d^2z/dxdy = d^2z/dydx$" ([Lindelöf,1873], p. 213).

f(x+h,y+k) = f(x,y). In order to prove the interchangeability theorem Lindelöf then considers a fixed point P with coordinates (x,y) and he concentrates on the expression

$$(9.51) \qquad \frac{\Delta(h,k)}{hk} = \frac{f(x+h,y+k)-f(x+h,y)-f(x,y+k)+f(x,y)}{hk}$$

as a function of h and k. When h and k are not equal to zero this expression is continuous in (h,k). If, moreover, f_x and f_y exist and are continuous in the neighbourhood of P, then according to Lindelöf the continuity of Δ(h,k)/hk is not interrupted when h=0 or k=0.

From the fact that the expression Δ(h,k)/hk is continuous in a reduced neighbourhood of P, Lindelöf apparently draws the conclusion that at P the expression is also continuous. Then no matter how (h,k) approaches (0,0) we will always get the same value for lim Δ(h,k)/hk. This immediately yields the interchangeability property.

Lindelöf's criticisms are very penetrating and with his explicit use of counter-examples, he represents a new attitude in the history of the theorem. Yet his new proof shows that he still underestimated the problem. In particular the last part of the proof is not satisfactory.

In retrospect, I think that Lindelöf was the victim of the geometric images that accompany his reasoning. He considered the expression (9.51) as a function of the variables h and k; he convinced himself of the fact that the function changes continuously in a reduced neighbourhood of (0,0) and he simply could not imagine that the value of lim Δ(h,k)/hk might depend on the manner in which the origin is approached.[46]

Schlömilch's 1874 proof

In the third 1868 edition of his Compendium, Schlömilch simply repeated the 1862 proof. In 1874, however, Schlömilch came up with a modification of the proof that is no longer based upon the assumptions that Lindelöf had

---

[46]Lindelöf's use of the word "monodromic" to denote the fact that a function is single-valued is, although understandable, exceptional. In the nineteenth century the word "monodromic" is almost exclusively used for complex functions. Compare: "He [Cauchy-T.K.] defines, for example, as *monodromic* a (single-valued) function defined in a finite portion S of the complex plane when the function always takes the same value at one and the same point P in S, whatever path z may take within S to arrive at P" ([Bottazzini,1986], p.170).

criticized.[47] Perhaps he had read Lindelöf's paper. On the other hand, he must have considered the matter of secondary importance, because nowhere does he explicitly point out that he modified the 1862 proof.
Schlömilch reasons as follows. We can apply the mean-value theorem to f(x+h,y)-f(x,y) as a function of y on [y,y+k].The result is

$$(9.52) \qquad \Delta(h,k)= k\{f_y(x+h,y+\theta k)-f_y(x,y+\theta k)\}$$

We can apply the mean-value theorem once more, this time to $f_y(x,y+k)$ as a function of x on [x,x+h]. We obtain

$$(9.53) \qquad \Delta(h,k)/hk = f_{yx}(x+\eta k,y+\theta k)$$

By first applying the mean-value theorem with respect to x and then to y, Schlömilch similarly derives:

$$(9.54) \qquad \Delta(h,k)/hk = f_{xy}(x+\mu k,y+\nu k)$$

Instead of immediately passing to the limit in (9.53) and (9.54) as in 1862, Schlömilch in 1874 first draws the conclusion

$$(9.55) \qquad f_{xy}(x+\mu k,y+\nu k) = f_{yx}(x+\eta k,y+\theta k)$$

and then allows h and k to converge to zero. This yields the desired result, and the dubious appeal to simultaneous convergence to zero of h and k is superfluous. The order of convergence is no longer important.
In 1874 Schlömilch finished the proof with a remark concerning the conditions under which the interchangeability theorem holds. He wrote:

> "In the present case this [the fact that the mean-value theorem holds conditionally -T.K.] implies that x and y are not allowed to have values for which one or another of the functions
> $f(x,y)$, $\partial f(x,y)/\partial x$, $\partial f(x,y)/\partial y$, $\partial^2 f(x,y)/\partial x\partial y$,
> *assuming the existence of the last three, becomes infinite* or discontinuous" (Emphasis is mine - T. K.).[48]

---

[47]Schlömilch's 1874 proof also occurs in [Jordan,1882], [Genocchi,1884] and elsewhere, without reference to Schlömilch.

[48]"Im vorliegenden Fall folgt hieraus [the fact that the mean-value theorem holds conditionally -T.K.], dass x und y keine solchen Werthe halten dürfen, für welche die eine

In this respect the only difference with the 1862 edition is that the words which I have emphasized were added in the 1874 edition.
It is remarkable that both Schlömilch and Blanchet, at the moment when they formulate the sufficient conditions, assume the equality of the mixed second order derivatives. As we have seen, Blanchet writes about "*the* second derivative" and Schlömilch only mentions f_{yx}, while his proof shows that he should write f_{xy} and f_{yx}.

IX.3.5 Schwarz: The conditions become subject of investigation

With his classic paper "Ueber ein vollständiges System von einander unabhängiger Voraussetzungen zum Beweise des Satzes $\frac{\partial}{\partial y}(\frac{\partial f(x,y)}{\partial x}) = \frac{\partial}{\partial x}(\frac{\partial f(x,y)}{\partial y})$", H. A. Schwarz inaugurated a new period in the history of the theorem. As we have seen, Schwarz's predecessors were in general aware of the fact that the theorem only holds conditionally, but they did not pay much attention to the conditions under which it actually holds.
Lindelöf is a transitional figure in this respect: in Lindelöf's work there is a shift of attention to the conditions under which the theorem holds, but he still underestimates the problem. Schwarz not only sought some set of necessary conditions, but he explicitly made the conditions the object of his investigation. He raised the problem of the necessary and sufficient conditions for the equality of the mixed partial derivatives and in a sense solved this problem. The rather careful use of what are in fact ε-δ definitions makes Schwarz's proofs quite acceptable from modern point of view. For example, Schwarz uses the following definition: a function f(x,y) of two variables is continuous in (x_0,y_0) if $|f(x,y)-f(x_0,y_0)|$ becomes arbitrarily small as soon as $|x-x_0|<h$ and $|y-y_0|<k$ for sufficiently small h>0 and k >0 (Cf.[Schwarz,1873], p.278).
Schwarz's work on the interchangeability theorem illustrates the important modification that the nineteenth century tradition underwent in the second half of the century, particularly under the influence of Weierstrass. Weierstrass and mathematicians influenced by him executed a successful foundational research project within nineteenth century analysis that led to new versions of all fundamental theorems in real analysis. It is characteristic that the conditions under which the theorems hold are all precisely defined and that the proofs are all given by means of explicitly used ε–δ definitions.

---

oder andere der Functionen f(x,y), $\partial f(x,y)/\partial x$, $\partial f(x,y)/\partial y$, $\partial^2 f(x,y)/\partial x \partial y$ die Existenz der drei letzten vorausgesetzt, unendlich oder discontinuierlich wird"([Schlömilch,1874], p.72).

Schwarz considers the question: precisely when do we have the interchangeability property for a continuous function defined on a continuous two-dimensional manifold:

$$f_{xy}(x,y) = f_{yx}(x,y)\ ?$$

His answer consists of three theorems:

1. $f_{xy}(x,y) = f_{yx}(x,y)$ holds on the entire manifold, if f_x, f_y, f_{xy}, f_{yx}, f_{xx}, f_{yy} (exist and) are continuous on the manifold.
2. $f_{xy}(x,y) = f_{yx}(x,y)$ holds on the entire manifold, if f_x, f_y, f_{xy}, f_{yx} (exist and) are continuous on the manifold.
3. $f_{xy}(x,y) = f_{yx}(x,y)$ holds on the entire manifold, if f_x, f_y, f_{xy} (exist and) are continuous on the manifold.

Schwarz consciously uses counter-examples in his attempt to find mutually independent and sufficient conditions. After having proved his first version of the theorem, he remarks (without giving examples) that one can easily define functions $f(x,y)$ which satisfy all conditions of that theorem except that $f_{xx}(x,y)$ and $f_{yy}(x,y)$ do not exist, and because the theorem only directly concerns $f_{xy}(x,y)$ and $f_{yx}(x,y)$, the question arises as to whether the requirements of the existence and continuity of f_{xx} and f_{yy} are really necessary. Schwarz's answer is: no. Next he proves the second version of the theorem. Then he comes up with the well-known example which shows that one cannot drop the requirement of the continuity on the manifold of at least one of the mixed partial derivatives:
$f(x,y) = y^2 \operatorname{arctg}(x/y) - x^2 \operatorname{arctg}(y/x)$; $f(0,0) = 0$. ([Schwarz,1873], pp.280-281). In this case f, f_x, f_y exist and are continuous for all $(x,y) \in \mathbb{R}^2$. The derivatives f_{xy} and f_{yx} exist for all (x,y), but both are discontinuous in $(0,0)$. Moreover, $f_{xy}(0,0) = 1$ and $f_{yx}(0,0) = -1$.
With the third version of the theorem Schwarz shows, however, that one need not require more than the continuity of one of the mixed partial derivatives. He gives two proofs of this third version of the theorem. I will describe the second proof.[49]
Let again

$$\Delta(h,k) = f(x_0+h,y_0+k) - f(x_0+h,y_0) - f(x_0,y_0+k) + f(x_0,y_0)\} \tag{9.56}$$

Then using the continuity of f_x, f_y and f_{xy}, Schwarz proves by means of the

[49]The first proof is based upon the interchangeability of the two integrals in a double-integral of a continuous function of two variables.

mean-value theorem that

(9.57) $\Delta(h,k) = hkf_{xy}(x_0+\theta_1 k, y_0+\theta_2 k)$

and

(9.58) $\Delta(h,k) = k\{f_y(x_0+h, y_0+\theta_3 k) - f_y(x_0, y_0+\theta_3 k)\}$

Then

(9.59) $$|f_y(x_0+h,y_0)-f_y(x_0,y_0) - f_{xy}(x_0,y_0).h| =$$
$$\left|\frac{\Delta(h,k) - f_{xy}(x_0,y_0)hk - \Delta(h,k) + f_y(x_0+h,y_0)k - f_y(x_0,y_0)k}{k}\right| =$$

(by means of (9.57) and (9.58))

$$| h\{f_{xy}(x_0+\theta_1 k,y_0+\theta_2 k) - f_{xy}(x_0,y_0)\}+$$
$$-\{f_y(x_0+h,y_0+\theta_3 k)-f_y(x_0+h,y_0)\}+$$
$$+\{f_y(x_0,y_0+\theta_3 k)-f_y(x_0,y_0)\}| \leq$$

$$| h\{f_{xy}(x_0,y_0+\theta_2 k) - f_{xy}(x_0,y_0)\}|+$$
$$| h\{f_{xy}(x_0+\theta_1 k,y_0+\theta_2 k) - f_{xy}(x_0,y_0+\theta_2 k)\}|+$$
$$|f_y(x_0+h,y_0+\theta_3 k) -f_y(x_0+h,y_0)|+$$
$$|f_y(x_0,y_0+\theta_3 k)-f_y(x_0,y_0)|$$

Let us define on a neighbourhood $O = \{(x,y) | \, |x-x_0| \leq h \;\&\; |y-y_0| \leq k\}$:

(9.60) $$h' = \max |f_{xy}(x,y_0) - f_{xy}(x_0,y_0)|$$
$$k' = \max |f_{xy}(x,y)-f_{xy}(x,y_0)|$$
$$k'' = \max |f_y(x,y)-f_y(x,y_0)|$$

Then (9.59) implies that on that neighbourhood O:

(9.61) $$|f_y(x_0+h,y_0)-f_y(x_0,y_0) - f_{xy}(x_0,y_0).h| <$$
$$(h.k.(h' + k') + k.k'' + k.k'')/k =$$
$$h.h' + h.k' + 2k''$$

The left-hand side of (9.60) is independent of k and also h' is independent of k. Moreover, we have for k->0 that Lim k' = Lim k'' = 0. So we obtain

(9.62) $|f_y(x_0+h,y_0)-f_y(x_0,y_0) - f_{xy}(x_0,y_0).h| < h.h'$

and, obviously, from (9.62) and Lim h' = 0 for h->0 follows

(9.63) $f_{xy}(x_0,y_0)= f_{yx}(x_0,y_0)$ Q.E.D.

Schwarz started with six conditions and attempted to drop as many of them as possible[50]. Whether or not Schwarz in this way reached his original goal, i.e. the determination of a complete system of independent conditions, depends on the way one defines the problem. If one reads the third theorem as follows: "If for a function f(x,y) that is defined and continuous on a manifold G:

i) f_x is continuous on the entire manifold G
ii) f_y is continuous on the entire manifold G
iii) f_{xy} is is continuous on the entire manifold G

(where continuity, of course, implies existence), then

$f_{xy}= f_{yx}$ on the entire manifold G.",

then Schwarz did reach his goal, because none of the three conditions implies one of the others. However, as soon as one does not want to prove the interchangeability property for the whole manifold, but only for one particular point (x_0,y_0), then one gets theorems of the kind proved later by Dini and Peano ([Dini,1877],pp.117-128, [Peano,1890] and [Dini,1907], pp.164-170).

Peano in 1890 and Dini in 1907 proved: "If for a function f(x,y) defined in a neighbourhood $O(x_0,y_0)$ of a point (x_0,y_0) f_x and f_{xy} exist in the entire neighbourhood $O(x_0,y_0)$ and, moreover i) f_y exists in a neighbourhood $O_x(x_0,y_0)$ [=(def.)$O(x_0,y_0)\cap\{(x,y)|y=y_0\}$] and ii) f_{xy} is continuous in (x_0, y_0), then holds the following $f_{xy}(x_0,y_0)= f_{yx}(x_0,y_0)$ ".

Because f_{yx} does not exist at all in (x_0,y_0) without i), Dini calls ii) the only essential condition for the validity of the theorem ("la sola condizione veramente inerente alla validità del teorema" ([Dini,1907], p.168).

[50]It is not clear whether the order of presentation in Schwarz's paper is paedagogical or whether it corresponds to the order of genesis of the results.

Cauchy (1823 textbook) The theorem concerns the relation (expressed in the form of a formula) between a dependent variable and two (or more) independent variables; the theorem expresses the equality of the mixed partial derivatives (defined in terms of the limit concept); the validity of the theorem is restricted by implicit conditions of continuity.
Cauchy-Moigno (1840 textbook) Idem
Schlömilch (1847 textbook / repeated in 1853 textbook) Idem
Bertrand (1864 textbook) Idem
Blanchet (1841 paper) Blanchet proves a hidden lemma in Cauchy's 1823 proof and brings out the condition of continuity of the first derivatives and the mixed derivatives.
Schlömilch (1862 textbook / 1868 textbook) Schlömilch gives a different proof; continuity of the function, its first derivatives and the mixed derivatives is now required.
Lindelöf (1867 paper) Using counter-examples (for the first time in the history of the theorem) Lindelöf points out weak spots in most of the earlier proofs. For example, with the exception of Blanchet, all above mentioned nineteenth century authors assume that the limit of a function of two variables is uniquely determined when the two variables vanish "at the same time" independently of the relation that exists between the two variables.
Schlömilch (1873 textbook) Schlömilch modified his 1862 proof in such a way Lindelöf's criticism was no longer applicable.
Schwarz (1873 paper) The theorem acquires basically its modern form. For all earlier authors the theorem played a secondary role - it had to be proved in order to be able to apply it. In the background is a notion of function that is very general and no longer tied to analytical expressions. Independently of the applicability of the theorem in calculations the conditions under which the theorem holds become subject of investigation.
Dini (1877 textbook) / *Peano* (1890 paper) / *Dini* (1907 textbook) Schwarz still restricted himself to necessary conditions for interchangeability in a whole neighbourhood; Dini and Peano gave necessary conditions for interchangeability in one point.

Table II: The nineteenth century proofs discussed in the text.

IX.4 Conclusion

Table II contains a list of the nineteenth century versions of the interchangeability theorem that I have discussed. The series of versions of the theorem supports the reconstruction in the previous chapter with respect to the existence of a nineteenth century tradition in analysis. They all represent the Conceptual Tradition. The break with the eighteenth century is obvious. For Cauchy a function was no longer a formula but a relation between variables expressed by a formula. Cauchy studied functions in terms of notions like continuity and differentiability that did not primarily concern formulae but the relation between variables expressed by those formulae. With Cauchy functions are not simply continuous, but continuous on intervals, and they are not simply discontinuous but discontinuous for particular values of the independent variable. With Cauchy the interchangeability theorem stopped being unconditionally valid. A change in the notion of function immediately led to a completely different interchangeability theorem that only holds conditionally. Blanchet soon pointed out hidden assumptions in Cauchy's proof concerning continuity. The series of different versions that I have discussed shows that nineteenth century mathematicians were not at first very much interested in a precise formulation of necessary conditions. Undoubtedly they did not expect any interesting, or in terms of the MMRT, weighty results from such investigations. Only Schwarz's different versions of the theorem and their proofs show a different attitude. These proofs are part of research projects that led to a precise formulation of (minimal) conditions for all theorems expressing relations between fundamental notions like continuity, differentiability etc.

Schwarz's work on the interchangeability theorem reflects the nineteenth century tradition in its mature form. The nineteenth century versions of the theorem discussed so far sufficiently illustrate the internal development of the nineteenth century tradition in analysis. We need not describe the history of the interchangeability theorem any further.[51]

---

[51] For a survey of part of the subsequent history of the theorem I refer to [Koetsier, 1990].

Chapter X Some concluding remarks

X.1 The rationality of mathematics

In Chapter I we saw that the Methodology of Proofs and Refutations (MP&R) captures important aspects of mathematical rationality (Chapter I). However, Lakatos's Methodology of Scientific Research Programmes (MSRP) which he defined for natural science, had to be modified in order to be applicable to mathematics (Chapter II). A critical discussion of Lakatos's "Cauchy and the continuum" and Spalt's *Vom Mythos der mathematischen Vernunft* made it clear that such a modification, and a demonstration of its applicability to mathematics by means of factual rational reconstructions, would require great methodological care in order to avoid vicious circularity (Chapters III and IV). I decided to base myself primarily on secondary sources and only to consider cases about which there exists considerable agreement among historiographers. The idea was to use the expertise of the historiographers, to look at history through their eyes, and to attempt to formalize their views in terms of a modified MSRP. A comparison of Giorello's descriptions of the work of Archimedes and Riemann in terms of the MSRP with the existing historiography on the development of the work of those great mathematicians, led to the notions of research project and research tradition (Chapter V).

The next step was made in Chapter VI, that of defining an outline of a modified MSRP - the Methodology of Mathematical Research Traditions (MMRT) - followed in the next chapters by the verification of the validity of this methodology by means of two rational reconstructions. In Chapter VII, on the basis of the existing historiography, I distinguished two traditions in Greek mathematics - the Demonstrative Tradition and its successor, the Euclidean Tradition - and I argued that the Euclidean Tradition replaced its predecessor in accordance with the MMRT. In Chapter VIII, I distinguished two traditions in analysis - the Formalist Tradition and the Conceptual Tradition. I also argued there that the transition from the one tradition to the other, in the beginning of the nineteenth century, took place along the lines of the MMRT. The Formalist Tradition and the Conceptual Tradition, and, in particular, their internal development, were studied in more detail in Chapter IX.

Chapters VII and VIII showed that historians are inclined to use notions like research tradition and research project in an informal way. Their descriptions of some major transitions can be easily translated into the language of MMRT. Using that language we can also say that in the two cases I considered, the appearance of the new tradition was accompanied by the occurrence of highly progressive research projects in the sense of the MMRT. In other words, factual rational reconstructions of major developments in mathematics on the basis of the MMRT are possible, and although historians might hesitate to accept them because of the all-encompassing character of the conceptual apparatus of the MMRT, the reconstructions represent a historiographically legitimate way of viewing these developments.

Concerning Lakatos's rationality thesis this study yields a view of the development of mathematics that can be summed up as follows. A mathematician working on a particular mathematical problem is by definition working at the *micro-level*. He studies a particular *domain* of mathematical entities and attempts to develop a *theory* for that domain, i.e. a set of statements considered to be true of that domain. At the micro-level a rational mathematician obeys Lakatos's MP&R. At that level mathematicians frame conjectures, then attempt to refute or prove these conjectures, and afterwards they subject proofs and refutations to critical analysis, etc.

However, micro-level work in general concerns a sub-domain of a much more general *fundamental domain* of mathematical entities that also encompasses other sub-domains that are being studied in other theories. And yet theories concerning different sub-domains of a fundamental domain are usually developed in accordance with the same *methodology of proof.* Such a fundamental domain, together with such a general methodology of proof, is a *mathematical research tradition.* Research traditions usually encompass large groups of mathematicians. At this *macro-level*, the methodology of mathematical research traditions (MMRT) describes the rationality of the development of mathematics. An important role is played by *research projects*, characterized by their heuristic unity - research goals together with sets of hints exemplified by one or more paradigmatic cases. In a research project conjectures and suggestions for ways to prove the conjectures are generated by analogy. At the micro-level, research projects add something essential to the MP&R - they suggest how and what to prove or to refute. The research projects are the "driving force" behind the development. New mathematics, when fitting into the framework of an existing tradition or representing a new tradition, is always the result of a research project. According to the MMRT a rational mathematical community makes choices between research projects and research traditions in such a way that it can expect to contribute to mathematics as much as possible in terms of the *weight of conjectures and theorems* that are going to be proved.

In brief, as regards the rationality thesis my claim is the following. The rationality of mathematics consists in the MP&R on the micro level and the MMRT on the macro level since the development of mathematics can be reconstructed successfully by means of these methdodologies.

X.2 Fallibility and continuity

X.2.1 The fallibility of mathematics in general

I will now consider the fallibility thesis in the light of my conclusions concerning the rationality thesis. In Chapters I and II, I discussed the possibility that mathematics is fallible in a way comparable to the fallibility of natural science in Popper's view. Such a thesis implies the possibility of considerable theoretical discontinuity - it implies the possibility that mathematical theories are refuted entirely. I called this the *strong fallibility thesis*. However, we did not find any evidence to support such a strong thesis. On the contrary, everything points to the validity of a much more moderate thesis, the *weak fallibility thesis*, implying that always considerable parts of a mathematical theory remain unrefuted.

Let us first consider the micro-level. In Chapter I, I investigated the history of Euler's formula for polyhedra. We saw that Euler proved the theorem V-E+F=2 with a notion of polyhedron in mind that appeared to possess an extension with an obscure borderline. An explicit restriction to convex polyhedra by Legendre shows an awareness of this obscurity. Distinguishing between convex and non-convex polyhedra is an example of conceptual refinement by dissociation. The examples of polyhedra found by L'Huilier and Hessel, for which the formula does not hold, revealed that the theorem in the form V-E+F=2 is indeed not universally valid for arbitrary polyhedra. It also became clear that the restriction to convex polyhedra was too severe. The problem was only satisfactorily solved after conceptual refinements by dissociation brought about by the introduction of the notion of connectivity. The pattern is as follows: potential unintended interpretations of notions appear and the theorems thus "falsified" are saved after suitable conceptual refinement. When we compare the beginning - Euler's version of the theorem - and the end of the development which we discussed - the versions of Listing, Jordan and Becker of the theorem - we could argue that Euler, without explicitly saying so, restricted himself to convex polyhedra. In that case Euler made no mistake as far as the truth of the theorem is concerned. In this interpretation the history of Euler's theorem hardly supports fallibilism. Let us, however, consider the most favourable interpretation if we want to defend fallibilism. In this interpretation we would argue that in Euler's view the theorem actually holds for *all* polyhedra. From the point of view of Listing, Jordan and Becker the conclusion is then clear: Euler made a mistake. Yet Listing e.a. did not reject Euler's theorem in its totality: all that happened was that the range of its validity was restricted. A universality claim, an aspect of the theorem, was falsified.

In the case of the history of the notion of uniform convergence we can argue in precisely the same way. When he proved that the limit function of a convergent series of continuous functions is continuous, Cauchy either thought only of uniformly convergent series of functions, or else he thought of all functions. The latter interpretation is the most favourable for fallibilism, but also in this case the original theorem was not rejected in its totality; by means of a suitable conceptual refinement the range of its validity was restricted.

The two examples, the history of Euler's formula for polyhedra and the history of the notion of uniform convergence, suggest the following weak fallibility thesis: *mathematical theories are weakly fallible in the sense that one can never exclude the occurrence of unintended possible interpretations of fundamental notions that require a restriction of universality claims by means of conceptual refinement. Because only the range of validity of theorems is restricted weak fallibility implies far going continuity.*

Let us also enter the macro-level perspective of the methodology of mathematical research traditions. Let us look at the way in which some traditions have replaced others. How fallible is the collection of theories that belongs to a mathematical research tradition? The two examples of competition between traditions in Chapters VII and VIII, but also the history of the interchangeability theorem for partial differentiation in Chapter IX, show that within the winning tradition the fundamental domain of the losing tradition is re-interpreted. In the Euclidean Tradition, for example, the fundamental domain of the Demonstrative Tradition is re-interpreted. This re-interpretation has the form of a more precise determination of the object in the form of definitions, postulates and axioms. The corrections involved are small in this case. From the point of view of the Euclidean Tradition, the Demonstrative Tradition perhaps lacked precision, but it is a fact that the bulk of the results reached in the Demonstrative Tradition was taken in by the Euclidean Tradition. The object of the Formalist Tradition was also re-interpreted in the Conceptual Tradition in analysis. When formulae were re-interpreted by Cauchy as relations between variables, some eighteenth century results, for example on divergent series, temporarily lost their meaning. Yet most of the theorems involved were not refuted in any way. Instead, as we have seen, after a break of half a century the nineteenth century tradition started absorbing many of the results rejected by Cauchy. However, most of the eighteenth century results were simply maintained by Cauchy and the other representatives of the Conceptual Tradition, although the results needed new proofs. In the course of time the new interpretation was more precisely defined and the eighteenth century results only survived conditionally in the end. The history of the interchangeability theorem for partial differentiation nicely shows this process. Roughly speaking one can say that the Formalist Tradition in fact restricted itself to a particular class of functions, the analytical functions. For such functions the properties derived by the eighteenth century mathematicians are also valid from the nineteenth century point of view. The shortcoming of eighteenth century analysis was that the results were less generally valid than was assumed. There is a striking similarity here with the history of Euler's formula and the history of the notion of uniform convergence. The macro-level perspective confirms our conclusion with respect to the micro-level: fallibility in mathematics seems to consist in the fact that results are fallible in the sense that after a process of conceptual refinement of fundamental notions, the domain of validity of those results may be restricted. As far as fallibility is concerned, it apparently does not make much difference whether a theorem was replaced by a successor theorem along the lines of the MP&R or whether it was replaced because of a change of tradition in accordance with the MMRT. In the latter case the theorem in question need not have been explicitly refuted in any way; it is simply replaced by its successor because a progressive new tradition requires new proofs of all relevant theorems.

It is also interesting to note what happened in the nineteenth century to the fundamental domain of the Euclidean Tradition with the appearance of hyperbolic geometry. Euclidean geometry was not refuted but was restricted in its validity. In this case the development also shows weak fallibility in the above-defined sense. Old results were not thrown away, but merely restricted in their domain of validity because originally unintended interpretations of the central notions had shown up. It is interesting that in this case, most of the conceptual refinement involved had already taken place in Antiquity when the Euclidean plane was more precisely defined by definitions, postulates and axioms during the formation of the Euclidean Tradition. For example, as soon as one attempts to prove Postulate 5 by reductio ad absurdum, one separates situations for which it holds from situations for which it does not hold on a logical level. However, because Euclidean space was considered to be the one and only space, other logically possible situations were not considered.

The unintended interpretations were present, but only on a logical level, and they were not taken seriously mathematically. Only in the nineteenth century did the originally unintended interpretations become acceptable when the crucial conceptual refinement took place: physical space was distinguished from geometrical space(s).

X.2.2 The fallibility of twentieth century mathematics

When we look at the development of mathematical knowledge over the entire period from Antiquity to the present, mathematical knowledge must be considered as (weakly) fallible in the sense described above. But then we are looking at the problem of the fallibility of mathematics in a rather global way, whereas the fallibility of mathematical knowledge within a tradition is undoubtedly related to the methodology of proof in that tradition. For example, a hypothetical tradition which would consider all conjectures of the form $\forall(x)P(x)$ as proven if after a certain effort the mathematical community did not come up with a counter-example, is more fallible than traditions that require more in their methodology of proof. It is quite possible to take a global point of view and to compare the methodologies of proof of different traditions with respect to their fallibility. The proofs of the Demonstrative Tradition for example were primarily based upon visualization. One can view Brouwer's intuitionism as an attempt to found a twentieth century tradition in mathematics also based upon mental visualization. If the "mental eye" is guided in the right way, this evidence carries considerable conviction. We have seen that in the proofs of the Formalist Tradition a considerable inductive "element" was present and with it a built-in uncertainty. The proofs of the Euclidean Tradition were based primarily upon deductive reasoning and the same holds for the Conceptual Tradition in analysis and the Structuralist Tradition. Deductive reasoning may also be fallible, particularly when new notions are being studied and when the reasoning is quite informal.

The Structuralist Tradition, however, has a special place in this respect when compared with other traditions. There has been extensive research this century on the foundations of mathematics. Before the twentieth century no tradition in the history of mathematics so successfully reflected upon its own fundamental assumptions. The result has been an extremely high standard of rigor in the form of a very precisely defined methodology of proof. Mac Lane has argued that nowadays we possess an exact notion of absolute rigor which rests upon the possibility of formalizing proofs completely ([Mac Lane, 1981], p.465). A great number of mathematicians undoubtedly share this view. Although absolutely rigorous proofs are rarely given explicitly, according to Mac Lane the sketches that we call mathematical proofs nowadays usually give enough detail to indicate how a full rigorous proof might be constructed. Does this mean that, leaving aside ordinary errors such as a mistake in a calculation, the Structuralist Tradition is infallible? Let us consider some points where that tradition is, or might be, fallible.

i) An informal proof within the framework of an established theory that is, or ought to be, in principle formalizable, may after critical analysis appear to be wrong. There may be an ordinary error involved, but, of course, it is also possible that along the lines of the methodology of proofs and refutations an interesting hidden lemma shows up. The possibility of this happening with classical results that have often been checked, that have been proved in different ways and that are reproduced repeatedly in textbooks is presumably small.

ii) It is to be expected, particularly when a new theory is being developed and new notions are being defined, that conceptual refinement along the lines of the methodology of proofs and refutations will occur.

iii) However, supposing that from now on we only accept completely formalized proofs, would the fallibility of mathematics in that case be restricted to errors in the manipulation of formulae? In other words, does the existence of Mac Lane's exact notion of absolute rigor guarantee infallibility of the results?

The answer to these last questions is no. We can see this by considering "Hilbert's programme". Hilbert's programme is an example of an attempt within the Structuralist Tradition to deal with the problem of fallibility. Hilbert distinguished finitistic mathematics from infinitistic mathematics. It was Hilbert's intention to reconstitute all mathematics as an elaborate formal system and to justify that system finitistically. Assuming the infallibility of finitistic mathematics, the infallibility of formal infinitistic mathematics would in such a way be guaranteed. There are good reasons to assume the infallibility of finitistic mathematics.[1] It is clear that finitistic mathematics is a part of mathematics which can only be considered as fallible from an extremely abstract point of view. Not only has no one been able to come up with

---

[1] Hilbert never precisely defined the limits of finitistic mathematics and there was at that time no need to do so. And if Hilbert had succeeded in giving an obviously finitistic consistency proof there would have been no need for such a definition. In the mean-time such definitions have been given. For example Tait argued in 1981 that Hilbert's finitism is captured by the formal system PRA of primitive recursive arithmetic, also known as Skolem arithmetic ([Simpson,1988], p. 352).

a refutation of any sort, but it is also completely unclear as to what a refutation would look like. However, Hilbert's attempt to show the consistency of infinitistic mathematics has failed. Gödel's theorem shows that it is impossible to justify all infinitistic mathematics by means of a formalist reduction to finitism.

The result is that infinitistic mathematics must be considered as fallible in two ways. The possibility of inconsistency of formal systems describing infinite structures cannot be excluded and, moreover, it cannot be excluded that within a formalization of an informal theory concerning infinite structures, a theorem can be derived that contradicts a theorem from the informal theory. For example, Wang has written with respect to the set-theoretic axioms:

> "Set theory presupposes a good deal of idealization and extrapolation. Extrapolation clearly implies an element of arbitrariness or at least a choice from different alternatives. [...] To eliminate alternatives involves a complex historical process of successions of experimentations, consensi, and rationalizations. We learn to consider certain consensi self-evident and we may be wrong even as to the truth of what is taken to be self-evident" ([Wang,1977], p.317).

It is clear: regarding the notion of (infinite) set we cannot exclude the possibility of conceptual refinement.

iv) Sometimes mathematicians consciously introduce fallible methods in mathematics. In this respect it is interesting to quote Chaitin, who suggested that

> "mathematics is perhaps more akin to physics than mathematicians have been willing to admit, and that perhaps a more flexible attitude with respect to adopting new axioms and methods of reasoning is the proper response to Gödel's theorem. Probabilistic proofs of primality via sampling[...] also suggest that the sources of mathematical truth are wider than usually thought. Perhaps number theory should be pursued more openly in the spirit of experimental science" ([Chaitin,1982], p. 301).

And indeed the ever growing role of computers in mathematics could also easily lead to an extensive "experimental number theory". In general it is to be expected that if interesting regularities in some area of mathematics were to show up - that would not be accessible, at least for the time being, to a deductive treatment - one would attempt to study them"experimentally".

Mac Lane is right in the sense that the Structuralist Tradition is characterized by a standard of rigor that is extremely high in comparison with earlier traditions. In Lakatos's work "twentieth century mathematics" on one hand and "mathematics over the centuries" on the other are sometimes insufficiently distinguished. One could argue that twentieth century mathematics is part of mathematics over the centuries, and that in precisely the same way as the fundamental notions of the Euclidean Tradition and the Formalist Tradition have been replaced more or less unexpectedly by more refined notions, the fundamental notions of the Structuralist Tradition will be also replaced. Such an argument is very suitable in teaching us modesty, but the fact remains that no tradition so successfully dealt with its foundations as did the Structuralist Tradition.

X.3 The tradition-independent subject matter of mathematics

X.3.1 Mathematical facts

In Chapter V, I mentioned Marchi's suggestion of making mathematical theorems analogous to scientific facts, and although I later avoided the use of the term "mathematical fact" and continued to speak of conjectures and theorems, I also pointed out that it is tempting to consider mathematical facts as being (to a certain extent) tradition-independent. In order to see this again, let us first consider a very simple example.

Ogilvy and Anderson tell us a story about a colonel leading an expedition into Ethiopia ([Ogilvy& Anderson,1966], pp.12-13). The colonel, wishing to buy 7 bulls costing 22 Maria Theresia thalers each, was confronted by a peculiar system of multiplication. The local priest and his assistant made two rows of parallel holes in the ground, and the interpreter told the colonel that the holes were called "houses". They put 7 pebbles in the first hole of the first row and 22 in the first hole of the second row. Then the two mathematical wizards repeatedly doubled the 7 pebbles in row one, putting 14 pebbles in the second house, 28 pebbles in the third house etc. Moreover, they repeatedly divided the 22 pebbles in the second row in half, putting 11 pebbles in the second house of that row, 5 pebbles in the third house etc. During the repeated division, fractions were neglected. The following situation was finally reached:

First row	7	14	28	56	112
Second row	22	11	5	2	1

In order to get the answer the priest, because of their being "evil", removed all pebbles in the first row corresponding to even houses in the second row, and he added the pebbles from the remaining houses in the first row corresponding to the odd houses in the second row:

$$14 + 28 + 112 = 154.$$

The colonel reported later that he was very much amazed by the fact that this method of multiplication always gives the correct answer, in spite of the "errors" that are made when the fractions are neglected during the division. The phenomenon that this peculiar method always works is a mathematical fact, which can be considered from the point of view of various traditions. These traditions can differ widely in the way they describe the fact and in the way they explain it. We don't know whether the priest and his assistant explained the success of their method of multiplication in some way or another. One can imagine a tradition in which the fact that certain methods work is explained by means of a myth.

Let us also consider the example of Pythagoras' theorem from the point of view of the Demonstrative Tradition, the Euclidean Tradition and the Structuralist Tradition. What do the three theorems involved have in common? One could argue that younger traditions always interpret the theorems that belong to older traditions, that the three theorems form an ordered sequence in that sense and that the three theorems only have in common that they belong to such an ordered sequence. If that were so, one would expect in the history of mathematics at least some similarly ordered sequences showing theorems being totally deformed over the course of time. Mathematical knowledge, however, appears to be remarkably stable. This is why I prefer to assume that the three theorems of Pythagoras view one particular mathematical fact from different perspectives. The three theorems of Pythagoras then all refer to one particular mathematical fact. The Euclidean perspective was in a way more precise than the perspective of the Demonstrative Tradition; the 20th century perspective is even more precise and includes the insight that there exist structures that are in several respects similar to the Euclidean plane, but for which Pythagoras' theorem does not hold.[2]

Let us look at the situation from another point of view. One of the remarkable things concerning our mathematical theories is that, as far as we know, they are usually consistent. Putnam has argued that the consistency of our mathematical knowledge implies the existence of an interpretation (in the sense of intended model). Putnam:

---

[2]Davis and Hersh give a sequence of seven different versions of the Chinese Remainder Theorem that illustrates my point very well ([Davis&Hersh,1983], pp.187-195). Although they object to Platonism (p.410), regarding the Chinese Remainder Theorem they write about "the changing nature of *the statement of the theorem*"(p.187 - emphasis is mine-T.K.). Apparently they do not doubt that in their sequence we meet with one and the same theorem seven times.

> "If there is *no* interpretation under which most of mathematics is *true*, if we are really just writing down strings of symbols at random, or even by trial and error, what are the chances that our theory would be consistent, let alone mathematically fertile?"([Putnam, 1975], p. 60).

Indeed, the discovery of a valid proof of the negation of an existing theorem, i.e. the discovery of inconsistency, is a relatively rare phenomenon in mathematics. New proofs of old theorems, however, are often found.[3] This is of course completely in accordance with the weak fallibility thesis. Applying Putnam's argument the weak fallibility of mathematics and the fact that the striving for mathematical knowledge in accordance with the MMRT has been so extremely successful imply that mathematics refers to a subject matter which makes the sentences of our mathematical discourse true or false. The frequent occurrence of independent discoveries and also the frequent occurrence of external successes - that is the unexpected application of a theory outside its own domain - can only be understood, from my point of view, as a confirmation of the fact that we are dealing in mathematics with a subject matter that enjoys an objective tradition-independent existence.

However, if one assumes that there exist mathematical facts that are to a certain extent tradition-independent, one immediately faces difficult questions. It is remarkable that our investigations concerning the rationality and the fallibility of mathematics inspired by Lakatos bring us back to the classical problems in the philosophy of mathematics. How should we characterize these mathematical facts? What is their relation to physical facts? We have access to physical facts through our senses. How do we have access to mathematical facts? These are difficult questions and satisfactory answers do not exist. In a paper published in 1981 the mathematician Mac Lane encouraged philosophers to renew the study of the philosophy of mathematics, a subject which he described as being "dormant since about 1931".[4] Other authors have expressed themselves in a similar way. For example Putnam, in a paper published in 1979 on recent trends in the philosophy of mathematics, came to the conclusion that none of the existing views on the nature of mathematics were valid.[5] And in 1979 Goodman argued that the four major views in the philosophy of mathematics - formalism, intuitionism, logicism and platonism - arise

---

[3]According to Eves, [Loomis,1968] contains 370 proofs of Pythagoras' theorem ([Eves, 1983], p.54).

[4][Mac Lane,1981], p.462

[5][Putnam,1979]

from "an oversimplification of what happens when we do mathematics".[6] From Goodman's point of view a more adequate philosophy of mathematics has yet to be formulated.[7] Although it is inevitable that my answer will be very tentative, at the end of this study I would like to attempt to formulate the beginning of a very global, tradition-independent answer to the question: What is it that mathematicians have studied over the ages and how did they study it? The answer should of course be compatible with the view of the history of mathematics developed in the preceding chapters.

X.3.2 Mathematics is not a natural science

Many philosophers view mathematics and natural science as being completely different, both in their subject matter and in their methods. Mathematics is then viewed as dealing with objects not given in experience and as dealing with truths derived independently of experience. Lakatos opposed such philosophies of mathematics and always emphasized the resemblance between mathematics and natural science. He pointed at the failure of all attempts made at giving a completely satisfactory a priori foundation for all mathematics, and he stressed the revisability of mathematics. Related views have been defended by others. For example, in a recent paper, "Mathematics as a Natural Science" ([Goodman,1990]), Goodman[8] describes mathematics as a *natural science of great generality*. He opposes the view that "mathematics consists of truths more secure than those of any empirical science" and argues that

> "central parts of mathematical analysis are among the most completely tested and most fully developed parts of our total science of nature. We should try to show that mathematics is as well founded as the most secure parts of physics, but not that it is better founded than physics" ([Goodman,1990], p. 185).

The following statements illustrate Goodman's argument:

> "The physicist studying systems satisfying Laplace's equation may talk about fields and forces and potentials. The mathematician studying systems satisfying Laplace's equation may talk about functions and vectors and numbers. [...] The difference is not one of subject matter, but one of level of generality" ([Goodman,1990], p.191).

---

[6][Goodman,1979], p.540

[7][Goodman,1979], p.550

[8]For philosophers: Nicholas D. Goodman, who should not to be confused with Nelson Goodman

Lakatos concentrated on the methodology of mathematics, leaving the status of its subject matter aside. In "A Renaissance of Empiricism" he argued that historico-critical case studies would have to answer the question as to whether or not constructivism, Platonic intuition, convention, experience, or a combination of them is the source of truth in mathematics. In this respect Goodman goes further than Lakatos and argues that the source of truth in mathematics is experience, and that mathematics is only more indirectly empirical than the other natural sciences. Goodman actually defends a view on the relation of mathematics and natural science that does not differ much from Quine's "holism". Quine has argued that science as a whole is one comprehensive explanatory theory, justified by its ability to explain sensations. Mathematics and logic are part of this theory, differing from natural science only in the sense that they assume a very central position. Since relinquishing logical or mathematical truths causes great upheaval in the network of our knowledge, they are not easily given up.[9]

Philosophies of mathematics are always attempts at reducing all of (true) mathematics to one particular aspect of the whole phenomenon. Logicists reduce mathematics to logic, (extreme) formalists concentrate on formula manipulation, intuitionists emphasize our capacity of mental construction. Unfortunately none of the existing philosophies of mathematics is completely satisfactory. The stress on one particular aspect always excludes other aspects that are nevertheless in practice considered as belonging to mathematics. A characterization of mathematics should do justice to mathematical practise. Philosophers like Goodman and Quine tend to reduce mathematics to natural science. Such a reduction, however, is not compatible with mathematical practice. Natural scientists study empirical reality; mathematicians, however, do not. To a certain extent the sciences can be ordered in terms of their generality: the socially possible is restricted by the biologically possible, the biologically possible is restricted by the chemically possible, the chemically possible by the physically possible and the physically possible by the mathematically possible. Such an ordering might create the impression that the difference between physics and mathematics is the same as the difference between biology and chemistry, and that mathematics is simply more general than physics but not essentially different. However, it is not so much the fact that it is the most general science that gives mathematics a qualitatively highly

---

[9]This view has been repeatedly expressed by Quine, in his well-known paper "Two Dogmas of Empiricism" for example (Reprinted in[Quine,1953]).

exceptional position. Mathematics owes its special position to the fact that the object under investigation need not exist in empirical reality. When Goodman writes

> "[people's] experience of the physical world is the main source of their mathematical knowledge" ([Goodman,1981], p.58)

he is undoubtedly right. However, when he continues without further arguments:

> "Thus there is no need for any profound separation between the epistemology of mathematics and the epistemology of empirical science" ([Goodman,1981], p.58)[10]

he is wrong. Mathematics and natural science are related in many different ways and it would be absurd to deny the important role of mathematical thinking in natural science. There is, however, a profound difference between the two questions: How do we study a particular mathematical structure, whether it exists in empirical reality or not? How do we find out whether a particular mathematical structure "occurs" in empirical reality? The former question concerns the epistemology of mathematics. The latter question belongs to the epistemology of empirical science.

X.3.3 A tentative global characterization of mathematics

In 1981 Mac Lane wrote that a sound philosophy of mathematics ought to begin with a description of the present state of mathematics ([Mac Lane,1981]). Regarding the present state of the subject, he argued that the different branches of mathematics arise from human activities that have a schematic structure made up of certain basic ideas. Arithmetic and number theory are related to counting, algebra to calculating, geometry to shaping, set theory and combinatorics to grouping, etc. In order to illustrate how elaborate the interlocking pattern of ideas is, he gave the following scheme for the origins of some of the concepts of abstract algebra:

---

[10] I do not always understand Goodman. On the one hand he does his utmost to assimilate mathematics to natural science. He writes a paper called "Mathematics as Natural Science" ([Goodman,1990]) in which he argues that mathematics is about the physical world. On the other hand he seems also inclined to give the object of mathematics a special status. The concluding statements of [Goodman,1981] are: "Still, the mathematician's theory is itself an abstract object, and it is an abstract object that is directly accessible to us. Mirrored in its structure, even if only dimly, are the outlines of that infinite complexity which we call mathematical reality".

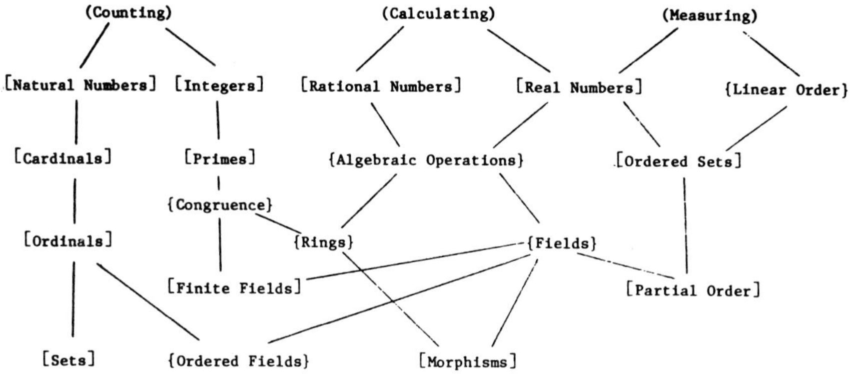

Figure 10.1 ([Mac Lane,1981], p.464)

Mac Lane writes:

> "Mathematics starts from a variety of human activities, disentangles from them a number of notions which are generic and not arbitrary, then formalizes these notions and their manifold interrelations" ([Mac Lane,1981],p.465).

and

> "Mathematics deals with the construction of a variety of formal models of aspects of the world and of human experience. On the one hand, this means that mathematics is not a direct theory of some underlying platonic reality, but rather an indirect theory of formal aspects of the world (or of reality, if there is such). On the other hand, our thesis emphasizes that mathematics involves a considerable variety of models" ([Mac Lane,1981], pp. 466-467).

Mac Lane placed great emphasis on the variety of models that mathematics deals with.[11] As a result of this complexity, simplistic descriptions of mathematics cannot be adequate. In a 1985 paper Knuth had a similar message. He studied page 100 of nine books on mathematics all covering different mathematical subjects, and he compared the types of thinking occurring on those

[11] Cf. also [Mac Lane,1986]

pages ([Knuth,1985]). The books included among others a calculus text, a standard topology text, a book on commutative algebra, a book on logic and a book on constructive mathematics. Although the categories that Knuth uses to compare the nine books are debatable and not precisely defined, his paper shows very well that there is no such thing as a simple definition of "mathematical thinking". "Formula manipulation" is different from "reduction to simpler problems", and "abstract reasoning" is different from "algorithmic thinking". Knuth also mentions the possibility of distinguishing different ways of "visualization": geometric, abstract, recursive, etc.[12] Knuth drew the following conclusion:

> "What insights do we get from these nine randomly selected examples of mathematics? In the first place, they point out something that should have been obvious to me from the start, that there is no such a thing as "mathematical thinking" as a single isolated concept; mathematicians use a variety of modes of thought, not just one" ([Knuth,1985], p. 180).

The object of mathematics

Mac Lane and Knuth considered modern mathematics. In the previous chapters I studied the development of mathematics over the centuries. The question that I would like to answer is: How should we characterize the subject matter that has been studied by mathematicians over the centuries in different ways in different research traditions? The answer should encompass both the synchronic variety described by Mac Lane and Knuth and also the diachronic variety that the successive research traditions reveal. Mac Lane defines mathematics as a "theory of formal aspects of the world" (see the above quotations). That definition is incomplete, because the forms studied within mathematics need not exist in the material world. It is enough that they are considered to be *possible* in the widest sense of the word. The difference between social and natural scientists on one hand and mathematicians and logicians on the other, is that social scientists and natural scientists study what is possible (and impossible) in empirical reality, i.e. under certain conditions given in experience, while mathematicians and logicians study what is *unconditionally possible (and impossible)*. Social scientists study reality under the aspect of the socially possible; biologists study reality under the aspect of the biologically possible; physicists study reality under the aspect of the physically possible. However, mathematicians and logicians do not study reality from the aspect of

---

12 [Knuth,1985],p.180

the mathematically and logically possible, but from the aspect of the possible as such. Mathematicians in particular, strive for knowledge of a *domain of ideal configurations from the point of view of the possible as such.*

The method of mathematics

In the epistemology of natural science, the role of perception is a major subject of investigation because we have access to empirical reality through our senses. How do we have access to the ideal configurations that mathematics studies? The following answer fits in very well with the MMRT. Actualizing an ideal configuration for ourselves in one way or another is the way to have access to that configuration. I will use the word "construction" for the way in which we create actual "images" of ideal configurations through our imagination. Of course one could ask what precisely is this capacity of constructing? Can one define it in a simple way? I think that there is no simple definition. Perception and in particular our observations of the different ways in which we can manipulate the world is a major source of constructions. Constructions are not subjected to restrictions. The papers by Mac Lane and Knuth show that there is a great variety of different ideal configurations and that a great variety also exists in the ways of thinking that are being used to study those configurations. "Anything goes" as long as the construction yields a configuration that is considered to be possible and which permits investigation. (Cf. Mac Lane's scheme in Figure 10.1).

A mathematical fact formulated within a particular tradition always refers to a configuration which is in some way constructed in that tradition. Different traditions have different ways of constructing, defining and studying configurations. In a way the limits of our collective imagination are at the same time the limits of what we consider to be possible. But just as empirical reality in natural science is studied in a very controlled way determined by a long history of competition, so does a long history of competition between research traditions and research projects in mathematics to a large extent determine how we use our imagination, our powers of idealization.

Kitcher's views on mathematical reality

The above remarks on object and method of mathematics are only a first approximation and they lack precision. Fortunately, in his book *The Nature of Mathematical Knowledge*, Kitcher develops a view of the nature of mathematical reality that nicely supplements the above characterizations ([Kitcher,1983], Ch. 6). Kitcher argues that *mathematics describes the ideal operational activity of an ideal subject.* An example of an elementary form of mathematical activity is *collecting. Correlating* is another one. Those activities can be tied to the manipulation of physical objects, but there are also higher levels of collecting or correlating. Kitcher, for example, views arithmetic as decribing the performances of an ideal agent who can execute a number of primitive operations. In the same way Euclidean geometry describes the performances of an ideal agent on ideal objects in an ideal space. In Kitcher's view we cannot know in advance what constructions the ideal agent can and cannot perform. Kitcher adopts a pragmatic attitude here: he adjusts his treatment of these issues to the manipulations that we appear to be able to perform. Kitcher then is not a constructivist in the strict sense of the word. He accepts classical logic on the ground that he wants to be able to generate classical mathematics and he also accepts statements referring to the finished product of a possibly infinite series of ideal operations if such statements can be interpreted as ultimately referring to ideal operations ([Kitcher,1983], p. 141-146). These views are very close to mine. The main difference is that Kitcher attempts to avoid all Platonist elements. I would argue, for example, that the product of the operations that are, on Kitcher's account, the subject matter of mathematics, refers to a possible ideal configuration, maybe or maybe not physically possible, but anyway possible in the widest sense of the word. Kitcher does not forgo the claim that mathematics studies finished ideal configurations but it is necessary, he says, "to regard that claim as ultimately interpreted in terms of ideal operations" ([Kitcher,1983], p.142). Kitcher seems to be a true constructivist in the sense that he ties the truth of mathematical statements to the possibility of a construction proving that truth. Yet Kitcher's position here is not completely clear. In order to obtain all of mathematics he has to require that the ideal agent be capable of performing operations that cannot be performed by a finite human being. If we accept this idealization, as Kitcher does, and if we accept also the principle of the excluded third as the best idealization that "construes the activity of the ideal subject as complete in a certain respect" ([Kitcher,1983], p. 142), then there are no good reasons to reject the Platonistic language of modern mathematics. As long as we are aware of the fact that this language is based upon a further idealization - the product of the ideal subject's activities is considered as a finished whole - there is no philosophical need to reduce the surface form of mathematical statements to some constructivist essence.

Mathematics is quasi-empirical
Possible configurations have first to be constructed in order to be studied - either mentally by means of our imagination, or by our imagination supported by (partial) material representations (on a blackboard for example). Moreover, considerable idealizations are involved in the constructions. If the construction lacks precision or if the idealizations are such that the result of the construction is very complex, there are no reasons to suppose that the mathematical claims that we make about them would be in every respect infallible. Fallibility, in particular the fallibility of universal claims, is to be expected. My analyses in the preceding chpaters confirmed this. On the other hand, the fact that the way in which we define a possible configuration is the point of departure of the theory about that configuration, implies in general that re-defining the configuration will either lead to a different theory concerning a different configuration, if the new definition deviates too much from the old definition, or else it will lead to a successor theory that encompasses the old theory, if the old configuration can be seen as a special case of a more general species represented by the new definition. In other words, weak fallibility is to be expected: there will be refutations in the sense that our imagination will sometimes yield unexpected, unintended interpretations of mathematical notions, but there will also be continuity and accumulation of truths.

With respect to both ontology and epistemology, there are clear differences between mathematics and empirical science. Yet in spite of the fact that the object and method of mathematics differ from the object and method of the empirical sciences, there are important analogies between mathematical and empirical knowledge. Mathematics, from my point of view, is definitely not an empirical science, but it makes good sense to call mathematics a quasi-empirical science. As a matter of fact my conclusions on Lakatos's rationality and fallibility theses are such that I agree with Putnam when he writes that

> "mathematical knowledge resembles *empirical* knowledge - that is, that the criterion of truth in mathematics just as much as in physic is success of our ideas in practice, and that mathematical knowledge is corrigible and not absolute" ([Putnam,1975], p.51).

Here I interpret "success in practice" in the sense of "success in terms of yielding weighty results". And a result may be weighty because of its usefulness in physics. Traditions are replaced by other more successful traditions and, to a certain extent, our knowledge thus undergoes corrections. When looking at the history of mathematics from the time of the Greeks until the present, one must conclude, however, that the corrections in mathematics, compared to the corrections in natural science since the Greeks, have been minor. There has been a considerable continuity and a considerable steady accumulation of truths. The rational enterprise of striving for mathematical knowledge has been remarkably successful in its own way.

Bibliography

[Abel,1826]: N. H. Abel, Untersuchungen über die Reihe $1+m.x/1 + m(m-1)x^2/1.2 + m(m-1)(m-2)x^3/1.2.3 + ...$, In [Wangerin,1895]

[Abel,1826b]: N. H. Abel, Lettre à Holmboe, le 16 janvier 1826, Oeuvres complètes de Niels Henrik Abel,Tome I, L.Sylow & S.Lie, (eds), Christiania,1881,pp.256-258

[Abel,1881]: N. H. Abel, Oeuvres complètes, 2 vols., L. Sylow & S. Lie (eds), Christiana, 1881

[Asquith&Hacking,1981]: P. D. Asquith, I. Hacking (eds), Proceedings of the 1978 Biennal Meeting of the Philosophy of Science Association, Vol. II, East Lansing, Michigan, 1981

[Asquith&Kyburg,1979]: P. D. Asquith, Henry E. Kyburg (eds), Jr., Current Research in Philosophy of Science, East Lansing Michigan, 1979

[Becker,1862]: J. K. Becker, Zur Polyedrometrie, Archiv der Mathematik und Physik **38**, 1862, pp. 345- 355

[Becker, 1934]: O. Becker, Die Lehre von Geraden und Ungeraden im Neunten Buch der Euklidischen Elemente, References are to the reprint in [Becker,1965], pp.125-145

[Becker, 1965]: O. Becker (Herausg), Zur Geschichte der Griechischen Mathematik, Darmstadt, 1965

[Becker, 1966]: O. Becker, Das Mathematische Denken der Antike, Göttingen, 1966

[Bell e.a.,1983]: A.W. Bell, J. Costello, D. E. Küchemann (eds), A Review of Research in Mathematical Education, Part A: Research on Learning and Teaching, Windsor, 1983

[Benacerraf&Putnam,1983]: P. Benacerraf, H. Putnam (eds), Philosophy of Mathematics, Selected Readings, Cambridge, 1983

[Van Bendegem,1988]: J. P. Van Bendegem, Non-Formal Properties of Real Mathematical Proofs, In PSA 1988, Proceedings of the 1988 Biennial Meeting of the Philosophy of Science Association Volume One, (A. Fine, J. Leplin eds), East Lansing, Michigan, 1988, pp. 249-254

[Van Bendegem,1988]: J. P. Van Bendegem (editor), Philosophica **42**, Recent Issues in the Philosophy of Mathematics I, 1988

[Van Bendegem,1989]: J. P. Van Bendegem (editor), Philosophica **43**, Recent Issues in the Philosophy of Mathematics II, 1989

[Berggren,1976]: J. L. Berggren, Spurious Theorems in Archimedes' Equilibrium of Planes: Book I, Archive for History of Exact Sciences **16**, 1976, pp.87-103

[Bernstein,1979]: D. L. Bernstein, The Role of Applications in Mathematics, American Mathematical Monthly **86**, 1979, pp. 245-253

[Beth,1959]: E. W. Beth, The Foundations of Mathematics: A Study in the Philosophy of Science, Amsterdam, 1959

[Blanchet,1841]: P. -H. Blanchet, Extrait d'une Lettre adressée à M. Liouville, Journal de Mathématiques pures et appliquées, Tome VI, 1841, pp. 65 - 68

[Bos,1974]: H. J. M. Bos, Differentials, higher-order differentials and the derivative in the Leibnizian calculus, Archive for History of Exact Sciences **14**, 1974/75, pp. 1-90

[Bos,1980]: H. J. M. Bos, Newton, Leibniz and the Leibnizian Tradition, In [Grattan-Guinness,1980], pp. 49-93

[Bos,1984]: H. J. M. Bos, Arguments on Motivation in the Rise and Decline of a Mathematical Theory; the "Constructions of Equations", 1637-ca.1750, Archive for History of Exact Sciences **30**, 1984, pp. 331-380

[Bossut&Lagrange,1799]: Bossut , J. -L. Lagrange, Rapport sur un mémoire présenté à la classe par le citoyen Callet, Mémoires de l'Institut National des Sciences et des Arts, Sciences Mathématiques et Physiques, Tome III, Prairial An IX(1799), pp.1-11

[Bottazzini,1986]: U. Bottazzini, The Higher Calculus: A History of Real and Complex Analysis from Euler to Weierstrass, Berlin etc., 1986

[Bourbaki,1950]: N. Bourbaki, The Architecture of Mathematics, American Mathematical Monthly **57**, 1950, pp. 221-232

[Bowen, 1984]: A. C.Bowen, Review of [Szabo,1978], Historia Mathematica **11**, 1984, pp.335-345

[Boyer,1959]: C. B. Boyer, The History of the Calculus and its Conceptual Development, New York, 1959 [First published in 1949]

[Boyer,1968]: C. B. Boyer, A History of Mathematics, New York etc., 1968

[Brill&Noether,1892-93]: A. Brill, M. Noether, Bericht über die Entwicklung der Theorie der algebraischen Functionen in älterer und neuerer Zeit, Jahresbericht der Deutschen Mathematiker - Vereinigung, Band 3, 1892-93, pp.107-566

[Brouwer,1905]: L. E. J. Brouwer, Leven,Kunst en Mystiek, Delft, 1905

[Browder,1976]: F. E. Browder (ed.), Mathematical Developments arising from Hilbert Problems, Providence, Rhode Island, 1976

[Brückner,1900]: M. Brückner, Vielecke und Vielflache, Theorie und Geschichte, Leipzig, 1900

[Buck&Cohen,1971]: R. C. Buck, R. S. Cohen (eds), PSA 1970 In Memory of Rudolf Carnap, Boston Studies in the Philosophy of Science , Vol. VIII, Dordrecht, 1971

[Burkhardt,1911]: H. Burkhardt, Über den Gebrauch divergenter Reihen in der Zeit von 1750-1860, Mathematische Annalen **70**, 1911, pp.169-206

[Cartan,1943]: H. Cartan, Sur le fondement logique des mathématiques, La Revue Scientifique **81**, Numéro 3216, Janvier 1943, pp. 3-11

[Cauchy,1813a]: A. - L. Cauchy, Recherches sur les polyèdres, Journal de l'École Polytechnique, Tome IX, Paris 1813, pp. 68-86 (Reprinted in [Cauchy,1882], 2nd series, Vol. 1, pp. 7-25)

[Cauchy,1813b]: A.-L. Cauchy, IIe Mémoire sur les polygones et les polyèdres, Journal de l'École Polytechnique, Tome IX, Paris 1813, pp. 87-98 (Reprinted in [Cauchy,1882], 2nd series, Vol. 1, pp. 26-38)

[Cauchy,1821]: Cours d'analyse de l'école royale polytechnique, 1er partie: analyse algébrique, Paris, 1821 (References are to the reprint in [Cauchy,1882], 2nd series, Vol.3)

[Cauchy,1823]: Résumé des leçons données à l'école royale polytechnique sur le calcul infinitésimal, Vol.1, Paris, 1823 (References are to the reprint in [Cauchy,1882], 2nd series, Vol.4, pp.5-261).

[Cauchy,1829]: Leçons sur le calcul différentiel, Paris, 1829 (References are to the reprint in [Cauchy,1882], 2nd series, Vol.4, pp. 263-609)

[Cauchy,1853]: A.-L. Cauchy, Note sur les séries convergentes dont les divers termes sont des fonctions continues d'une variable réelle ou imaginaire, entre des limites données, Comptes Rendus de l'Académie, Tome XXXVI, p. 454 etc. (References to the reprint in [Cauchy,1882], 1st series , Vol. 12, pp. 30 - 36)

[Cauchy,1882]: Oeuvres complètes d'Augustin Cauchy, 2 series, Paris, 1882-

[Chaitin,1982]: G. Chaitin, Gödel's Theorem an Information, International Journal of Theoretical Physics **21**, 1982, pp. 941-954. References are to the reprint in [Tymoczko,1986], pp. 300-311

[Clairaut,1739]: A.C.Clairaut, Recherches générales sur le calcul intégral, Histoire de l'Académie des Sciences de Paris, 1739, pp. 425-436

[Clairaut,1740]: A. C. Clairaut, Sur l'intégration ou la construction des équations différentielles du premier ordre, Histoire de l'Académie des Sciences de Paris, 1740, pp. 293-323

[Cohen&Wartofsky,1974]: R. S. Cohen, M.W. Wartofsky (eds), Methodological and historical essays in the natural and social sciences, Dordrecht-Holland / Boston-U.S.A., 1974

[Cohen e.a.,1976]: R.S. Cohen, P.K. Feyerabend, M.W.Wartofsky (eds), Essays in Memory of Imre Lakatos, D. Reidel Publishing Company, Dordrecht-Holland / Boston-U.S.A., 1976

[Coxeter,1973]: H. S. M. Coxeter, Regular Polytopes, New York, 1973 (Corrected reprint of the original 1963 publication)

[Czwalina,1972]: A. Czwalina (Herausg.), Archimedes Werke, Darmstadt, 1972

[Daval&Guilbaud,1945]: R. Daval, G.-T. Guilbaud, Le Raisonnement Mathématique, Paris, 1945

[Davis&Hersh, 1983]: Philip J. Davis, Reuben Hersh, The Mathematical Experience, Penguin Books, 1983 (Reprint of the first 1980 edition)

[Dedekind,1876]: R. Dedekind, Bernhard Riemann's Lebenslauf, [Weber,1876], pp. 507-526

[De Vries,1981]: G. de Vries, Lakatos' Bewijs, Kennis en Methode **4**, 1981, pp. 4-21

[Dieudonné,1978]: J. Dieudonne e.a., Abrégé d'histoire des mathématiques, Tome II, Paris, 1978

[Dini,1877]: U. Dini, Analisi Infinitesimale, Lezioni dettate nella r. università di Pisa, Parte I, Calcolo differenziale, Anno academico 1877-78, Autografia Bertini, Pisa

[Dini,1878]: U. Dini, Fondamenti per la teorica delle funzioni di variabili reali, Pisa, 1878

[Dini,1907]: U. Dini, Lezioni di Analisi Infinitesimale, Vol. I, Calcolo Differenziale, Pisa, 1907

[Donovan e.a., 1988]: A. Donovan, L. Laudan, R. Laudan (eds), Scrutinizing Science, Empirical Studies of Scientific Change, Dordrecht -Boston-London, 1988

[Duhamel,1847]: J. M. C. Duhamel, Cours d'analyse, première partie, seconde édition, Paris, 1847

[Van Dormolen, 1977]: J. van Dormolen, Learning to understand what giving a proof really means, Educational Studies in Mathematics **8**, 1977, pp. 27-34

[Dijksterhuis, 1929]: E. J. Dijksterhuis, De Elementen van Euclides I, Groningen, 1929

[Dijksterhuis,1956]: E. J. Dijksterhuis, Archimedes, Copenhagen, 1956

[Engelsman,1984]: S. B. Engelsman, Families of Curves and the Origins of Partial Differentiation, Amsterdam-New York-Oxford, 1984

[Euler,1748]: L. Euler, Introductio in analysin infinitorum, Vol. I.,Lausanne, 1748 (Reprinted in [Euler,1911], 1st series, Opera Mathematica, Vol. 8)

[Euler,1755] L. Euler, Institutiones calculi differentialis, St. Petersburg,1755, (Reprinted in [Euler,1911], 1st series, Opera Mathematica, Vol.10)

[Euler,1758a]: L. Euler, Elementa doctrinae solidorum, Novi commentarii academiae scientiarum Petropolitanae, Tom. IV (ad annum 1752 et 1753), 1758, pp. 109 - 140 (Reprinted in [Euler,1911], 1st series, Opera Mathematica, Vol. 26, pp. 71-93)

[Euler,1758b]: L. Euler, Demonstratio non nullarum insignium proprietatum quibus solida hedris planis inclusa sunt praedita, Novi commentarii academiae scientiarum Petropolitanae, Tom. IV (ad annum 1752 et 1753), 1758, pp. 140-160, (Reprinted in [Euler,1911], 1st series, Opera Mathematica, Vol. 26, pp. 94-108)

[Euler,1911]: L. Euler, Opera Omnia, ed. Societatis Scientiarum Naturalium Helveticae, 4 series, Leipzig, Berlin, Zürich, Basel, 1911-

[Eves,1983]: H. Eves, An Introduction to the History of Mathematics, Philadelphia etc., 1983 (5th edition)

[Fine,1978]: A. Fine, Conceptual Change in Mathematics and Science: Lakatos' Stretching Refined, [Asquith&Hacking,1981], pp. 328-341

[Federico,1982]: P. J. Federico, Descartes on Polyhedra, A Study of the De Solidorum Elements, New York, etc., 1982

[Feit&Thompson]: W. Feit, J. G. Thompson, Solvability of Groups of Odd Order, Pacific Journal of Mathematics **13**, 1963, pp. 771-1030

[Feferman,1978]: S. Feferman, The Logic of Mathematical discovery Vs. the Logical Structure of Mathematics, [Asquith&Hacking,1981], pp. 309 - 327

[Fisher,1978]: G. M. Fisher, Cauchy and the Infinitely Small, Historia Mathematica **5**, 1978, pp. 313 - 331

[Flett,1980]: T. M. Flett, Differential Analysis, Differentiation, differential equations and differential inequalities, Cambridge etc., 1980

[Fontaine,1732]: A. Fontaine, Nouvelle Méthode pour la Solution des Problèmes de Maximis & Minimis, References are to [Fontaine,1764], pp. 1-5

[Fontaine,1734]: A. Fontaine, Sur les Courbes Tautochrones, References are to [Fontaine,1764], pp. 15-23

[Fontaine,1738]: A. Fontaine, Le Calcul Intégral, Première Méthode, References are to [Fontaine,1764], pp. 24 - 83

[Fontaine,1764]: A. Fontaine, Mémoires données à l'Académie Royale des Sciences, non imprimés dans leur temps, Paris, 1764

[Fontaine,1767]: A. Fontaine, Addition à la Méthode pour la Solution des Problèmes de Maximis et Minimis, Histoire de l'Académie Royale des Sciences, Année 1767, Paris,1770, pp. 588-613

[Fourier,1822]: J. Fourier, Théorie analytique de la chaleur, Paris, 1822 [References are to the reprint in Oeuvres de Fourier (G. Darboux ed.), Vol. 1, Paris 1888

[Frajese,1974]: Opere di Archimede, A cura di Attilio Frajese, Torino, 1974, (References are to the reprint of 1988)

[Franklin,1987]: J. Franklin, Non-deductive Logic in Mathematics, British Journal for the Philosophy of Science **38**, 1987, pp. 1-18

[Fremont,1969]: H. Fremont, How to Teach Mathematics in Secondary Schools, Philadelphia, 1969

[Freudenthal, 1973]: H. Freudenthal, Mathematik als pädagogische Aufgabe, Band I, Stuttgart, 1973

[Von Fritz, 1952]: K. von Fritz, Der gemeinsame Ursprung der Geschichtsschreibung und der exakten Wissenschaft bei den Griechen, Philosophia Naturalis **2**, 1952, pp.200-223 & Berichtigung auf pp.376-379

[Von Fritz,1971]: K. von Fritz, Grundprobleme der Geschichte der antiken Wissenschaft, Berlin-New York, 1971

[Frobenius,1880] F.G.Frobenius, Über die Leibnitzsche Reihe, Journal für die reine und angewandte Mathematik **89**, 1880, pp.262-264. (References are to the reprint in G.F.Frobenius, Gesammelte Abhandlungen, Band II, Berlin,1986, pp.8-10)

[Fuys,1985]: D. Fuys, Van Hiele Levels of Thinking in Geometry, Education and Urban Society **17**, 1985, pp. 447-462

[Genocchi,1884]: A. Genocchi, Calcolo differenziale e principii di calcolo integrale, Torini, 1884

[Gergonne,1813]: identical with [L'Huilier,1813]

[Giere,1985]: R. N. Giere, Philosophy of Science Naturalized, Philosophy of Science **52**, 1985, pp. 331 - 356

[Giusti,1984]: E. Giusti, Gli "errori" di Cauchy e i fondamenti dell' analisi, Bolletino di Storia delle Scienze Matematiche **4**, 1984, pp. 24-54

[Glas, 1986]: E. Glas, On the Dynamics of Mathematical Change in the Case of Monge and the French Revolution, Studies in the History and Philosophy of Science **17**, 1986, pp. 249-268

[Glas, 1989^{I}]: E. Glas, Testing the Philosophy of Mathematics in the History of Mathematics, Part I: The Sociocognitive Process of Conceptual Change, Studies in the History and Philosophy of Science **20**, 1989, pp. 115-131

[Glas, 1989^{II}]: E. Glas, Testing the Philosophy of Mathematics in the History of Mathematics, Part II: The Similarity between Mathematical and Scientific Growth, Studies in the History and Philosophy of Science **20**, 1989, pp. 157-174

[Giorello,1975]: G. Giorello, Archimedes and the Methodology of Research Programmes, Scientia **110**, 1975, pp. 125-135 (Translation of G. Giorello, Archimede e la metodologia dei programmi di ricerca, Scientia **110**, 1975, pp. 111-123

[Giorello,1981]: G. Giorello, Intuition and Rigor: Some Problems of a 'Logic of Discovery' in Mathematics, pp. 113-135 in Maria Luisa Dalla Chiara (ed.), Italian Studies in the Philosophy of Science, Boston Studies in the Philosophy of Science, Vol. 47, Dordrecht, 1981

[Gödel,1947]: K. Gödel, What is Cantor's continuum problem?, In [Benacerraf&Putnam,1983], pp. 470-485

[Goodman,1979]: N. D. Goodman, Mathematics as an Objective Science, American Mathematical Monthly **86**, 1979, pp. 540-551

[Goodman,1981]: N. D. Goodman, The Experiential Foundations of Mathematical Knowledge, History and Philosophy of Logic **2**, 1981, pp. 55-65

[Goodman,1990]: N. D. Goodman , Mathematics as Natural Science, The Journal of Symbolic Logic **55**, 1990, pp. 182-193

[Grabiner,1974]: J.V. Grabiner, Is Mathematical Truth Time-Dependent?, American Mathematical Monthly **81**, 1974,pp.354-365

[Grabiner,1981]: J.V. Grabiner, The Origins of Cauchy's Rigorous Calculus, Cambridge-London, 1981

[Grabiner,1981a]: J.V.Grabiner, Changing Attitudes Towards Mathematical Rigor: Lagrange and Analysis in the Eighteenth and Nineteenth Centuries, [Jahnke&Otte,1981], pp.311-330.

[Grattan-Guinness,1972]: I. Grattan-Guinness, Joseph Fourier 1768-1830, M.I.T., 1972

[Grattan-Guinness,1980]: I. Grattan-Guinness (ed.), From the Calculus to Set Theory 1630-1910, An Introductory History, London, 1980

[Greenberg,1981]: J. L. Greenberg, Alexis Fontaine's Fluxio-differential Method and the Origins of the Calculus of Several Variables, Annals of Science **38**, 1981, pp. 251 -290

[Greenberg, 1984]: J. L. Greenberg, Alexis Fontaine's Route to the Calculus of Several Variables, Historia Mathematica **11**, 1984, pp. 22 - 38

[Hallett,1979]: M. F. Hallett, Towards a Theory of Mathematical Research Programmes , British Journal for the Philosophy of Science **30**, 1979, pp. 1-25 and pp. 135-159 (This paper is based on parts of [Hallett,1979a])

[Hallett,1979a]: M. F. Hallett, The Nature of Progress in Mathematics: Case studies in the Early History of Set Theory and topology, Ph. D. Thesis 1979, University of London

[Hankel, 1965]: H. Hankel, Zur Geschichte der Mathematik in Altertum und Mittelalter, Hildesheim, 1965 [Reprografic reprint of 1874 edition]

[Heath,1921]: T. L. Heath, A History of Greek Mathematics, Vols I and II, Oxford, 1921

[Heath,1926]: T. L. Heath, The Thirteen Books of Euclid's Elements, Three Volumes , Second revised edition, Cambridge, 1926

[Heath,1949]: T. L. Heath, Mathematics in Aristotle, Oxford, 1949

[Hessel,1832]: J. F. Hessel, Nachtrag zu dem Eulerschen Lehrsatze von Polyëdern, Crelle Journal der Mathematik **8**, 1832, Hft. 1, pp. 13-20, Tafel I.

[Van Hiele, 1973]: P. M. van Hiele, Begrip en Inzicht, Purmerend, 1973

[Higgins,1940]: Th. J. Higgins, A Note on the History of Mixed Partial Derivatives, Scripta Mathematica **7**, 1940, pp. 59-62

[Hilbert,1900]: D. Hilbert, Mathematical Problems, in [Browder,1976], pp. 1-34

[Hintikka e.a., 1981]: J. Hintikka, D. Gruender, E. Agazzi (eds.), Theory Change, Ancient Axiomatics, and Galileo's Methodology, Proceedings of the 1978 Pisa Conference on the History and Philosophy of Science, Vol. 2, Dordrecht: Holland/Boston: U.S.A. / London: England

[L'Huilier,1811]: S. A. J. L'Huilier, Démonstration immédiate d'un théorème fondamental d'Euler sur les polyèdres, et exceptions dont ce théorème est susceptible, Mémoire de l'Académie Impériale des sciences de Saint-Pétersbourg, Tome IV (Publié 1813), pp. 271-301

[L'Huilier,1813]: S. A. J. L'Huilier, Mémoire sur la polyédrometrie; contenant une démonstration directe du *Théorème d'Euler* sur les polyèdres, et un examen des diverses exceptions, auxquelles ce théorème est assujetti, Annales de Mathématiques pures et appliquées, rédigé par J.-D. Gergonne **3**, 1812 - 1813, pp. 169-191

[Hopf,1932]: H. Hopf, Differentialgeometrie und topologische Gestalt, Jahresbericht der Deutschen Mathematiker Vereinigung **12**, 1932, pp. 209 -229

[Jahnke&Otte,1981]: H. N. Jahnke, M. Otte, Epistemological and Social Problems of the Sciences in the Early Nineteenth Century, Dordrecht-Boston-London, 1981

[Jahnke,1987]: H. N. Jahnke, Motive und Probleme der Arithmetisierung der Mathematik in der ersten Hälfte des 19. Jahrhunderts - Cauchys Analysis in der Sicht des Mathematikers Martin Ohm, Archive for History of Exact Sciences **37**, 1987, pp. 101 - 182

[Jongeling 1988]: B. Jongeling, Over rationaliteit en vooruitgang in de wetenschap, Krisis **33**, december 1988, pp. 66-88

[Jonquières,1890a]: de Jonquières, Note sur un point fondamental de la théorie des polyédres, Comptes Rendus hebdomadaires des séances de l'Académie des Sciences **110**, janvier-juin 1890, pp. 110-115

[Jonquières,1890b]: J. P. E. de Fauque de Jonquières, Note sur le théorème d'Euler dans la théorie des polyédres, Comptes Rendus hebdomadaires des séances de l'Académie des Sciences **110**, janvier-juin 1890, pp. 169-173

[Jordan,1882]: C. Jordan, Cours d'analyse de l'école polytechnique, tome I, Paris, 1882

[Juškevič&Winter,1965]: A. P. Juškevič, E. Winter, Leonhard Euler und Christian Goldbach, Briefwechsel 1729-1764, Berlin, 1965

[Kalmar,1967]: L. Kalmár, Foundations of Mathematics -Whither Now?, in [Lakatos, 1967], pp. 187 -207 (including discussion).

[Kitcher,1983]: Ph. Kitcher, The Nature of Mathematical Knowledge, Oxford, 1983

[Klein,1979]: F. Klein, Vorlesungen über die Entwicklung der Mathematik im 19.Jahrhundert (Ausgabe in einem Band), Reprint, Berlin, etc, 1979

[Klein, 1894]: F. Klein, Riemann und seine Bedeutung für die Entwicklung der modernen Mathematik, Text of a 1894 lecture, In Klein, Gesammelte math. Abhandlungen, III, pp. 482-497

[Kline,1972]: M. Kline, Mathematical Thought from Ancient to Modern Times, New York, 1972

[Knorr,1975]: W. R. Knorr, The Evolution of the Euclidean Elements, A Study of the Theory of Incommensurable Magnitudes and Its Significance for Early Greek Geometry, Dordrecht-Holland/Boston-U.S.A., 1975

[Knorr,1978]: W. R. Knorr, Archimedes and the Elements: Proposal for a Revised Chronological Ordering of the Archimedean Corpus, Archive for History of Exact Sciences, **19**, 1978, pp. 211-290

[Knorr,1981]: W. R. Knorr, On the Early History of Axiomatics: The Interaction of Mathematics and Philosophy in Greek Antiquity, In [Hintikka e.a.,1981], pp. 145 - 186

[Knorr,1981a]: W. R. Knorr, On the Early History of Axiomatics: a Reply to Some Criticisms, In [Hintikka e.a., 1981], pp. 193-196

[Knuth,1985]: D.E. Knuth, Algorithmic Thinking and Mathematical Thinking, American Mathematical Monthly **92**, 1985, pp. 170-181

[Koetsier,1986]: T. Koetsier, Joseph Louis Lagrange (1736-1813), His Life, His Work and His Personality, Nieuw Archief voor Wiskunde, Vierde Serie, Deel 4, No. 3, 1986, pp. 191-205

[Koetsier,1990]: T. Koetsier "Etappen in der Entwicklungsgeschichte des sogenannten Vertauschbarkeitssatzes von Schwarz", In D. D. Spalt (Herausg.), Rechnen mit dem Unendlichen, Beiträge zur Entwicklung eines kontroversen Gegenstandes, Birkhaüser Verlag, 1990, pp. 59-78.

[Kuhn,1970]: Th. S. Kuhn, The Structure of Scientific Revolutions (Second enlarged edition), Chicago, 1970

[Lacroix,1810]: S. F. Lacroix,Traité du calcul différentiel et du calcul intégral, Tome I, seconde édition, revue et augmentée, Paris, 1810

[Lacroix,1814]: S. F. Lacroix,Traité du calcul différentiel et du calcul intégral, Tome II, seconde édition, revue et augmentée, Paris, 1814

[Lacroix,1819]: S. F. Lacroix,Traité du calcul différentiel et du calcul intégral, Tome III, seconde édition, revue et augmentée, Paris, 1819

[Lagrange,1772]: J. L. Lagrange, Sur une espèce de calcul relatif à la différentiation et à la intégration des quantités variables, References are to the reprint in [Lagrange,1867], pp. 441-476

[Lagrange,1797]: J.-L. Lagrange, Théorie des fonctions analytiques, Paris, 1797, (Reprinted in [Lagrange,1867], Vol. 9)

[Lagrange,1806]: J.-L.Lagrange, Leçons sur le calcul des fonctions, Paris 2ième édition, 1806, (Reprinted in [Lagrange,1867], vol. 10)

[Lagrange,1867]: J.-L. Lagrange, Oeuvres, 14 Vols., Paris, 1867-1892

[Lakatos,1963-64]: I. Lakatos, Proofs and Refutations, British Journal for the Philosophy of Science **14**, 1963-64, pp. 1-25, 120-139, 221-245, 296-342. References are to the only slightly different reprint in [Lakatos,1976], pp.6-105

[Lakatos,1967]: I. Lakatos (ed.), Problems in the Philosophy of Mathematics, Proceedings of the International Colloquium in the Philosophy of Science, London, 1965, Vol. 1, Amsterdam, 1967

[Lakatos,1971]: Lakatos' "History of Science and Its Rational Reconstructions", Comments on the paper by Kuhn, Feigl, Hall and Koertge, followed by Lakatos's "Replies to Critics", In [Buck&Cohen,1971]

[Lakatos,1976]: J. Worrall & E. Zahar (eds.), Imre Lakatos, Proofs and Refutations, The Logic of Mathematical Discovery, London, 1976

[Lakatos,1978a]: J. Worrall & G. Currie (eds.) Imre Lakatos, Philosophical Papers, Volume 1, The Methodology of Scientific Research Programmes, London, 1978

[Lakatos,1978b]: J. Worrall & G. Currie (eds.) Imre Lakatos, Philosophical Papers, Volume 2, Mathematics, Science and Epistemology , London, 1978

[Laudan,1977]: L. Laudan, Progress and Its Problems, Towards a Theory of Scientific Growth, Berkeley etc., 1977

[Laugwitz,1987]: D. Laugwitz, Infinitely Small Quantities in Cauchy's Textbooks, Historia Mathematica **14**, 1987, pp. 258 - 274

[Lautman,1937a]: A. Lautman, L'axiomatique et la méthode de division, Recherches philosophiques, 8 septembre 1937, References are to the reprint in [Lautman,1977], pp.291-304

[Lautman,1977]: A. Lautman,1977, Essai sur l'unité des mathématiques et divers écrits, Collection 10|18, Paris, 1977

[Legendre,1794]: A. - M. Legendre, Éléments de géométrie, Paris, 1794

[Leibniz,1971]: G. W. Leibniz, Mathematische Schriften, Herausgegeben von C. I. Gerhardt, Band IV, 1971, Hildesheim, New York

[L'Huilier,1811]: S. A. J. L'Huilier, Démonstration immédiate d'un théorème fondamental d'Euler sur les polyèdres, et exceptions dont ce théorème est susceptible, Mémoire de l'Académie Impériale des sciences de Saint-Pétersbourg, Tome IV (Publié 1813), pp. 271-301

[L'Huilier,1813]: S. A. J. L'Huilier, Mémoire sur la polyédrometrie; contenant une démonstration directe du *Théorème d'Euler* sur les polyèdres, et un examen des diverses exceptions, auxquelles ce théorème est assujetti, Annales de Mathématiques pures et appliquées, rédigé par J.-D. Gergonne **3**, 1812 - 1813, pp. 169-191

[Lindelöf,1867]: L. Lindelöf, Remarques sur le théorème concernant l'ordre des différentiations, Acta Societas Scientiarum Fennicae, Tomus VIII, 1867, pp.207-213

[Listing,1862]: J. B. Listing, Der Census räumlicher Complexe oder Verallgemeinerung des Eulerschen Satzes von den Polyedern, Aus dem zehnten Bande der Abhandlungen der königlichen Gesellschaft der Wissenschaften zu Göttingen, Göttingen, 1862

[Loomis,1968]: E. S. Loomis, The Pythagorean Proposition, Washington, 1986 (Reprint of the 2nd 1940 edition)

[Mac Lane, 1981]: S. Mac Lane, Mathematical Models: a Sketch for the Philosophy of Mathematics, American Mathematical Monthly **88**, 1981, pp. 462-472

[Mac Lane,1986]: S. Mac Lane, Mathematics Form and Function, New York etc., 1986

[Marchi,1976]: P. Marchi, Mathematics as a Critical Enterprise, in [Cohen e.a.,1976], pp. 379-393

[Matthiesen,1863]: L. Matthiesen, Ueber die scheinbaren Einschränkungen des Euler'schen satzes von den Polyedern, Zeitschrift für Mathematik und Physik **8**, 1863, pp. 449-450 und Tafel IV.

[Maull,1976]: N. L. Maull, Reconstructed Science as Philosophical Evidence, In PSA 1976, Proceedings of the 1976 Biennial Meeting of the Philosophy of Science Association, Vol. 1, 1976, East Lansing, Michigan, pp. 119-129

[Mehrtens,1976]: H. Mehrtens, T. S. Kuhn's Theories and Mathematics: a Discussion Paper on the 'New Historiography' of Mathematics, Historia Mathematica **3**, 1976, pp. 297-320

[Mittelstraß, 1965]: J. Mittelstraß, Die Entdeckung der Möglichkeit von Wissenschaft, Archive for History of Exact Sciences **2**,1962-1966, pp.410-435

[Moigno,1840]: M. l'abbé F. N. M. Moigno, A.-L. Cauchy, Leçons de Calcul Différentiel rédigées d'après les méthodes et les ouvrages publiés ou inédits par M. l'abbee Moigno, Paris, 1840

[Monna,1975]: A. F. Monna, Dirichlet's principle, A mathematical comedy of errors and its influence on the development of analysis, Utrecht, 1975

[De Morgan,1842] A. De Morgan, The Differential and Integral Calculus, London, 1842

[De Morgan,1844]: A. De Morgan, On Divergent Series, and various Points of Analysis connected with them, Transactions of the Cambridge Philosophical Society , Vol. VIII, 1844, pp.182-203

[Mulder, 1988]: H. M. Mulder, To See the History for the Trees, Congressus Numerantium **64**, 1988, pp. 25-43

[Murphy&Hendrick,1984]: A. C. Murphy, R.E.Hendrick, Lakatos, Laudan and the Hermeneutic Circle, Studies in the History and Philosophy of Science **15**, 1984, pp.119-130

[Neuenschwander,1972]: E. Neuenschwander, Die ersten vier Bücher der Elemente Euklids, Archive for History of Exact Sciences **9**, 1972/73, pp.325-380

[Neuenschwander,1981]: E. Neuenschwander, Über die Wechselwirkungen zwischen der französischen Schule, Riemann und Weierstraß. Eine Übersicht mit zwei Quellenstudien, Archive for History of Exact Sciences **24**, 1981, pp. 231-255.

[Peano,1890]: G. Peano, Sur l'interversion des dérivations partielles, Mathesis, recueil mathématique à l'usage des écoles spéciales et des établissements d'instruction moyenne **10**, 1890, S. 153-154

[Ogilvy&Anderson,1966]: C. S. Ogilvy, J. T. Anderson, Getallentheorie, Utrecht-Antwerpen, 1969 (Dutch translation of C. S. Ogilvy, J.T. Anderson, Excursions in Number Theory, New York, 1966)

[Poinsot,1810]: L. Poinsot, Mémoire sur les polygones et les polyèdres, Journal de l'École Polytechnique **4**, 1810, pp. 16-48

[Polya,1954]: G. Polya, Mathematics and Plausible Reasoning, Vol. 1, Induction and Analogy in Mathematics, Princeton, 1954

[Polya,1968]: G. Polya, Mathematics and Plausible Reasoning, Vol.2, Patterns of Plausible Inference, Princeton, 1968 (second edition)

[Popper,1981]: K. R. Popper, Objective Knowledge, An Evolutionary Approach, Oxford, 1981 (Revised Edition)

[Putnam,1975]: H. Putnam, What is Mathematical Truth?, In [Tymoczko,1986], pp. 49 -65 (Reprinted from H. Putnam, Philosophical Papers, Vol. 1, Cambridge: Cambridge University Press, 1975)

[Putnam,1979]: H. Putnam, Philosophy of Mathematics: A Report, In [Asquith & Kyburg, 1979], pp. 386-398

[Quine,1953]: W. V. Quine, From a Logical Point of View, Cambridge: Harvard, 1953

[Resnik&Kushner,1987]: M. D. Resnik, D. Kushner, Explanation, Independence and Realism in Mathematics, British Journal for the Philosophy of Science **38**, 1987, pp. 141-158

[Richards,1987]: J. L. Richards, Augustus De Morgan, the History of Mathematics, and the Foundations of Algebra, ISIS **78**, 1987, pp. 7-30

[Riemann,1851]: B. Riemann, Grundlagen für eine allgemeine Theorie der Functionen einer veränderlichen complexen Größe, doctoral dissertation, [Weber,1876], pp. 3-47

[Riemann, 1854]: B. Riemann, Ueber die Hypothesen, welche der Geometrie zu Grunde liegen", in [Weber,1976], pp. 254-269

[Riemann,1857]: B. Riemann, Theorie der Abel'schen Functionen, In [Weber,1876], pp. 81-135

[Riemann,1876a]: B. Riemann, Gleichgewicht der Electricität auf Cylindern mit kreisförligen Querschnitt und parallelen Axen, in [Weber,1876], pp. 413-416

[Riemann,1876b]: B. Riemann, Fragmente philosophischen Inhalts, [Weber,1876], pp. 473 - 506

[Robinson,1966]: A. Robinson, Non-standard Analysis, Amsterdam, 1966

[Salmon,1967]: W. C. Salmon, The Foundations of Scientific Inference, Pittsburgh, 1967

[Schlömilch,1847]: O. Schlömilch, Handbuch der Differenzial- und Integralrechnung, Erster Theil, Differenzialrechnung, Greifswald 1847

[Schlömilch,1853]: O. Schlömilch, Compendium der höheren Analysis (Erste Auflage), Braunschweig 1853

[Schlömilch,1862]: O.Schlömilch, Compendium der höheren Analysis (Zweite Auflage), Erster Band, Braunschweig 1862

[Schlömilch,1868]: O.Schlömilch, Compendium der höheren Analysis (Dritte Auflage), Erster Band, Braunschweig 1868

[Schlömilch,1874]: O.Schlömilch, Compendium der höheren Analysis (Vierte Auflage), Braunschweig 1874

[Seidel,1847]: Ph. L. Seidel, Note über eine Eigenschaft der Reihen welche discontinuirliche Functionen darstellen, In Ostwald's Klassiker der Exacten Wissenschaften, Nr. 116, Herausg. H. Liebmann, pp. 35 - 45

[Seidenberg,1959]: A. Seidenberg, Peg and Cord in Ancient Greek Mathematics, Scripta Mathematica **24**, 1959, pp.107-122

[Seidenberg,1972/73]: A. Seidenberg, On the Area of a Semi-Circle, Archive for History of Exact Sciences **9**, 1972/73, pp.171-211

[Schwarz,1873]: H. A. Schwarz, Ueber ein vollständiges System von einander unabhängiger Voraussetzungen zum Beweise des Satzes $\frac{\partial}{\partial y}(\frac{\partial f(x,y)}{\partial x}) = \frac{\partial}{\partial x}(\frac{\partial f(x,y)}{\partial y})$", [Schwarz, 1890], pp. 275-284

[Schwarz,1890]: H. A. Schwarz, Gesammelte Abhandlungen, Band 2, Berlin, 1890

[Simpson,1988]: S. G Simpson, Partial Realizations of Hilbert's Program, The Journal of Symbolic Logic **53**, 1988, pp. 349-363

[Spalt,1981]: D. D. Spalt, Vom Mythos der Mathematischen Vernunft, Darmstadt, 1981

[Spalt,1983]: D. D. Spalt, Eine längst fällige, wenngleich unnötige Rehabilitation Cauchys, Mathematikunterricht **4** , 1983, pp. 60 - 76

[Spalt,1988]: D. D. Spalt, Das Unwahre des Resultatismus. Eine historische Fallstudie aus der Analysis, Mathematische Semesterberichte **35**,1988, pp. 6-37

[Staudt,1847]: G. K. C. von Staudt, Geometrie der Lage, Nürnberg, 1847

[Steele, 1936]: A. D. Steele, Über die Rolle von Zirkel und Lineal in der Griechischen Mathematik, Teil I und II, Reprint in [Becker,1965], pp.146-202

[Stegmüller,1977]: W. Stegmüller, The so-called circle of understanding, In Collected Papers on Epistemology, Philosophy of Science and History of Philosophy, Volume II, Dordrecht-Holland/Boston-U.S.A., 1977, pp.1-25

[Steiner,1826]: J. Steiner, Leichter Beweis eines Stereometrischen Satzes von Euler, Journal für die Reinen und Angewandte Mathematik **1**, ,1826, pp. 364 - 367. Reprint in [Steiner,1881], pp. 95-100

[Steiner,1881]: K. Weierstrass (Herausg.), Jacob Steiner's Gesammelte Werke, Band I, Berlin, Verlag von G. Reimer, 1881

[Steinitz,1916]: E. Steinitz, Polyeder und Raumeinteilungen, in W.F. Meyer, H. Mohrmann (Herausg.), Encyklopädie der Mathematischen Wissenschaften, Band 3, Teil 1, zweite Hälfte, Leipzig, 1914-1931

[Struik,1967]: D. J. Struik, A Concise History of Mathematics,Third Revised Edition, New York, 1967

[Szabó,1960]: A. Szabó, Anfänge des euklidischen Axiomensystems, Archive for History of Exact Sciences **1**, 1960-62, pp.37-106

[Szabó,1978]: A. Szabó, The Beginnings of Greek Mathematics, Dordrecht-Boston, 1978

[Tarski,1964]: A. Tarski , Inleiding tot de logica en tot de methodeleer der deductieve wetenschappen, Amsterdam, 1964

[Taton,1951]: R. Taton, L'Oeuvre Scientifique de Monge, Paris, 1951

[Tucciarone,1973]: J. Tucciarone, The Development of the Theory of Summable Divergent Series from 1880 to 1925, Archive for History of Exact Sciences **10**, 1973, pp.1-40

[Tóth,1965]: I. Tóth, Das Parallelenproblem im Corpus Aristotelicum, Archive for History of Exact Sciences **3**, 1966/67, pp.249-422

[Tymoczko,1986]: T. Tymoczko (ed.), New Directions in the Philosophy of Mathematics, An Anthology, Boston-Basel-Stuttgart, 1986

[Van Bendegem,1988]: J. P. Van Bendegem, Non-Formal Properties of Real Mathematical Proofs, In PSA 1988, Proceedings of the 1988 Biennial Meeting of the Philosophy of Science Association Volume One, (A. Fine, J. Leplin eds), East Lansing, Michigan, 1988, pp. 249-254

[Van Bendegem,1988]: J. P. Van Bendegem (editor), Philosophica **42**, Recent Issues in the Philosophy of Mathematics I, 1988

[Van Bendegem,1989]: J. P. Van Bendegem (editor), Philosophica **43**, Recent Issues in the Philosophy of Mathematics II, 1989

[Ver Eecke,1948]: P. Ver Eecke, Proclus de Lycie, les commentaires sur le premier livre des Éléments d'Euclide, Bruges,1948

[Van Fraassen,1980]: B. C. Van Fraassen, The Scientific Image, Oxford, 1980

[Van der Waerden,1961]: B. L. van der Waerden, Science Awakening, Vol. I, Groningen, 1961

[De Vries,1981]: G. de Vries, Lakatos' Bewijs, Kennis en Methode **4**, 1981, pp. 4-21

[Wang,1977]: H. Wang, Large Sets, In R.E.Butts, J Hintikka, Logic, Foundations of Mathematics, and Computability Theory, Part I, Dordrecht-Holland/Boston-U.S.A., 1977, pp. 309-333

[Wangerin,1895]: A. Wangerin (Herausg.), Ostwald's Klassiker der Exakten Wissenschaften, Nr. 71, Leipzig, 1895

[Weber,1876]: H. Weber (Herausg. unter Mitwirkung von R. Dedekind), Bernhard Riemann's gesammelte Mathematische Werke und wissenschaftlicher Nachlass, Leipzig, 1876

[Weil,1978]: A. Weil, History of mathematics: Why and how, In [Weil, 1979], pp. 434-442

[Weil, 1979]: A. Weil, Œuvres Scientifiques, Collected Papers, Vol. III (1964-1978), New York-Heidelberg-Berlin, 1979

[Youschkevitch,1976/77]: A. P. Youschkevitch, The Concept of Function up to the Middle of the 19th Century, Archive for History of Exact Sciences **16**, 1976/77, pp.37-85

Index

Eleatic philosophy 191